Autopoiesis
**A Theory of
Living Organization**

THE NORTH HOLLAND SERIES IN
General Systems Research
Dr. George Klir, *Editor*

Autopoiesis

A Theory of Living Organization

Series Volume 3

Edited by
Milan Zeleny

Columbia University
New York, New York

North Holland
New York • Oxford

Elsevier North Holland, Inc.
52 Vanderbilt Avenue, New York, New York 10017

Sole Distributors outside the USA and Canada:

Elsevier Science Publishers B.V.
P.O. Box 211, 1000 AE Amsterdam, The Netherlands

Library of Congress Cataloging in Publication Data

Main entry under title:

Autopoiesis, a theory of living organization.
   (North Holland series in general systems research; 3)

   Includes bibliographies and index.
   1. System theory—Addresses, essays, lectures. 2. Order (Philosophy)—
   Addresses, essays, lectures. 3. Evolution—Addresses, essays,
   lectures, I. Zeleny, Milan, 1942–
Q295.A88   003   80-27038
ISBN 0-444-00385-1

Desk Editor   John Haber
Design   Edmée Froment
Art Editor   Glen Burris
Art rendered by   Vantage Art, Inc.
Production Manager   Joanne Jay
Compositor   Maryland Composition
Printer   Haddon Craftsmen

Manufactured in the United States of America

Contents

Foreword

When the general historians (a discipline still to be invented) write the history of the 1970s, they may well identify an intellectual mutation which produced two new words, autopoiesis and allopoiesis, and the ideas which they symbolize, as the most significant mutation of the decade from the point of view of its long-run impact. The mutation originated, as many do, on the edge of the world's scientific habitat, if my Chilean friends will forgive me. The principle that important new mutations are more likely to take place toward the edges of the ecosystem in which they will eventually flourish is by no means well established in evolutionary theory; nevertheless, there is a good deal to be said in its support. One recalls an earlier remark, "Can any good thing come out of Nazareth?"

There is something about this whole enterprise that is slightly reminiscent of the voyage of Columbus. Its pioneers, F. J. Varela, H. R. Maturana, and M. Zeleny, like Columbus's three boats, probably thought they were heading for the Indies, that mysterious, still largely unknown, though spicy realm of human knowledge that studies the overwhelmingly mysterious and complex phenomenon of life. What they may well have sighted is a whole new continent, which no Amerigo has yet named, but the hazy outlines of which are now visible to the early explorers. This is nothing less than the study of the whole developmental process of the universe, that is, the general theory of evolution.

One of the great puzzles of the universe, as we now visualize it, is that it seems to contain two time arrows: one is the famous second law of thermodynamics, that entropy increases and potential decays. The other time arrow is that of evolution, which segregates entropy and builds up increasingly complex structures of order, no doubt at the cost of creating more disorder elsewhere. What we are looking at in this volume is nothing less than the great principle of the instability of chaos and the unattainability of equilibrium. The universe emerges as a disequilibrium system from the "big bang" on. The study of equilibrium, which is a remarkable amount of what goes by the name of science, now seems concerned with intellectual constructs unknown in the real world, though very useful in learning about it.

No matter who discovers a new continent, we are almost sure to find somebody who has been there before, and this is certainly the case with autopoiesis. The Irish monks and the Viking adventurers were there before Columbus. I would certainly argue as an economist that Adam Smith in *The Wealth of Nations* discovered the concept of autopoiesis in the "invisible hand." This is the concept of the development of social order through the unplanned interactions of large numbers of individuals, each seeking their own gain through production and exchange, coordinated through the development of the system of relative prices. This is a considerable conceptual advance beyond the rather one-dimensional mechanics of Newton. I would even credit Adam Smith with having perceived the evolutionary significance of positive feedback in expanding "hypercycles," as E. Jantsch and M. Eigen refer to them, in his perception that the division of labor and the rise in productivity depend on the extent of the market but also themselves promote the expansion of the market. It has taken us perhaps 200 years to perceive the extraordinary general implications for science of Adam Smith's insights.

The concepts of autopoiesis and allopoiesis are now undergoing such rapid expansion and development that it is a little hard to say what they are. My own inclination, again no doubt as an economist, is to identify autopoiesis with the processes which develop an unplanned order out of randomness, that is, out of essentially stochastic or probabilistic systems, simply because of the principle that whenever an uncertain or probabilistic event actually comes off, this changes the probabilities of all potential events around it. This change itself, of course, has to have a certain order, and where this order comes from is something of a mystery. We might call it order of the second kind.

Order of the first kind would be allopoiesis, which I tend to identify with the production of order according to some previous plan. An obvious example of this is the production of the house from a blueprint, or of any human artifact from the image of it which exists in the minds of its producers. Similar processes would be the production of a research report from the original proposal, a nation from a constitution, or any organization from its charter.

One of the most fascinating and difficult problems is that of the relative roles of autopoiesis and allopoiesis in the transformation of genotypes into phenotypes, for instance, of the egg into the chicken. This is still one of the great mysteries of the world. In the production of human artifacts the genetic structure, in the shape of blueprints or mental images, bears a strong one-to-one relationship with the final product. In the case of biological production, this is not so. An animal does not look at all like the double helix of the DNA that presumably contained the information that produced it. One suspects, therefore, that there is a good deal of order of the second kind in the development of the egg into the animal, but how much is auto- and how much is allopietic I do not have the slightest idea.

Evolution, however, is clearly primarily autopoietic. A chicken may be a centrally planned economy, at least in some degree originating from its egg of the plan. An ecosystem, however, is free private enterprise beyond the wildest dreams of Milton Friedman. It is not a "community," whatever the biologists say. It has no mayor; it has no perceptible plan. It emerges by the invisible hand of ecological interaction. The same would seem to be true of the whole evolutionary process.

The critical question that is unfolding here could well be the most important question about the universe. This is the nature of potential and the processes by which potential is realized. It is clear that there must have been potential for this book in the "big bang," otherwise the book would not be here. But the nature of this potential and the processes by which it is eventually realized are a deep mystery. It is certainly very hard to believe that there was a "plan" for this book in the "big bang," so one suspects that autopoietic processes have been extremely important in producing it. This book, indeed, represents an event of extreme improbability that actually came off; having come off, it will affect all subsequent history, perhaps not very much, or perhaps more than we think.

These essays raise far more hares that can possibly be caught. The whole problem of the autopoietic nature of language, for instance, is touched on in one or two papers and is a huge field for inquiry. One wonders also if there is not a large field open for what might almost be called order of the third kind, that is, how "know-what" is translated into "know-how." Even within the scientific community it is the rare economist who gets rich, the rare political scientist who wins an election, the rare linguist who is comprehensible, and perhaps the rare student of conversations who can converse. These reflections open up some delightful explorations for the future.

This book is an excellent supplement to a previous volume,[1] and indeed the unfamiliar reader should probably read the previous volume first, as a good deal of this volume is in a sense commentary on earlier work. The wide geographical and disciplinary range of its authors is a tribute to the explosive quality of these ideas.

Kenneth E. Boulding

---

[1] Milan Zeleny, ed., *Autopoiesis, Dissipative Structures and Spontaneous Social Order* (Boulder, Colorado: Westview Press, AAAS Selected Symposium 55, 1980).

Preface

Autopoiesis, as originally conceived by Maturana and Varela, belongs to the newly emerging category of paradigms addressing the issues of self-organization and spontaneous phenomena within physical, biological, and social systems: *order through fluctuation* and *dissipative structures* (Prigogine), *ultra- and hyper-cycles* of precellular self-organization (Eigen), and *spontaneous social orders* (von Hayek). Autopoiesis is the most recent of these concepts, and perhaps the most radical and controversial, but it is also the most general and has the potential of encompassing all the others.

The reader will find this book challenging, disturbing, uneven, brilliant, naive, and ambitious—all at the same time. He will have the rare opportunity of experiencing a new paradigm in the process of becoming. Regardless of his final verdict, because of the experience he will never be the same.

Editing this kind of book presented a series of extraordinary challenges. First, how do you corral strongly individualistic (none of the papers is co-authored), creative thinkers, scattered all around the world (the authors' backgrounds and origins are noted in the introduction to their chapters) and through a large variety of disciplines (included are biologists, economists, philosophers, cyberneticians, biophysicists, logicians, systems and computer analysts, engineers, operations researchers, and linguists) and induce them to appear within the confines of a single volume?

Second, how do you reconcile their varying levels of discourse, different styles, and different worldviews? These are people largely unaccustomed to editorial domination—a totally unpliable material for any editor. They are like sharp, multicolored pieces of glass: you can make them fit into a possibly beautiful mosaic, but you cannot mold them into grey clay and make a statue.

Third, how do you convince editors, reviewers, and publishers that there is a need for such a book, that it is commercially viable, and that a mosaic can be as important in science as another monument of grey clay?

This volume was conceived in Vienna in 1976. Preliminary versions of most of the papers were presented at the NATO International Conference on Applied General Systems Research in Binghamton, New York, August 1977. The next two years were spent on writing, revision, and exchange of opinions. Now the book is ready.

One of the early reviewers compared the book to a *zoo*, "inhabited not by spirited, graceful, and unique animals, but by their human counterparts." How does one present a zoo, and how does one deal with such a potentially rich experience? As the editor, I have prepared a guided tour through our "zoo": an introductory tutorial and a set of comments presenting the individual "cages." Now, this first "pass through" is obviously a personal view, and neither the readers nor the "animals" will necessarily be inclined to agree with it. But it is integrated, complete, and reasonably informative.

But, dear reader, please do come back and roam the "zoo" on your own! Get rid of the guide, wander freely, take long pauses—enjoy the experience. It might be difficult or even incomprehensible at first, but then the pieces may start falling together and the mosaic of order and beauty emerge. And don't despair if they do not: It was still worth the entry, and, in only a few decades or so, there will be another show.

The above metaphors simply express the fact that this is a *transdisciplinary* book—not because the authors come from different disciplines, and not because they cross so many disciplinary boundaries in their expositions, but because the concept and methodology of autopoiesis are unclassifiable and transcend discipline-constrained ways of thinking. It is therefore likely that the book will be comprehended most readily by generally minded, adventurous thinkers, while it might cause the greatest difficulty to narrowly defined specialists.

What are some of these possible difficulties? Several of the papers are just too abstract and vague for an experimentally and empirically trained mind. Such concepts as autonomy, unity, organization, and structure might be uncomfortably general for those devoted to the experimental verification or refutation of ideas. But one has to build an abstract model of reality, a framework of seeing, before being able to see anything. It is this editor's opinion that grasping a few such abstractions would represent a good intellectual investment on the part of the readers. Many new terms (including "autopoiesis" itself), new assumptions, new propositions, and unfamiliar styles appear and there is a relative nonreliance on previously published literature. All this is potentially disturbing to a smooth and uncomplicated process of understanding. But truth is rarely easy to acquire and never self-evident in the beginning.

Kenneth E. Boulding, then the President of the American Association for the Advancement of Sciences, characterized autopoiesis as a "breakthrough in science" at the 1979 AAAS Meeting in Houston. An AAAS Symposium volume on the topic quickly followed in 1980, with Boulding writing its preface. The original monograph of Maturana and Varela on autopoiesis was translated into English and appeared in the Boston Studies in the Philosophy of Science, also in

1980. Elsevier North Holland adds this volume to a previously published book by Varela, *Principles of Biological Autonomy*. The process of leaving shadow and obscurity is often satisfying, and I wish all the contributors to this volume, both individually and collectively, success with their readers. They are on their own now, in the bright and burning sun of scientific scrutiny and judgment.

As the editor, I would like to acknowledge the involvement and encouragement received from several colleagues during the process of working on this volume: George J. Klir, William Gray, Heinz von Foerster, Joel N. Morse, Stafford Beer, Richard H. Howe, Howard Pattee, Hugo Uyttenhove, and Manfred Kochen.

Erich Jantsch once observed, ''Nothing will fall into place in this book.'' But then he added, ''Nothing should be *made* to fall into place, perhaps.'' And so it is up to the reader to gather the proper pieces, to fit them together, to create a pattern of his own. The reader should be actively involved in the process of becoming, not a passive observer of the state of being: That was my intention.

Milan Zeleny

Villa Serbelloni
Bellagio, Italy

Autopoiesis
**A Theory of
Living Organization**

PART I
PROPOSITION

Life is a wave, which in no two consecutive moments of its existence is composed of the same particles.

John Tyndall

The first part includes the papers that present the main concepts and implications of autopoiesis—hence the title Proposition. They are mostly positive in their renditions of autopoiesis, with perhaps the exception of Jantsch's paper.

The contributions in the second part of the volume explore the interfaces of autopoiesis with the existing paradigms of thought, offer criticism, or take a generally more skeptical view of the concept. Where the first group attempts an independent, autonomous, intellectual "take-off," the second group strives to incorporate or to embed autopoiesis within the existing framework of thought. Both groups perceive autopoiesis as a fruitful and powerful idea—they differ only with respect to assessing the extent of its influence and potential impact.

The reader will soon see why Jantsch's paper was adjoined with the first group: it appears to be more "self-critical" than "critical" with regard to autopoiesis.

The positive character of this section is partially implied by the domination of a rather homogeneous group of four Chileans: Maturana, Varela, Uribe, and Guiloff. The remaining contributors to this section are already taking autopoiesis from its "parents" and sending it out into the world, on its own. All such "children" ultimately do so, even in defiance of their parents' protectiveness.

So in the first part we present a thesis, in the second part the antithesis, and we leave the synthesis to the reader. Well, not entirely. There is the "guide" of sort, taking the reader through the book as a whole—a template for a synthesis.

It would be a very incomplete and poor introduction to autopoiesis that did not expose the reader to the writing and argumentation of its chief progenitors: Humberto R. Maturana and Francisco J. Varela. A perceptive reader will also have a chance to peek into the creative "kitchen" of this pair and confront the human side of science: the two authors decided to present their own personal views of autopoiesis, although they did agree to a joint introduction.

The fundamental question is the following. Is autopoiesis the necessary *and* sufficient condition for all (or most) biological phenomenology, or is it only a necessary condition? Where Maturana still adheres to the first notion, Varela cannot accept the sufficiency part of the argument and calls for a development of complementary symbolism based on the notions of information and purpose.

The "sufficiency proposition" of autopoiesis is revolutionary and even outrageous in the current scientific sense. Both Maturana and Varela have started from that assumption. Later, after both had been exposed to the challenges of computerism, empiricism, and formalism, it became a matter of who "blinks" first. Varela blinked first.

The net benefactor of this dialectical controversy is the theory of autopoiesis itself, as well as the reader entering at this stage of its development. Autopoiesis is not a dogma anymore: it has become a living, flexible, adaptable philosophy. A philosophy?

Autopoiesis is capable of unifying the three traditional parts of philosophy through its comprehensive treatment of identity (logic), autonomy (ethics), and their (dialectical) relationship (aesthetics). The interactions among autonomous identities are the subject of aesthetics (viewing beauty as a social relationship), where the unity of philosophy is to be found. The three domains of phenomena—identity, autonomy, and interactions—are irreducible to each other and are often treated as separate, "nonintersecting" domains. Yet, such a distinction is entirely observer dependent and the mental separation of domains, although often useful, has nothing to do with the phenomena proper. The fact that one may choose, purposefully, to view an autopoietic unity as a nonautonomous, input–output controlled, environmentally driven behavioral system has no effect on the necessary autopoiesis of that system. One may of course also choose to view an autopoietic unity as being autonomous (self-controlled) in its environment (i.e., focus on the interactions of its components). The environment then ceases to be a source of "inputs" but merely generates perturbations affecting the system's autopoietic existence. The "outputs" and the properties of the system as a whole thus emerge from the interactions of its components.

It appears that one can consider these two points of view as equally valid and fully complementary, as is implied by Varela, or recognize their natural complementarity while upholding the primacy of the observer-independent autopoiesis over the observer-dependent autonomy control perspective, as is implied, although not explicitly stated, by Maturana.

This editor has attempted to become a surrogate for the reader of a self-respecting intelligence. A reader's understanding is often crushed by such responses as, "That is not what I am saying" from more "complicated" authors. The only answer I know is, "Why not say it clearly?" Their "That is not what I mean" is even easier: "Why do you say it, then?" I shall attempt to carry out this sort of dialogue for the reader.

Chapter 1
What Is Autopoiesis?

Milan Zeleny

1.1 Introduction

Readers working their way through this volume will learn about auto-poiesis from 15 different expositions, including those of the very creators of the concept: Maturana, Varela, and Uribe. But experience shows that a careful tutorial orientation, before a plunge into the articles them-selves, can go a long way toward providing a framework for understand-ing. One acquires a template, a point of reference, and the subsequent reading and study can take place in a directed, selective, and therefore creative way.

Autopoiesis means literally "self-production." We should be careful about the variety of expressions with similar connotation: self-organi-zation, -renewal, -creation, -generation, -maintenance, -perpetuation, and the like. Similarly, there are also such terms as biopoiesis, heteropoiesis, and allopoiesis characterizing different aspects of the processes of pro-duction. The audience may worry why we use the Greek "autopoiesis" instead of the English "self-production." "Self-production" could mean many different things to different people: it could be interpreted in a va-riety of ways. "Autopoiesis" is not a translation but a label for a partic-ular, clearly defined interpretation of "self-production."

We observe self-production phenomena intuitively in living systems. The cell, for example, is a complex production system, producing and synthesizing macromolecules of proteins, lipids, and enzymes, among others; it consists of about 10^5 macromolecules on the average. The entire macromolecular population of a given cell is renewed about 10^4 times during its lifetime. Throughout this staggering turnover of matter, the cell maintains its distinctiveness, cohesiveness, and relative autonomy. It produces myriads of components, yet it does not produce only something else—*it produces itself.* A cell maintains its identity and distinctiveness even though it incorporates at least 10^9 different constitutive molecules during its life span. This maintenance of unity and wholeness, while the components themselves are being continuously or periodically disassembled and rebuilt, created and decimated, produced and consumed, is called "autopoiesis."

1.2 Basic Concepts

In order to achieve a more precise characterization of autopoiesis, only intuitively introduced above, it is necessary to define some of the basic concepts that enter into its definition.

Unity. An entity distinguished from its background by the observer, either as a whole, without referring to its components (simple unity), or through identifying its components (composite unity).

Production process. Any process of synthesis, transformation, or destruction realized in the space of components. For example, disintegration of macrocomponents "produces" substrate on which other processes of production can "act."

Organization. A complex of relationships among components and component-producing processes that must remain invariant in order to constitute a unity distinguishable within its identity class.

Structure. A particular spatiotemporal arrangement of particular components through which the underlying organization is realized in a given space and at a given point in time.

Closed organization. A particular (circular) organization of processes that recursively depend on each other for their maintenance and realization; they form a recursive closure. Compare with linear or serial chains or "trees" of processes, forming an open organization.

Autonomous unity. A unity distinguished (or described) as a composite unity integrating its components. A nonautonomous (controlled) unity is then distinguished as a simple unity (i.e., as a component of a larger system within which it operates). The autonomy or control mode of description depends on the cognitive preference of the observer.

REMARK. It is fair to note Varela's *closure thesis* (Varela 1979) at this point: "Every autonomous system is organizationally closed." In our view, autonomy being a derived and observer-dependent notion, it is organizational closure that is sufficient, but not necessary, for perceived autonomy. Thus, *every organizationally closed system is autonomous* would be more in line with our current exposition. Equating autonomy with organizational closure (or autopoiesis) leads to mixing two non-intersecting domains of discourse and would make the concept of autonomy redundant. □

System. A composite unity characterized by its organization *and* structure. Referring to its organization identifies only the class of unities to which a system belongs; describing its structure identifies its concrete space of components. Neither mode is sufficient for fully describing a system.

Topological boundary. The part of a system's structure that allows the observer to identify it as a unity.

Many concepts, for example, existence, purpose, observer, and reproduction, need to be adequately defined. The reader is referred to Maturana's paper in this volume and to references at the end of this chapter (Varela 1979; Maturana and Varela 1980) for a more complete treatment. It is one of the contributions of autopoiesis that these concepts are now being more precisely defined and their prevailing *metaphorical* usage recognized.

1.3 Definition

We are ready to define an *autopoietic system:*

Autopoietic system. A unity realized through a closed organization of production processes such that (a) the same organization of processes is generated through the interaction of their own products (components), and (b) a topological boundary emerges as a result of the same constitutive processes.

In this definition, the organization of components and component-producing processes is maintained invariant through the interactions and flux of components. This invariance follows from the definition: should the organization change we would have a change in system's identity class. What changes is the system's structure and its parts, as, for example, a topological boundary in response to the perturbations in the environment of the system's autopoiesis.

The type of self-producing and self-maintaining closed organization described above is referred to as *autopoietic organization.* The nature of the components themselves and their spatiotemporal relations are secondary and refer to the structure of the system. Consequently, the system's topological boundary is a structural manifestation of the underlying organization, subject to change and compensatory adaptations. It does not constitute the system's organization; it represents its structural manifestation under certain conditions in a particular field of components. This does *not* mean that the topological boundary (and structure) is not conducive to or even necessary, through creating a favorable environment of components, for the maintenance of an autopoietic organization. Both organization and structure are mutually interdependent.

REMARK. The reader should be aware that the definition and interpretation of autopoiesis presented above differs from those of Maturana and Varela (see the papers in this volume and the reference list at the end of this chapter). Namely, "production" is defined in more general terms, "autonomy" is treated as being fully observer-dependent, and "topological boundary" is recognized as pertaining to structure. We cannot engage in an extended discussion here, but the works of others may be consulted (Zeleny 1978, 1980). □

An open organization of components and component-producing processes (linear, treelike, or other noncyclical concatenations) leads to allopoiesis; that is, the organization is not recursively generated through the interactions of its own products. In this sense, the system is not *self-producing*; it produces something other than "itself." This particular (allopoietic) concatenation of processes is capable only of production, not self-production. Allopoietic organizations are still invariant and can be spontaneously concatenated (under favorable conditions).

A particular concatenation of production processes can be assembled by humans through a purposeful design. We then speak of *heteropoiesis*. Man-made machines and contrivances, and their own productions as well, are heteropoietic—they are produced by another system. A machine, for example, is characterized by an organization of components produced by other processes (a person or another machine), and of processes of production whose products do not constitute the machine itself. So far, all heteropoietic systems are allopoietic (i.e., nonliving).

1.4 Computer Model of Autopoiesis

The best understanding of the previously defined concepts can be achieved by contemplating computer simulation outcomes of modeling the simplest autopoietic organization of production processes.

Consider Figure 1, which summarizes the essential building blocks of

<u>Components</u>

  * <u>catalyst(s)</u>, represented by an asterisk

 ( ) <u>holes</u>, represented by a blank space

  o <u>substrate</u>, represented by a small circle

  ▯ <u>free link</u>, represented by an empty square

 −▯ <u>singly-bonded link</u>, represented by a square with
        a quota inside and a single dash

 −▣− <u>fully-bonded link</u>, represented by a square with an
        APL division sign inside and two dashes

<u>Organization of Components</u>

PRODUCTION (P): * + o + o→▯ + * + ( )

> A catalyst and two units of substrate produce one free link
> and a hole as a byproduct. The catalyst is neither affected,
> nor does it change its position. This stipulation can be re-
> laxed. Production can take place only when the two units of
> substrate are within a predetermined neighborhood of the
> catalyst.

BONDING (B): ▯ + ▯→▯-▯   -▣-▯ + ▯→ -▣-▣+▯

> Two free links can be bonded together to start a chain. One
> free link can be bonded with an existing chain of bonded
> links, making the chain longer.

```
        -▣-▣              ▯      -▣-▣            ▯
           \        +     |  →      \            |
          ▣-▯            ▯-▣       ▣-▣-▣
```

> Two existing chains of bonded links can be bonded into
> a longer chain.

                    ▯ + ( )→ o + o

DISINTEGRATION (D): -▣-▣-▯ + ( )→ -▣-▯ + o + o

> Any link, free or bonded, can disintegrate into two units
> of substrate provided there is a hole (space) available in
> the immediate neighborhood of such link so that the addition-
> al unit of substrate can occupy the space.

Figure 1. APL-Autopoiesis: A Model of the Cell

the so-called APL-AUTOPOIESIS model. The three processes—production (P), bonding (B), and distintegration (D)—are postulated as being
feasible because of the specific properties of components moving randomly over a two-dimensional tesselation grid. These processes, in turn,
determine the interactions that take place in the space of components.

  The processes can be organized in a number of ways, acting independently, in series, and so on. Consider, for example, production acting
alone: all substrate would be simply transformed into links and the process
of production would cease. Or, we can concatenate production and bonding: all substrate is transformed into links, links subsequently bond, form-

ing a linearly growing structure(s), a crystal. The underlying organization is open, the system allopoietic.

The reader can continue with these "mental exercises." For example, consider the environment of *links* and concatenate the processes of bonding and disintegration: all the links are ultimately transformed into the substrate's environment, with some intermittent bonded chains emerging and disappearing in the process. In neither of these cases were the terminal products necessary for fueling the processes themselves—there was no organizational closure.

Consider the simplest organizational closure: production and disintegration. Each process, in order to maintain itself, is dependent on the products of the other process. The processes are interdependent and concatenated in a circular manner. Depending on the relative rates of both processes, the result would be a mixture of substrate and links oscillating around some equilibrium proportion.

All three processes, concatenated in a circular fashion, are necessary for autopoiesis. In Figure 1 such a closed organization is implied. Putting aside the notion of origin and examining an ongoing system, observe that disintegration "produces" the substrate necessary for production, production "produces" the catalyst necessary for itself and the links necessary for bonding, and bonding "produces" the stuff necessary for disintegration.

Yet, the circular organization of processes is still not sufficient. The rates of "production" must be in harmony. A too "vigorous" rate of disintegration would never allow bonded chains to form a viable topological boundary; a too low disintegration rate would lead to "crystallization," and so forth. In Figure 2 we present "snapshots" from a particular history of a balanced autopoietic organization. The topological boundary emerges at about TIME 40 (fourth frame), and it allows the unity to maintain its autopoiesis indefinitely. In this particular run we allow the rates to get out of balance and the unity gradually loses its autopoiesis and ultimately disappears.

A series of interesting experiments can be performed with the APL-AUTOPOIESIS program. In Figure 3 observe how the same autopoietic organization manifests itself in a different structure, in response to simulated environmental perturbation. In Figure 4 observe that the same organization (no additional complexity interjected into the rules of interaction is needed) is capable of self-reproduction if a component (second catalyst) is "placed" within the original enclosure. Autopoiesis itself, under favorable conditions, begets self-reproduction, heredity (similar structure and components), and ultimately evolution.

It is not our intention to describe all the variety and details pertaining to the APL-AUTOPOIESIS model. The program is commercially available and interested readers can duplicate and creatively extend all of the

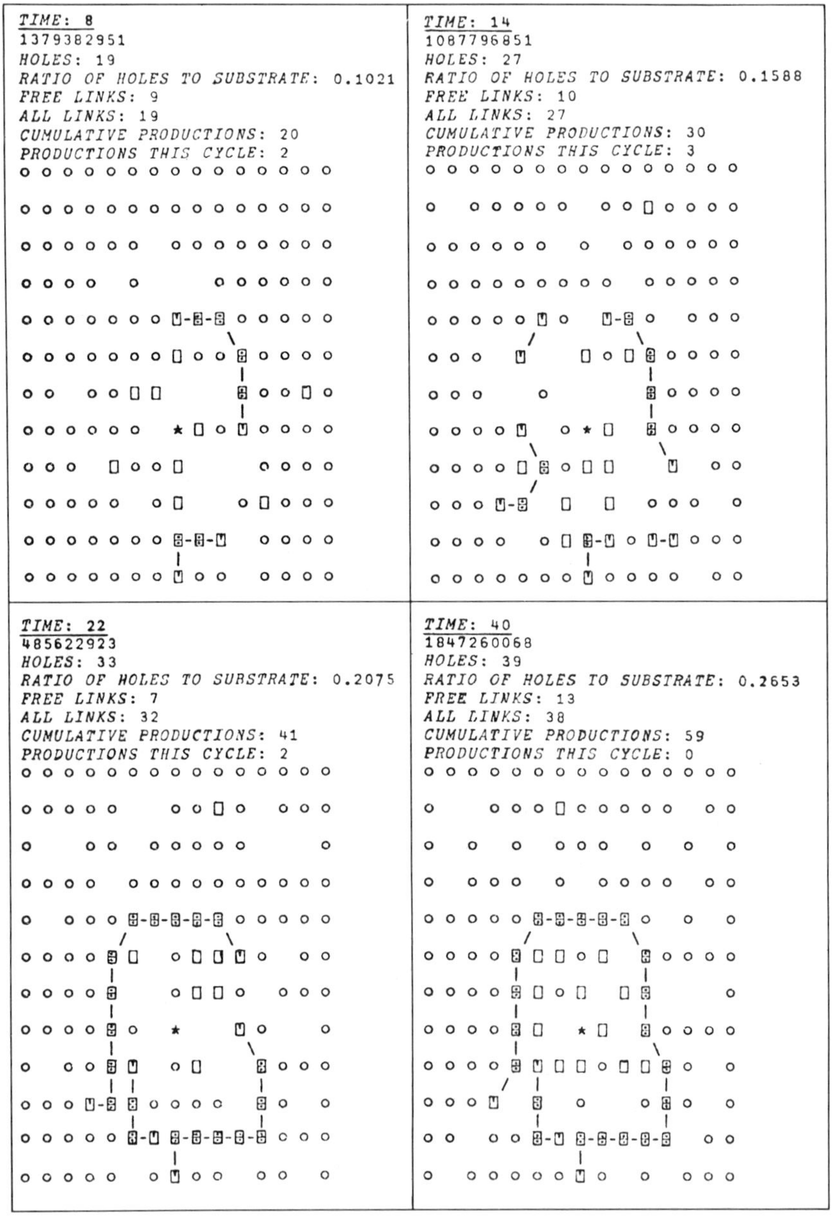

Figure 2

```
TIME: 46                                        TIME: 50
1403767852                                      1917167081
HOLES: 37                                       HOLES: 34
RATIO OF HOLES TO SUBSTRATE: 0.2450             RATIO OF HOLES TO SUBSTRATE: 0.2165
FREE LINKS: 13                                  FREE LINKS: 9
ALL LINKS: 36                                   ALL LINKS: 33
CUMULATIVE PRODUCTIONS: 59                      CUMULATIVE PRODUCTIONS: 60
PRODUCTIONS THIS CYCLE: 0                        PRODUCTIONS THIS CYCLE: 0
o o o o o o o o o o o o o o o                   o o o o o o o o o o o o o o o

o    o o o o o   o o o o o o                    o    o o o o o   o o o o o o

o    o o o o o o o o o o o o                    o    o o o o o o   o o   o o

o    o o o o o   ▯ o o o     o                  o o o o o o o o o ▯ o o o   o
                                                                  |
o    o o o ▣-▣-▣-▣-▣ o o o   o                  o   o   o ▣-▣-▣-▣-▣ o o o o
          /           \                                /
o o o o ▣ ▯ ▯ ▯ ▯ ▣   o o                      o o o ▣ o ▯ ▯   o o   o o o
        |             |                               |
o o o o ▣ o o       ▣   o o                    o   o o ▣ ▯ o     ▯ ▯ o o   o
        |             |                               |
o o   o ▣ ▯ ▯ ★ ▯ ▯ ▣ o   o                    o o o o ▣ ▯ o ★ o     ▣ o o o o
        |             \                                |                \
o   o   ▣ ▯ ▯ o   ▯   ▯ o o o                  o o   o ▣ ▯   ▯ o ▯ ▯ ▯ o o o
      / |                                            / |
o o o ▯ o ▣ o ▯ o   o o   o o                  o o o ▯   ▣   o   ▯ o o o   o
      |                                              |
o o   o o ▯ o ▣-▣-▣-▣-▯   o                    o o o o o ▣-▯ ▣-▣-▣-▣-▯   o o
          |                                                  |
o   o o o o o ▯   o o   o   o                  o o o o   o o ▯   o o o o   o


TIME: 64                                        TIME: 70
1877143764                                      385435594
HOLES: 30                                       HOLES: 28
RATIO OF HOLES TO SUBSTRATE: 0.1818             RATIO OF HOLES TO SUBSTRATE: 0.1656
FREE LINKS: 8                                   FREE LINKS: 7
ALL LINKS: 29                                   ALL LINKS: 27
CUMULATIVE PRODUCTIONS: 65                      CUMULATIVE PRODUCTIONS: 67
PRODUCTIONS THIS CYCLE: 0                        PRODUCTIONS THIS CYCLE: 0
o o o o o o o o o o o o o o o                   o o o o o o o o o o o o o o o

o o     o     o o o o   o o o                  o     o o o   o o o o o o o

o   o o o o o o o o o o o o o o                 o   o   o o o   o o o o o o

o o     o o ▯-▣-▣-▣-▣ ▯ o o o o                o   o o o ▯-▯ o ▯-▣-▣ o o   o
                                                         |
o o o o   o o   o o o o o o o                  o o o o o   o o o o o o o
                                                         |
o o o o ▯   ▯   ▯ o ▯   o   o                  o   o o ▯ o o ▯ ▯   ▯   o o
        |       |                                      |
o o   o ▣ o o ★ ▯ ▯ ▣   o o o                  o o o o ▣ o o ★ o o o o o o o
        |       |                                      |
o   o o ▣ ▯ o       ▯ o o o                    o   o o ▣ ▯   ▯ o o ▯   o
      / |                                            / |               |
o o o ▯ o ▣ ▯ ▯ o o       o o                  o o o ▯ o ▣ ▯ o   ▯ ▯ o o o
      |                                              |
o o o   o ▣-▣-▣ ▯-▯ o o   o o                  o o o   ▣-▣-▣ ▯-▯ o o o o
      |                                                      |
o o o o o o o ▯ o o o   o   o                  o o o o o o   ▯ o o   o o ▯ o
```

Figure 2 (*continued*)

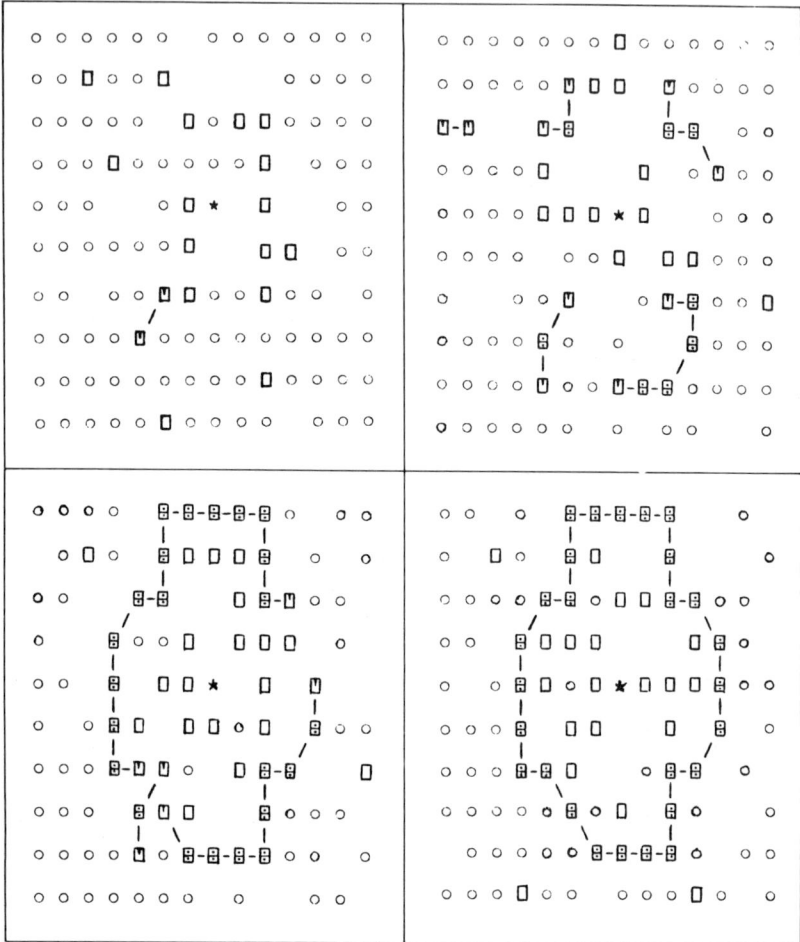

Figure 3

experiments described.[1] For more complex "nesting" of rules and for a full mathematical description of the model the reader should consult Zeleny (1977).

---

[1] APL-AUTOPOIESIS interactive modeling package and User's Manual are available from Computing and Systems Consultants, Inc., P.O. Box 1551, Binghamton, N.Y. 13902. This package is fully equipped with standard special purpose subprograms and allows users to create and implement their own programs and experimental variations through an unlocked APL control program AUTO. Inquire also about a video record of autopoietic simulation, on Sony videocassette, U-matic, KC-30.

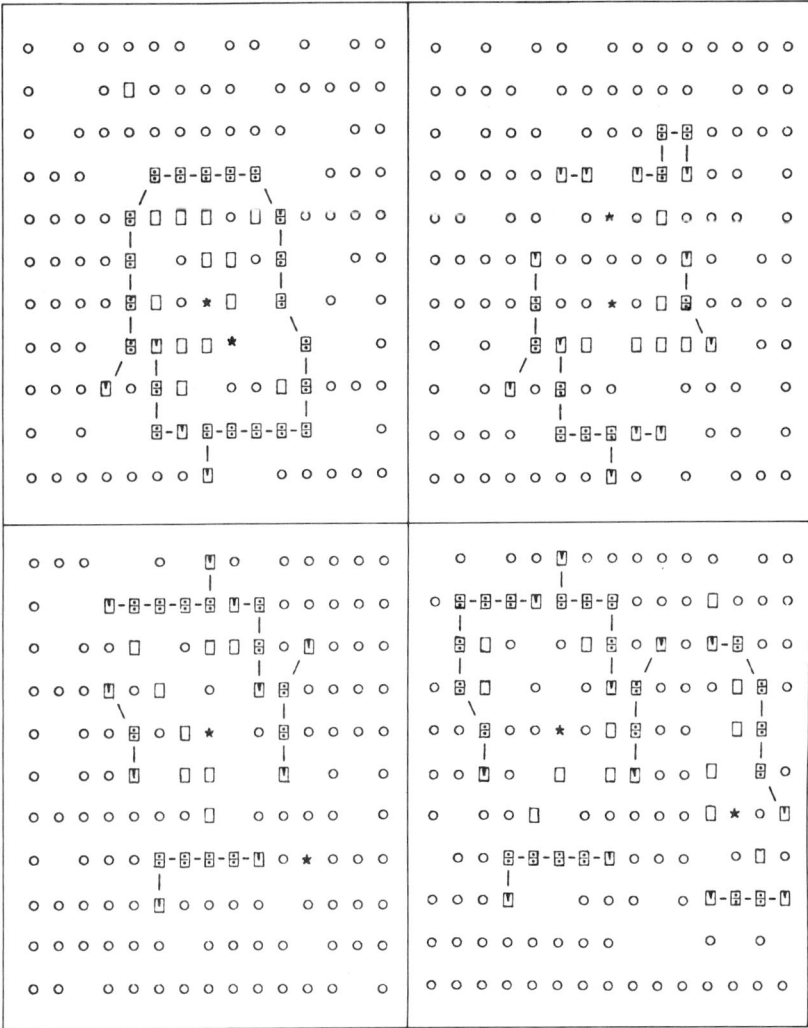

Figure 4

One more note. We mentioned the random movement of components. By contemplating the development of the space of autopoiesis in Figure 2, observe that the space becomes progressively restructured, "deformed," and the processes, although controlled through a random number generator, do not take place in a random fashion. They become constrained in space and time, they become "canalized" and predictable from a macroscopic viewpoint although not microscopically. The remarkable structural coupling of the unity and its environment, mutually affecting each other and adapting to each other, supports most beautifully our intuitive (though not scientific-rational) understanding of living systems.

1.5 Summary of Implications

The implications and conjectures that can be derived from the presented framework of autopoiesis are neither complete nor consensual, as witnessed by the remarkable variety of views and interpretations presented in this volume. This introduction presents only one view, intended as a template against which all other views can be tested, understood, and re-created.

Autopoietic systems are characterized by a closed organization of production processes and by an internally produced topological boundary.

Organizational closure is necessary but not itself sufficient for autopoiesis. The closure of processes must be "balanced" so that a topological boundary can be formed.

A topological boundary belongs to the domain of structural adaptations and is independent of the underlying organization. It is necessary, under certain conditions, for the maintenance of balanced organizational closure.

Organizational closures that do not form a topological boundary, although they might be self-producing in a broader sense, are not recognizable by the observer and not classified as autopoietic at this stage.

Structure is not simply emergent from the functioning of an autopoietic organization, but it affects the maintenance and viability of such an organization in a given environment.

Organizationally closed processes defined in a different space of components (e.g., social systems) could be autopoietic as they define their topological boundaries in the same domain. Such boundaries would of course be different from physical membranes, compartments, and the like. Their recognition requires different sets of criteria appropriate to the given domain.

In the domain of social interactions, the close coexistence and interspersion of autopoietic, allopoietic, and heteropoietic organizations makes their identification and separation more difficult than in a purely physical space.

Autonomy is an observer-dependent concept. All autopoietic systems can be viewed as being controlled within the larger system they constitute.

All living systems are autopoietic. All autopoietic systems are organizationally closed. The reverse implications cannot be asserted at this stage.

References

In addition to these references, directly related to the preceding remarks, an annotated bibliography of autopoiesis as of 1980 is appended.

Maturana, H. R., and Varela, F. J., (1980), *Autopoiesis and Cognition: The Realization of the Living,* Boston Studies in the Philosophy of Science, Vol. 42. Reidel, Boston.

Varela, F. J., (1979), *Principles of Biological Autonomy,* Elsevier, New York, pp. 55–60.

Zeleny, M. (1977), Self-Organization of Living Systems: A Formal Model of Autopoiesis, *International Journal of General Systems* 4 (1), 13–28.

Zeleny, M. (1978), APL-AUTOPOIESIS: Experiments in Self-Organization of Complexity, in *Progress in Cybernetics and Systems Research,* Vol. 3, edited by R. Trappl *et al.,* Hemisphere, Washington, D.C., pp. 65–84.

Zeleny, M., ed. (1980), *Autopoiesis, Dissipative Structures and Spontaneous Social Orders,* Westview, Boulder, Colo.

Autopoiesis 1980: An Annotated Bibliography

In addition to the papers this volume comprises, a sizable bibliography of works on autopoiesis has emerged over the past decade. Although the concept of autopoiesis has many intellectual precursors, here we list only the works related *directly* to its modern embodiment, thus employing the very term "autopoiesis" as a discursive vital core. The order of the listing is roughly chronological.

Maturana, H. R. (1970), The Neurophysiology of Cognition. in *Cognition: A Multiple View,* edited by P. Garvin, Spartan Books, New York, pp. 3–23. Possibly the first time autopoiesis is actually proposed (although the term itself is not used yet). The nervous system is viewed as a closed, state-determined system for the first time. In the same year, Maturana (The Biology of Cognition, Biological Computer Laboratory, BCL Report 9.0, University of Illinois, Urbana, 1970) discusses the observer-dependent operation of distinction.

Maturana, H. R., and Varela, F. J. (1973), *De Máquinas y Seros Vivos,* Editorial Universitaria, Santiago. The first, and now largely inaccessible, exposition of autopoiesis in Spanish. Work on this manuscript started in 1970–1971 through a close interface of both authors.

Varela, F. J., Maturana, H. R., and Uribe, R. B. (1974), Autopoiesis: The Organization of Living Systems, Its Characterization and a Model, *Biosystems* 5(4), 187–196. This is the first exposure of the principles of autopoiesis in English; it also contains the first computer tesselation automaton of autopoiesis. This paper was probably the most influential in communicating autopoiesis to Western science.

Maturana, H. R. (1975), The Organization of the Living: A Theory of the Living Organization, *International Journal of Man-Machine Studies* 7(3), 313–332. A rather philosophical treatment of autopoiesis, rich in new insights, pure and undiluted Maturana. Discussions of the nervous system, structural coupling, ontogeny and evolution, self-reproduction, and linguistic domains.

Maturana, H. R., and Varela, F. J. (1975), Autopoietic Systems, Biological Computer Laboratory, BCL Report 9.4, University of Illinois. Urbana. The first English translation of the original Maturana–Varela monograph (1973). Its subtitle: "A Characterization of the Living Organization." It also contains the famous Preface by Stafford Beer.

Beer, S. (1973), Preface to Autopoietic Systems, in Maturana and Varela (1975), pp. 1–16. Beer states, What I am now sure about is that they are right. Nature is not about codes: we observers invent the codes in order to codify what nature is about. These discoveries are very profound." And elsewhere, "yes, human societies *are* biological systems."

Zeleny, M. and Pierre, N. A. (1975), Simulation Models of Autopoietic Systems, *Proceedings of the 1975 Computer Simulation Conference,* Simulation Councils, Inc., La Jolla, Calif., pp. 831–842. Mathematical formalization of a computer model of autopoiesis, exploring its relationship to Conway's "Game of Life," and introducing the notions of bioalgebra and social autopoiesis.

Zeleny, M. and Pierre, N. A. (1976), Simulation of Self-Renewing Systems, in *Evolution and Consciousness,* edited by E. Jantsch and C. H. Waddington, Addison-Wesley, Reading, Mass., pp. 150–165. Popular and simple exposition, including the principles of human systems management.

Zeleny, M. (1977), Self-Organization of Living Systems: A Formal Model of Autopoiesis, *International Journal of General Systems* 4 (1), 13–28. Complete formalization of the APL simulation model of autopoiesis. Also, the questions of the origin of life and social autopoiesis.

Zeleny, M. (1977), Organization as an Organism, *Proceedings of the SGSR Annual Meeting,* Denver, Colo., SGSR, Washington, D.C., pp. 262–270. Basic notions of social autopoiesis and an institutional example.

Zeleny, M. (1978), APL-AUTOPOIESIS: Experiments in Self-Organization of Complexity, in *Progress in Cybernetics and Systems Research, Vol. 3, edited by R. Trappl et al.,* Hemisphere, Washington, D.C., pp. 65–84. Includes detailed descriptions of autopoiesis simulation experiments: growth, circadian rhythms, morphogenesis, mitosis, etc. Full APL-AUTOPOIESIS modeling package and user's manual is now available from Computing and Systems Consultants, Inc., P.O. Box 1551, Binghamton, N.Y. 13902.

Varela, F. J. and Goguen, J. A. (1978), The Arithmetic of Closure, in *Progress in Cybernetics and Systems Research,* Vol. 3, edited by R. Trappl et al., Hemisphere, Washington, D.C., p. 125. Complementing the concept of autopoietic organization by a broader notion of organizational closure. Reprinted in *Journal of Cybernetics* 8(3,4) (1978), 1–34.

Varela, F. J. (1979), *Principles of Biological Autonomy,* Elsevier, New York. Varela's summary and interpretation of autopoiesis, organizational closure, and the calculus of indications.

Faucheux, C., and Makridakis, S. (1979), Automation or Autonomy in Organizational Design, *International Journal of General Systems* 5(4), 213–220.

Andrew, A. M. (1979), Autopoiesis and Self-Organization, *Journal of Cybernetics* 9(4), 359–368.

Maturana, H. R. and Varela, F. J. (1980), *Autopoiesis and Cognition: The Realization of the Living*, Boston Studies in the Philosophy of Science, Vol. 42, Reidel, Boston. First and revised commercial publication of the original Maturana-Varela monograph (1973 and 1975), including S. Beer's preface.

Zeleny, M., ed. (1980), *Autopoiesis, Dissipative Structures and Spontaneous Social Orders*, AAAS Selected Symposium 55, Westview, Boulder, Colo. Includes contributions of Boulding, Maturana, Jantsch, Düchting, Allen, Gierer, and Zeleny. It covers a wide range of methodological approaches and strives for their transdisciplinary unification.

Boulding, K. E., Foreword, in Zeleny (1980), pp. xvi–xxi. Boulding characterizes autopoiesis's implication that chaos is unstable as somewhat shocking for a generation raised on the concept of entropy. Yet, he foresees that "the idea is likely to have a considerable impact on a good many disciplines, from thermodynamics to sociology."

Jantsch, E. (1980), The Unifying Paradigm behind Autopoiesis, Dissipative Structures, Hyper- and Ultracycles. in Zeleny (1980), pp. 81–87. An attempt at unification of a group of "new paradigms" employed in the study of self-organizing and spontaneously generated phenomena in physical, biological, and social systems.

Maturana, H. R. (1980), Autopoiesis: Reproduction, Heredity and Evolution, in Zeleny (1980), pp. 45–79. Maturana's most recent and undiluted views on the phenomenology arising from the assumption of autopoiesis. A challenging presentation: "I claim that nucleic acids do not determine hereditary and genetic phenomena in living systems."

Zeleny, M. (1980), Autopoiesis: A Paradigm Lost? in Zeleny (1980), pp. 3–43. A long-needed attempt at a historical embedding of autopoiesis, tracing the thoughts of Trentowski, Bernard, Leduc, Menger, Bogdanov, Smuts, von Hayek, Weiss, and others.

Zeleny, M., ed. (1981), *Cybernetics Forum*, Special Issue on Autopoiesis; contains reprints of articles by Varela, Maturana and Uribe, and Maturana, and Zeleny.

Zeleny, M. (1981), Autopoiesis Today, in Zeleny (1981). Review of the current state of autopoiesis research models, literature, and interest groups.

Zeleny, M. (forthcoming), *Autopoiesis of Life: Rebellious Thoughts*.

INTRODUCTORY REMARKS

Chapters 2 and 3 have in common our intention of reviewing and clarifying the notion of autopoiesis as we see it today. In this attempt we both have reached essentially the same central conclusions, yet we differ in our way of expressing them. These introductory remarks point to the central agreement of our views and to the difference in the way each has chosen to express them.

What most requires further clarification is the relation between the mechanisms that define the identity of an individual and the phenomenology that this entity, once constituted, can generate in its interactions. Individuality and the phenomenology of unities in their interactions are distinct domains that should not be confused. In fact, these domains are nonintersecting to the extent that one cannot be reduced to or deduced from the other. To consider either, the observer has to effect a change in perspective. However, the observer also has the capacity to move to a metalevel where both phenomenologies are considered simultaneously. This gives the tempting impression that the mechanism of individuality can be *identified* with the ontogeny and phylogeny of a unity and that the components of the unity can be used as explanatory terms for the unity's interactions. An example of this sort of epistemological confusion is the adscription of an "informational" quality to a structural component, a hormone molecule, for example. The "informational" quality of a hormone actually "means" that the observer can follow certain regularities in the behavior of the unity (the cell), considered as a simple entity, when it interacts with the hormone; the molecular quality of the hormone "means" that we can consider the changes that it effects in the components and the structure of the cell.

Autopoiesis is the mechanism of identity characteristic of living systems. From what we said above, it follows that the characterization of a living system and the biological phenomenology are distinct, nonreducible domains. The former is the necessary condition for biological phenomena to occur at all (the existence of unities in some space). Yet with autopoiesis alone we cannot characterize living systems as they appear. Their phenomenology depends on the interactions of unities in the media in which they exist; it depends on the kinds of structural couplings living systems undergo and on the constraints on these couplings. These contingent factors are what biologists observe in changing to a metadomain where the unity is treated as a simple entity endowed with properties that are observed through time and capable of generating the familiar events of evolution and development.

Both of us point in our respective chapters to the need to separate clearly the two phenomenological domains. There are at least two ways of dealing with this subtle epistemological situation, and we have each chosen a different one. We are not quite sure which is the best route to take; by and large, it is quite beyond us to decide. It is at least possible, however, to point out where the problem is and what the alternatives are.

Humberto R. Maturana
Francisco J. Varela

Maturana's paper is a good example of highly original and independent thought of a deeply reflective experimental biologist. Even the style of his presentation is unorthodox and unique, often disturbing, sometimes difficult to follow. Maturana does not compromise; he is confident enough to let his writing (structure) reflect his thinking (organization). He does not polish the structure.

To appreciate Maturana's paper, one should adopt a rather elaborate strategy. First, one must know some basic notions of autopoiesis prior to reading, preferably from other sources. Second, after reading sections 2.1–2.3.1 lightly, without attempting an immediate and complete understanding, one should *study* carefully the basic notions of Section 2.3.1 (the definitions of unity, organization, structure, space, etc.). *Then* one should turn back and read the paper anew, this time proceeding also to Sections 2.3.2 and 2.3.3. These sections deal with the biological phenomena proper: reproduction, heredity, and adaptation. Sections 2.1 and 2.2 present a summary and conclusions; reading them without preparation could be inefficient.

A critic might be tempted to pick a statement here and there, a sentence or a paragraph out of context, and point to its imprecision, vagueness, or even linguistic deficiency. That would not be appropriate. Maturana's strength is in relating the components of thought in novel and exciting ways, not in refining and polishing the components themselves. This paper should not be read in a "reviewer's fashion" (i.e., linearly, with a pencil in hand). This is writing that demands reflection on itself as a whole, not an analysis of its components.

Maturana maintains that autopoiesis implies autonomy (not vice versa) and, more problematically, that autopoiesis implies life (although the reverse implication seems to be self-evident). His major thesis is that the *organization* of an autopoietic system is *invariant;* that is, should the system undergo a change of its organization, it would become something else, some other system. In this sense, because of the organizational invariance, the system cannot be "self-organizing."

A more relaxed notion would admit the organizational invariance *within certain limits,* possibly assuming a series of interdependent, albeit different, organizations as being capable of characterizing a unity through their interrelatedness. What would remain invariant is their autopoiesis. Also, the question of organizational redundancy and memory should be raised.

In paragraph 5 of his Implications (p. 22), Maturana asserts that there are no restrictions on the space in which autopoiesis may take place. Only the systems exhibiting autopoiesis in physical space are called "living." This is an important statement not only because it suggests the possibility of contemplating autopoiesis in other than physical spaces (such as language, learning, or social interactions), but also because it raises the question of further specification of "physical" space (organic, inorganic, Earth, Mars, etc.). The implication is that there could be autopoietic systems, even in physical space, that would not be conventionally designated as "living."

It appears that Maturana carefully and elegantly resolved the autonomy–control dilemma through his concept of simple and composite unities. A simple unity exists in the space in which it can be distinguished by an observer; it "defines" its own boundaries of existence. A composite unity exists in the space defined by its components. Because the observer can exercise a discretion in viewing a system either as a simple or as a composite unity, he can also regard the system as being either controlled (with inputs and outputs) or autonomous, depending on the purposes of his inquiry.

Humberto R. Maturana was born in Santiago, Chile, in 1928. He studied medicine in Chile and then anatomy at the University College in London under J. Z. Young. Dr. Maturana's research interests have been in neuroanatomy and neurophysiology. At M.I.T., in cooperation with Lettvin, McCulloch and Pitts, he was involved in experimental work on neurophysiology of perception in the frog's eye, exploring the functional relationship between the perceptual sorting in the retina and the anatomy of the retinal cells. While teaching biology at the Medical School in Santiago, he developed the theory of the closed organization of living systems. This was first formulated as the theory of "autopoiesis" in collaboration with Francisco Varela in 1973. Professor Maturana's recent concerns are mostly with problems of cognition. *Address:* Facultad de Ciencias, Departamento de Biología, Universidad de Chile, Casilla 653, Santiago, Chile.

Chapter 2
Autopoiesis

Humberto R. Maturana

A system is autonomous if the relations that characterize it as a unity involve only the system itself, and not other systems. Thus defined, autonomy can be viewed as a central characteristic of living systems. Yet, since autonomy is not necessarily a feature exclusive to living systems, any attempt to explain the organization of living systems must show how they are autonomous and how all the phenomena proper to them arise as a result of their autonomy. It is in this context that I maintain that the notion of autopoiesis fully characterizes living systems as autonomous entities in physical space. It is also in this context that I maintain that the notion of reproduction does not enter into the characterization of living systems as unities and that the phenomenon of sequential reproduction by cellular division (or any other kind of partition into equivalent unities) is necessary only for the phenomenon of phylogenic evolution.

F. Varela and I have published these notions in a small book and in several articles (Maturana and Varela 1973; Varela, Maturana and Uribe 1974; Maturana 1975; Varela 1978). I now wish to review them and clarify some aspects of the theory of autopoiesis as I see it today.

2.1 Autopoiesis ($\alpha\upsilon\tau\acute{o}\sigma$ = self; $\pi o\iota\epsilon\nu\iota\nu$ = production)

We maintain that there are systems that are defined as unities as networks of productions of components that (1) recursively, through their interactions, generate and realize the network that produces them; and (2) constitute, in the space in which they exist, the boundaries of this network as components that participate in the realization of the network. Such systems we have called autopoietic systems, and the organization that defines them as unities in the space of their components, the autopoietic organization.

We also maintain that an autopoietic system in physical space (i.e., an autopoietic system whose components we define as physical, such as molecules) is a living system, and, therefore, that a living system is an autopoietic system in physical space.

2.2   Implications of Autopoiesis

The autopoietic organization of an autopoietic system is necessarily an invariant. This is obvious, of course, because these systems are defined and realized as unities by being autopoietic. What is not immediately obvious, however, is that all that happens to them must happen while they are autopoietic and through their being autopoietic; otherwise they disintegrate. Autopoiesis, therefore, results in the stabilization of auto-poiesis through its operation as the configuration of relations that must remain invariant through the history of change of an autopoietic system.

An autopoietic system is, from the point of view of its dynamics of states, a system that, while autopoietic, only generates states in auto-poiesis; that is, with respect to its states, an autopoietic system is a closed system that only generates one kind of states—states in autopoiesis. Obviously, this is a reformulation of the previous point. However, it is nec-essary to restate it in these terms because the notion of closure is essential for the understanding of the operation of living systems as systems.

Nothing is said in the characterization of living systems as autopoietic systems about the operational constraints under which their autopoiesis must be realized. This is because whatever constraints must be satisfied, they are determined by the properties of the components, and they are implied when it is said that an autopoietic system exists in the space in which its components exist. Thus, autopoietic systems in the physical space must satisfy thermodynamics and must be materially and energet-ically open, even though they are necessarily closed in their dynamics of states.

The notion of autopoiesis also says nothing about the nature of the components that realize the system as a network of productions. In fact, the components of an autopoietic system can vary infinitely so long as they have the properties that permit them to constitute it in the space that they define. Furthermore, the components of an autopoietic system are specified by its autopoietic organization, which determines which prop-erties must have the entities that, as components, realize it in a given space; an observer cannot identify the components independently of the autopoietic system that they integrate.

There is no restriction on the space in which an autopoietic system may exist. The physical space in which living systems exist is only one of many. In fact, living systems exist in the physical space as the space defined by their components. Accordingly, we have chosen to identify living systems with only autopoietic systems in the physical space because this is the space in which we exist, and because for that reason this space constitutes for us a peculiar limiting cognitive space. Otherwise, the properties of autopoietic systems as autopoietic systems must be isomorphic in every space.

An autopoietic system exists as a system in the space of its components, but as a unity it defines a space through its operation as a whole.

The unity of an autopoietic system is the result of the neighborhood relations and interactions (interplay of the properties) of its components, and in no way the result of relations or interactions that imply the whole that they produce. In other words, nothing takes place in the operation of the autopoietic network with reference to the unity of the network. Therefore, notions of regulation and control that may be used to describe what may take place in an autopoietic network, pertain to the metadomain in which the observer describes the system as a whole; they do not characterize the interactions of its components.

Reproduction is not a constitutive feature of autopoietic systems in general, nor of living systems in particular. Reproduction is secondary to the constitution of the unity to be reproduced; therefore, reproduction does not enter into the characterization of living systems.

2.3 Comments

My comments will center on the following:

1. The basic notions of unity, organization, structure, space, and the domains of perturbations.
2. The biological phenomena of reproduction, heredity, and adaptation.
3. The distinction from other systems.

2.3.1 Basic Notions

In what follows, the word "entity" refers in the most general manner, and without further qualifications, to anything that may be distinguished.

Unity. The basic cognitive operation that we perform as observers is the operation of distinction. By means of this operation we define a unity as an entity distinct from a background, characterize both unity and background by the properties with which this operation endows them, and define their separability. A unity thus defined is a simple unity that specifies through its properties the space in which it exists and the phenomenic domain that it may generate through its interactions with other unities.

If we recursively apply the operation of distinction to a unity, so that we distinguish its components, we redefine it as a composite unity that exists in the space that its components define; it is through the properties of its components that we observers can distinguish it. Yet we can always treat a composite unity as a simple unity that exists not in the space of its components, but in a space that it defines through the properties that characterize it as a simple unity. In this context, then, if an autopoietic system is treated as a composite unity, it exists in the space defined by its components; but if an autopoietic system is treated as a simple unity, the distinctions that define it as a simple unity characterize its properties as a simple unity and define the space in which it exists as such a simple unity.

Organization and Structure. The relations between components that define a composite unity (system) as a composite unity of a particular class constitute its organization. In this definition of organization the components are viewed only in relation to their participation in the constitution of the unity (whole) that they integrate. For this reason nothing is said about the properties that the components of a particular unity may have, other than those required by the realization of the organization of the unity.

The actual components (all their properties included), together with the actual relations that concretely realize a system as a particular member of the class of composite unities to which it belongs by its organization, constitute its structure. Therefore, the organization of a system, as the set of relations between its components that define it as a system of a particular class, is a subset of the relations included in its structure. It follows that any given organization may be realized through many different structures, and that different subsets of relations included in the structure of a given entity may be abstracted by an observer (or its operational equivalent) as the organizations that define different classes of composite unities.

The organization of a system, then, specifies the class identity of the system, and must remain invariant if the class identity of the system is to remain invariant: if the organization of a system changes, then its identity changes and becomes a unity of another kind. Yet since a particular organization can be realized by systems with otherwise different structures, the identity of a system may stay invariant while its structure changes within the limits determined by this same structure. If these limits are overstepped—that is, if the structure of the system changes so that its organization is destroyed—the system becomes something else, defined by another organization.

It is apparent that only a composite unity has both structure and organization. A simple unity does not; it only has properties that are defined by the operations of distinction through which it becomes separated from a background. It is also apparent that as soon as a composite unity is treated as a simple unity any question about the origin of its properties becomes inadequate because the properties of a simple unity are given through its distinction as a simple unity. Yet it is also apparent that although the properties of a composite unity arise from its organization, they are realized through the properties of its components. Accordingly, while two simple unities interact through the simple interplay of their properties, two composite unities interact in a manner determined by their structure through the interplay of the properties of their components.

Space. Operationally, a simple unity defines its space, that is the domain in which it can be distinguished as a unity. A simple unity, therefore, exists in a space that it defines. A composite unity, however, exists in a space defined by its components because it is through the properties of its components that it can be distinguished as a unity. Yet a composite unity treated as a simple unity defines a space as a simple unity and exists as a simple unity in such a space. According to this, although we have said that living systems are autopoietic systems in the physical space, strictly speaking the physical space is defined as the space in which living systems exist as autopoietic systems and interact as composite unities. Therefore, the physical space is necessarily a limit space for living systems because they cannot undergo interactions that are not mediated through their components, and the components define the physical space.

The specification of a space goes together with the specification of a phenomenic domain. As soon as a unity is defined, a phenomenic domain is defined. Accordingly, if a composite unity operates as a simple unity, it operates in a phenomenic domain that it defines as a simple unity, and that is necessarily different from the phenomenic domain in which its components operate. Therefore, the emergence of a phenomenic domain, as the result of the operational distinction of a composite unity as a simple unity, makes phenomenic reductionism (and, hence, explanatory reductionism) impossible. Furthermore, the dynamics of the establishment of unities through operational distinctions that specify their properties results in all phenomenic domains being necessarily realized through the operation (interplay) of the properties of the unities that generate them; that is, through relations of contiguity. Given that a component A interacts with another component B in such a way that the changes in B through its interactions with C result in the reduction of the production of D, an observer may say, by considering the whole, that A controls the production of D. A, B, C, and D, interact through relations of contiguity, but the relation in which A controls the production of D is not a relation of contiguity in the phenomenic domain defined by the components. Relations such as regulation, control, or function, therefore, are not relations of contiguity, but referential relations specified by the observers putting themselves in a metadomain of descriptions, by using their view of the whole as the reference for their description of the participation of the components that they describe in the constitution of the composite unity.

"Everything said is said by an observer . . ." (Maturana 1970), so everything said is a description in the observer's domain. Yet by defining a unity in this domain of descriptions, the observer specifies a reference description that may constitute the basis for a metadomain of descriptions of descriptions, and can do this recursively. Thus, although a characterization of a system as a composite unity, without reference to the whole in terms of neighborhood relations only, and a functional description in terms of relations between the components and the whole, are both descriptions made by an observer; they are operationally different because they take place in different descriptive domains. The first points to a system that would operate in the described manner if its components existed as described; the second points to how the relations and interactions of the components of such a system would appear to an observer who considers them in relation to the whole that they are observed to constitute. These two descriptions are complementary in the cognitive domain of the observer.

Domains of Perturbations. All that happens to a composite unity, whether in relation to its internally generated dynamics of structural change or in relation to its interactions and the structural changes that these trigger in it, is determined by its structure. Or, in other words, in every instance the structure of a composite unity determines both (1) its domains of structural changes without loss of identity (domain of states) and with loss of identity (domain of disintegrations) and (2) its domains of interactions that trigger its changes of state (domain of perturbations) and that trigger its disintegration (domain of destructive interactions). Or, in still other words, at every instance, the structure of a composite unity specifies which structural configuration of the medium in which it operates may perturb it, and which may trigger its disintegration. This is the case regardless of the composite unity considered; it therefore applies to autopoietic systems in any space. As every biologist knows, this is obviously the case for living systems.

Yet, although the structure of a composite unity specifies which configuration of the structure of the medium may perturb it, the actual perturbations that take place are determined by the structure of the medium. However, since the domain of perturbations of a composite unity such as an autopoietic system may change along its ontogeny (individual history) as a result of the structural changes triggered in it by the perturbations or by its internal dynamics, the actual sequences of perturbations and changes of state that a given composite unity actually undergoes is always a function of both the structure of the unity and the structure of the medium. As a consequence, the sequence of perturbations that a system undergoes selects along its ontogeny a path of structural changes that result in its structural coupling to its medium. If the composite unity is a living system, then this structural coupling appears revealed to an observer as a behavioral complementarity in which the conduct of the system is congruent with the changes of state of the medium in a manner that permits it to continue in its autopoiesis. Of course, if a destructive interaction takes place, then the process is interrupted and the system disintegrates.

2.3.2 Biological Phenomena

We maintain that a given phenomenon is a biological phenomenon only to the extent that it implies the realization of the autopoiesis of at least one living system. In other words, we maintain that all biological phenomena necessarily arise as a result of the autopoietic operation of a living system or of a group of them, and that this is their only peculiarity.

Reproduction. Reproduction takes place whenever a composite unity gives origin to another unity of the same class (same organization) through a process of fragmentation, and not through processes of construction or mapping, as would be involved in the phenomena of production or copy. Therefore, in reproduction the new unity should originate as an operationally independent entity. Defined in this manner, reproduction is a frequent phenomenon in nature. It takes place whenever a composite unity whose organization and the components that realize it are uniformly distributed throughout its expanse, with no component in a single dose or compartmentalized, undergoes a fragmentation that does not exclude any of the necessary components or processes from any of the fragments. This obviously takes place in the fragmentation of a crystal when the plane of fracture separates collections of unit cells. The same thing happens, in principle, in the fragmentation of an autopoietic unity, regardless of how the fragmentation is triggered, if its components are uniformly distributed and are not compartmentalized or in a single dose with respect to the plane of fracture. In contemporary eucaryotic cells the process of mitosis transforms a compartmentalized autopoietic unity that has some of its components exclusively in the nucleus, into a noncompartmentalized unity in which a plane of fracture can separate two unities with equivalent (but not necessarily identical) sets of components and the same organization. In procaryotic cells where there is no compartmentalization but some of the components are in single dose, reproduction takes place only after these components have become multiple.

It follows from all this that reproduction is independent of whether the fragmentation arises triggered by the interactions of the unity or through its internal dynamics or both. Accordingly, the fact that present-day mitosis is the result of evolution does not contradict the notion that in cellular reproduction there is nothing else but the fragmentation of an autopoietic unity in a process that is not exclusive to the living systems. What is peculiar to the reproduction of autopoietic systems is that the organization of the unities that reproduce is the autopoietic organization and that the process takes place, in principle, without interruption of their autopoiesis and through the realization of their autopoiesis.

Heredity. Heredity is also a universal phenomenon in nature and takes place whenever reproduction takes place regardless of the class of unities reproduced. In other words, heredity is a necessary result of reproduction, and as such it is a trivial consequence of the distribution of the components when the fragmentation takes place. Furthermore, it is obvious that similarity and difference in the unities produced in reproduction is determined by the same process of distribution of components. In fact, to the extent that the autopoietic organization allows for structural plasticity (structural change without loss of identity), variation is possible through the differential distribution of components, and, to the extent that certain components are necessary for the realization of the autopoietic organization of the unities resulting from reproduction, variation is restricted. If some components have properties that determine a particular feature of their participation in the production of other components, the former may determine some particular restriction in the domain of structural variability compatible with the autopoiesis of the unities involved in reproduction. Yet all of this occurs within the set of phenomena we already described.

Modern nuclear genetics is therefore not the study of heredity; rather it is the study of hereditary phenomena associated with the particular structures (components and relations) and productions that depend on the properties of the nucleic acids. In fact, all of the components of the cell participate in the cellular phenomena of heredity and genetics, albeit with different penetration in the succession of generations.

I have not used such notions as coding, message, information, or transmission of information, because they do not refer to the processes that generate the phenomena of reproduction, heredity, or genetic variation. They refer to relations in a metadomain of descriptions, and do not determine relations of contiguity between the components of the composite unity described. Their value, therefore, pertains to a metacognitive domain where the observer beholds the cell simultaneously as a simple and as a composite unity, and where he establishes a relation between these two otherwise independent entities.

Adaptation. In the history of interactions of a composite unity in its medium, both unity and medium operate as independent systems that, by triggering in each other a structural change, select in each other a structural change. If the organization of a composite unity remains invariant while it undergoes structural changes triggered and selected through its recurrent interactions with its medium, and if there are structural configurations of the medium that participate as recurrent selective features in the history of interactions of the unity, then the outcome of this history of interactions is the selection by the medium of a sequence of structural changes in the composite unity that results either in its adaptation to the recurrent features of the medium (as its operation without loss of class identity in relation to them) or in its disintegration. In other words, if a composite unity is structurally plastic, then adaptation as a process of structural coupling to the medium that selects its path of structural change is a necessary outcome. In this process the configuration of constitutive relations that remain invariant in the adapting composite unity determine the matrix of possible perturbations that the composite unity admits at any instance and hence operates as a reference for the selection of the path of structural changes that takes place in it in its history of interactions. Defined in this manner, adaptation is not peculiar to living systems. On the contrary, adaptation is a universal phenomenon that takes place whenever a plastic composite unity undergoes recurrent interactions with structural change but without loss of organization, and may arise in relation to any recurrent structural configuration of its domain of interactions. Furthermore, this domain of interactions could be anything with which the composite unity interacts as if with an independent entity or system, including its own configuration of internal states. Accordingly, all that is unique with respect to adaptation in living systems is that in them the autopoietic organization constitutes the invariant configuration of relations around which the selection of their structural changes takes place during their history of interactions.

If adaptation takes place during the individual history of one autopoietic unity, the phenomenon is ontogenic adaptation and corresponds to what is usually called learning. If adaptation takes place through a succession of generations in which each reproductive step offers an additional dimension of variability through the diversity produced in the offspring in each generation, the phenomenon is phylogenic adaptation, and the outcome is evolution.

2.3.3 Distinction from Other Systems

An autopoietic system is defined as a unity through relations of production of components, not through the components that compose it, whichever these may be. An autopoietic system is defined as a unity through relations of form (relations of relations), not through relations of energy transformation. An autopoietic system is defined as a unity through the specification of a medium in its realization as an autonomous entity, not through relations with a medium that determines its extension or boundaries. An autopoietic system is defined as a unity as a closed network of productions of the components that recursively, through their interactions, realize the network that produces them and constitute its boundaries by realizing the surfaces of cleavage that separate it as a composite unity in the space in which they exist.

It is the configuration of relations that defines the class of autopoietic systems, not the processes or components through which this configuration is realized. The only thing peculiar to autopoietic systems is their autopoietic organization. Yet since there are many systems that may have components of the same kind as the components of a particular subclass of autopoietic systems, or that may be defined by relations of form, or that may be realized as networks of productions—but are not autopoietic—the recognition of an autopoietic system may be difficult. This is particularly so if one forgets that the system that one intends to consider is realized in the space of its components, and, hence, one loses view of its boundaries.

In general, an observer who beholds a composite unity, and thus defines it as such, may assort its components in several different manners and claim that the composite unity may at the same time belong to a manifold of different classes of unities: an apple may at the same time appear to an observer as an apple and as a projectile. However, in the strict operational sense, different unities are defined by different operations of distinction that specify which relations define their different organizations and, hence, which relations are necessary and which are superfluous. Strictly, then, an apple is not a projectile, even though the same components may be organized as an apple or as a projectile. This, of course, is not new, but a mistake can be produced if one forgets that a unity is defined by its organization and not by its components. Thus, for example, in an autopoietic system identified as a separable entity in the physical space through the identification of its boundaries and components, an observer may describe dissipative processes, both from the energetic and material points of view, and claim that an autopoietic system is a dissipative system. However, a dissipative system is not an autopoietic system. A dissipative system is defined by relations of stability under flow, and, therefore, it is defined as a system by relations to another entity or system with respect to which it is supposed to exhibit stability. Accordingly, although the observer who calls an autopoietic system a dissipative system may direct another observer to an entity that includes the boundaries of an autopoietic unity, the direction is to a different system; the other observer is directed to a system defined by relations different from those that determine the boundaries of an autopoietic system. The structures of the two systems intersect, but they are operationally entirely different systems because they are defined by different organizations.

2.4 Epistemology

"Everything said is said by an observer to another observer that could be himself" (Maturana 1970).

The fundamental cognitive operation that an observer performs is the operation of distinction. By means of this operation the observer specifies a unity as an entity distinct from a background, and a background as the domain in which an entity is distinguished. An operation of distinction, however, is also a prescription of a procedure that, if carried out, severs a unity from a background, regardless of the procedure of distinction and regardless whether the procedure is carried out by an observer or by another entity. Furthermore, the prescriptiveness of an operation of distinction implies a universal phenomenology of distinctions which, through the specification of new procedures of distinction or through their recursive application in the reordering of the distinguished entities, can, in principle, endlessly give rise to new simple and composite unities, and, hence, to new nonintersecting phenomenic domains. In these circumstances, although a distinction performed by an observer is a cognitive distinction and, strictly speaking, the unity thus specified exists in the observer's cognitive space as a description, the observer defines a metadomain of descriptions from the perspective of which a reference is established allowing speech to occur as if a unity, simple or composite, existed as a separate entity that can be characterized by denoting or connoting the operations that must be performed to distinguish it as a separate entity.

From the perspective of a metadomain of descriptions, the distinction between the characterization of a unity and the observer's knowledge of it should be clear. In fact, knowledge always implies a concrete or a conceptual action in some domain, and the recognition of knowledge always implies an observer that beholds the action from a metadomain. Therefore, an observer who claims knowledge of a system also claims the ability to define a metadomain from the perspective of which the observer can simultaneously behold the system as a simple unity and describe its interactions and relations as a simple unity. In these circumstances, it is legitimate to distinguish between the characterization that an observer makes of a unity—by pointing either to its properties, if it is a simple unity, or to its organization, if it is a composite one—and the knowledge about a unity that the observer reveals, by describing either its operation as a simple unity, if it is a simple unity, or both its operation as a simple unity and the operation of its components as components, if it is a composite entity. In either case, however, the knowledge an observer has of the unities so distinguished consists in handling these unities in a metadomain of descriptions with respect to the domain in which they are characterized. In other words, an observer characterizes a unity by stating the conditions in which it exists as a distinguishable entity, but perceives it only to the extent that a metadomain is defined in which it can be treated as the characterized entity. Thus, autopoiesis in physical space characterizes living systems because it determines the distinctions that we can perform in our interactions with them, but we know them only as long as we can both operate with their internal dynamics of states as composite unities and interact with them as simple unities in the environment in which we behold them. The fact that the characterization of an entity is also a description made by the observer, and as such also belongs to the observer's descriptive domain (Maturana 1970), does not invalidate the operational effectiveness of the distinction of distinctions in the metadomain of descriptions in which the cognitive statements are made. The entity so characterized is a cognitive entity, but once it is characterized the characterization is also subject to cognitive distinctions valid in the metadomain in which they are made by treating the characterization as an independent entity subject to contextual descriptions. Therefore, complementarities such as system–environment, autonomy–control, and so on (Goguen and Varela 1977; Varela 1978) are complementarities in our cognition of the system that we observe in a context that allows us to establish such relations, but they are not constitutive features of the system because they do not participate in its organization through the interplay of the properties of its components. Accordingly, that one should not be able to account for or deduce all biological phenomena from the notion of autopoiesis alone is not a shortcoming of such a notion. On the contrary, this is to be expected because such a notion only refers to the characterization of a system in a domain of descriptions in which it is distinguished as a composite unity. In order to have a biological phenomenon a background must be involved and, hence, a metadomain of observations must be generated so that the phenomenon may be distinguished and described. For a biological phenomenon to take place, an autopoietic system must operate in a context; the processes that take place in the realization of the autopoietic network of productions are not biological phenomena. What is involved here is the dynamics of constitution of a composite unity and the cognitive distinction of a unity. A composite unity is constituted when a set of relations between components specifies a surface of cleavage that operationally defines a background with respect to which it delimits the related components as a simple unity. The unity thus constituted does not participate in its own constitution, because it is only with respect to a background that it has operational existence. The components and the unity that they compose exist in nonintersecting spaces.

2.5   Summary

My fundamental claims are the following. (1) Autopoiesis in the physical space is the necessary and sufficient condition that makes a system a living system, and as such, an autonomous entity. (2) A given phenomenon is a biological phenomenon only to the extent that its realization involves the realization of the autopoiesis of at least one living system. (3) Although everything takes place in living systems through the realization of their structurally determined autopoiesis, the actual occurrence of any biological phenomenon is always a function of the historical contingencies under which the participating living systems realize their autopoiesis. (4) Any phenomenon that may be involved in the autopoiesis of an organism may participate without contradiction in the domain of biological phenomena.

References

Goguen, J. A., and Varela, F. J. (1978). Systems and distinctions; duality and complementarity, *Int. J. Gen. Systems.* 5.

Maturana, H. R. (1970), *The Biology of Cognition,* Biol. Computer Lab. Res. Report 9.0, Univ. of Illinois, Urbana.

Maturana, H. R. (1975), The organization of the living: a theory of the living organization, *Int. J. Man–Machine Studies* 7, 313–332.

Maturana, H. R., and Varela, F. J. (1973), *De Máquinas y Seros Vivos,* Ed. Universitaria, Santiago de Chile. English version: *Autopoietic Systems,* Biol. Computer Lab. Res. Report 9.4, Univ. of Illinois, Urbana.

Varela, F. J. (1979), *Principles of Biological Autonomy*, Elsevier, New York.

Varela, F., Maturana, H., and Uribe, R. (1974), Autopoiesis: The organization of living systems, its characterization and a model, *Biosystems* 5, 187–196.

INTRODUCTORY REMARKS

Varela asserts that the autonomous (autopoietic?) character of living systems does take precedence, both logical and "functional," over the genetic and reproductive understanding of the individual unity. However, Varela attempts to place autopoiesis in a wider framework by introducing the concept of *organizational closure*. He views autopoiesis as being only a special case in a much larger class of organizations (characterized by a circular concatenation of constitutive processes): *organizationally closed* (i.e., recursively or circularly organized) systems. Thus, not all organizationally closed systems are necessarily autopoietic; the notion of the production of components should not be "stretched" beyond the notion of *chemical* productions. Varela thus limits the concept of autopoiesis to a particular class of biological productions taking place at the level of the cell. Dissipative structures, social groupings and orders, language, and so on are excluded from Varela's version of autopoiesis. In this sense, Varela ties autopoiesis intimately to a *particular* physical domain, with a particular type of material components. Systems that renew themselves either too quickly or too slowly, relative to the vantage point of the cell, are characterized only as metaphors of autopoiesis.

Section 3.3, on Descriptive Complementarity, is the most interesting: it marks Varela's transition from process-oriented to structure-oriented thinking. He also parts with Maturana, perhaps under the influence of Pattee and Goguen, and admits the complementarity and validity of using the notions of information, program, purpose, and the like as different but still symbiotic modes of explanation. He upholds the duality of the autonomy–control view of systems. But his arguments are still rather tentative; the issue is far from being resolved. It is this editor's feeling that his ultimate return to Maturana's less compromising position is more than probable.

This duality of modes of inquiry is further exemplified through the invoked notions of *Erklärung* and *Verstehen* (i.e., "explanation" and "understanding" in English, or the "how" and the "why" of science). Varela uses the word "explanation" for both *Erklärung* and *Verstehen,* but that is hardly what he means. He is after the complementarity of causal (operational) and teleological descriptions. Or. in Jantsch's interpretation, after the complementarity of chance and necessity, stochastic and deterministic descriptions. In any case, the complementarity of both modes of inquiry is emphasized, although admitting the post hoc nature of purpose and directiveness would not contradict the notion.

Varela looks at the question of whether or not dissipative structures (using the Belousov–Zhabotinsky reaction as an example) could be considered as being autopoietic. Where Maturana's answer is a resounding *no*, Varela considers them as "serious candidates." Eigen's hypercycles (i.e., organizational closure) are viewed as possible precursors of autopoietic cell-like units. An interesting discussion of the relationship between "structural" and "informational" components (molecules) then follows: there is no difference between them, except that due to the framework of our observation.

The main conclusion of Varela's paper, after all the detours, is the claim that autopoiesis is necessary and sufficient to characterize living systems, but not sufficient to explain living phenomena as perceived by an observer. For that, some sort of symbolic description is necessary. But to view information as a "thing" to be transmitted, taking its symbols at their face value, is too much even for Varela. He is bound to call it nonsense, thus closing the gap with Maturana again.

Admittedly, both Maturana and Varela have reached identical conclusions. To what extent it is necessary to emphasize the difference between the ways of their epistemological expression, and thus to introduce a cleavage in the "paternity claim" for the concept of autopoiesis, is an issue best left for both authors to grapple with. As Jantsch put it, there is a more powerful and beautiful paradigm hidden in the fog than any part of it, pulled out into the sun, can be. Therefore, let's not pull, but work on thinning the fog!

Francisco J. Varela was born in Chile in 1946. He studied biology with Humberto Maturana at the Universidad de Chile in Santiago. Later he received a Ph.D. in Biology from Harvard University, in 1970. His main interests lie in what McCulloch called "experimental epistemology," that is, the natural history and possible mechanisms of knowing. F. J. Varela published many papers in the areas of cybernetics, neurobiology, and philosophy of science. In addition to writing the original monograph on autopoiesis, co-authored with H. R. Maturana, he also published *Principles of Biological Autonomy,* in 1979. Currently he serves as an Associate Professor of Biology at the University of Chile. *Address:* Facultad de Ciencias, Universidad de Chile, Casilla 653, Santiago, Chile.

Chapter 3
Describing the Logic of the Living
The Adequacy and Limitations of the Idea of Autopoiesis

Francisco J. Varela

3.1 Introduction

The notion of autopoiesis was proposed by Humberto Maturana and myself (Maturana and Varela 1973)[1] with the intention of redressing what seemed to us to be a fundamental imbalance in the understanding of the living organization. This imbalance was reflected mainly on two fronts of current biological discourse: (1) The unitary organization of living systems was not properly taken into account, and the emphasis was mostly put on genetic analysis and reproductive capabilities. (2) The characterization of the logic of life was construed on the basis of certain chemical species coexisting with "new" added qualities referred to as information (regulation, control, program, and so on).

Our efforts were directed toward showing the following.

(1) The importance of the individual organization is fundamental, and the autonomous character of the living system takes precedence, both logical and functional, over the genetic understanding of the individual as a member of the species. The individual organization can be shown to be one of self-construction through recursive production of components, and it is this specific organization, autopoiesis, which is at the base of the autonomy of living systems. The most clear paradigm of this autopoiesic organization is the cell and its metabolic net. Once the individual organization is clearly defined, one can attempt to analyze the added complexities that autopoietic systems have undergone in the history of Earth, including their reproductive capacities and higher order aggregations.

---

[1] This publication is hereinafter referred to as AS (Autopoietic Systems).

(2) Informational and functional notions need not enter into the characterization of the living organization, as they belong to a domain different from the relations that define the system. Thus we proposed a critique to the current use of such notions as unnecessary for the definition of the logic of life, and claimed autopoiesis as necessary and sufficient to define the living organization, and, a fortiori, the phenomenology of the living.

In this paper I intend to review the idea of autopoiesis several years after its conception and to evaluate what I still consider central and what seems to need further development and modification. I have received a very positive influence from conversations with Howard Pattee; in fact, much of the discussion here is directed at dealing with questions raised by his own work vis-à-vis the idea of autopoiesis.

3.2 From Autopoiesis to Organizational Closure

In retrospect, I still see the notion of autopoiesis and its underlying epistemology as a very valuable step. It pointed to a neglecting of *autonomy* as basic to the living individual and rooted this elusive quality in a *mechanism for its identity*. This link between definitory relations in a system's organization, and its identity in the domain in which the components exists, is at the base of the phenomenology of autonomy. It brings out the essential connection between self-production and the domain of interactions defined by the system's functioning, so that whatever the domain of interactions is, it is not separable from the system's workings. This has a number of consequences for the understanding of *cognitive domains* in biological systems, which were partly analyzed in our original book (Maturana and Varela 1973) and have been extensively explored in other writings (Maturana 1974, 1975, 1979; Varela 1979).

A second significant point brought to the foreground by the notion of autopoiesis was the connection between unitary autonomy and the indefinite recursive or circular nature of the autopoietic organization. In other words, it made it evident that the "self," in self-production, pointed to a circular concatenation of process that attained coherence only because of and through its interdependence. Thus, autopoiesis is a particular case of a larger class or organizations that can be called *organizationally closed,* that is, defined through indefinite recursion of component relations. Much of my own work has been directed toward dealing with closure more explicitly (Varela 1975, 1977, 1979; Varela and Goguen 1976; Goguen and Varela, 1979).

This brings me to the first significant point regarding the idea of autopoiesis as I see it now. It is tempting to confuse autopoiesis with organizational closure and living autonomy with autonomy in general. The fact is that the definition of autopoiesis has some precision because it is based on the idea of *production* of components, and this notion of production cannot be stretched indefinitely without losing all of its power. The image that motivated this definition was, of course, that of chemical productions; such an image can be generalized to some extent. An example of this is the computer simulation of a two-dimensional simple autopoietic system reported elsewhere (Varela et al. 1974; see also Zeleny and Pierre 1976). But in order to say that a system is autopoietic, the production of components in some space has to be exhibited; further, the term production has to make sense in some domain of discourse.

Frankly, I do not see how the definition of autopoiesis can be *directly* transposed to a variety of other situations, social systems for example. It seems to me that the kinds of relations that define units like a firm (Beer 1975) or a conversation (Pask 1977) are better captured by operations *other* than productions. Such units are autonomous, but with an organizational closure that is characterizable in terms of relations such as instructions and linguistic agreement. The basic consequences of closure remain the same as in the particular case of autopoiesis, but the specifics change, that is, the kinds of defining relations. Unless a careful distinction is made between the particular (autopoiesis and productions) and the general (organizational closure and general computations), the notion of autopoiesis becomes a metaphor and loses its power. This is what has happened, in my view, with the attempts to apply autopoiesis directly to social systems. It seems both more precise and parsimonious simply to preserve the emphasis on autonomy and indefinite recursion (or self-reference) and to define subclasses of organizationally closed systems corresponding to the phenomenological domain under consideration; autopoiesis seems mostly adequate to the domain of cells and animals.

3.3 Descriptive Complementarity

In AS we argued that the notions of information and purpose are dispensable. This is because the living organization could be defined without resorting to such notions, and thus, the explanation underlying the living phenomena need not include them as constitutive components. Further, we argued, such notions cannot enter into the definition of a system's organization because they pertain to the domain of discourse between observers. Information and purpose can only enter for pedagogical purposes. They do not enter into an operational explanation, for which autopoiesis is complete, that is, based on distinctions of component properties that generate a phenomenic domain.

In retrospect, I believe this question needs further development. I still hold to be valid the criticism of the *naive* use of information and purpose as notions that can enter into the definition of a system on the same basis as material interactions [e.g., in Miller's (1965) definition of living systems].

But there are limitations on our original presentation, stemming from the fact that we did not take our criticism far enough to *recover* a non-naive and useful role of informational notions in the descriptions of the living phenomena. It criticized without a corresponding *Aufheben*.

The analysis is AS was based on the assumption that operational explanations are, in some sense, intrinsically preferable and sufficient. This now seems to me wrong in two senses that I shall try to make clear.

(1) It gives the operational explanation an epistemological status not compatible with the very intention of the criticism leveled against the naive use of information. (2) It neglects the fact that informational terms, although belonging to a different category of explanation than the operational terms used in autopoiesis, can still be used as valid explanatory terms, and that furthermore, different modes of explanation can coexist.

There was, evidently, a need in AS to overemphasize a neglected side of a polarity. Similarly, I have argued elsewhere that autonomy cannot, in fact, be conceived without a complementary consideration of how the system is also controlled in a dual context; in particular, autopoiesis and allopoiesis are complementary rather than exclusive characterizations for a system (Varela 1977; Goguen and Varela 1977). What I argue here is that an operational explanation for the living phenomenology needs a complementary mode of explanation to be complete, a mode of explanation that I shall refer to a *symbolic*.[2] A presentation of this expansion from the views expressed in AS has to begin with a very brief consideration of what is at the base of our tendency to prefer purely operational explanations and to relegate informational terms to the category of "purely pedagogical."

3.3.1 Modes of Explanation

Our preference for (purely) operational explanations is precisely that— a preference. Such preferences come from a community of inquiring individuals, a scientific community that, through the inheritance of a tradition, comes to agree on certain criteria of validity relative to certain values or intentions (Radnitzky 1973). The preference for "casual" explanations in the common sense of contemporary science comes from a predominantly manipulative and technological orientation present in science over the last 150 years. Given such a preference, operational explanations came to be *the* explanations: nonoperational explanations hold no power for manipulation and prediction. Prediction, in fact, is the symp-

---

[2] There are some obvious similarities with what Pattee calls the complementarity between dynamic and linguistic modes of description (Pattee 1977). There are also some significant differences.

tom of a successful explanation in this kind of philosophy, so that causality and prediction are a triumphant duo that characterizes modern science.

This triumphant duo, however, has to be viewed in a historical perspective. Alongside operational explanations, another equally outstanding tradition has existed that asserts the validity of finalistic or teleological explanation, where the terms of explanation are not "whys" but "what fors." These two traditional modes of explanation are best characterized with the German nouns *Erklärung* and *Verstehen,* usually translated as explaining and understanding (von Wright 1971).

Now, the intention behind a *Verstehen*-type explanation is not manipulation, but understanding, communication of intelligible perspective with regard to a phenomenic domain. Typical examples are teleological explanation in Aristotle or the vitalist explanations of the 18th century. To the extent that their main orientation is to understand and communicate this understanding, such explanations are fundamentally different in orientation from the operational explanation. It is, as we said, a historical fact that Western science took a very strong stand preferring operational explanation since the time of Galileo, and in fact, made *Verstehen*-type explanations into an enemy, to be banned forever from science.

From our perspective, at the end of this 20th century, sufficiently removed from the Age of Enlightenment, things look rather different. As I see it, four major developments have contributed to altering the preference for purely operational explanation. First, the great renovation inside physics, the model for logical empiricism, after the constitution of quantum mechanics and its variegated epistemological problems, that makes both naive causality and naive objectivity completely inadequate. Second, the rise of biological science that introduced into science the need to consider phenomena of unbounded complexity relative to physical sciences. The epitome of this development is the history of genetics, the Watson–Crick model, intertwining both structural components and the apparent need for a "coding" description. Third, the extensive development, linked to the use of biological concepts, of cybernetics and systems theory in the area of design and prescription of systems, where the notions of communication and purpose are at the core of what is not only the main subject, but many of its daily consequences such as computers and complex systems of regulation in human services. Finally, and in much more subdued form for the world of science, the reawakening in the European schools of thought, of the importance of the *Verstehen*-type explanation in human affairs.

All these developments in the span of time elapsed since the end of the 19th century forces us willy-nilly to a reevaluation of what we mean by our preference for operational explanations and, in fact, what we mean when we intertwine such explanatory modes whether talking about a computer program or an animal dance.

What, I submit, is essential to understand in this relationship is that both forms of explanation refer to modes of description relative to some perspective of the observer—or rather, we should say, of an inquiring community. In the causal description the fundamental assumption is the phenomena occur through a *network of nomic* (lawlike) relationships that follow one another. In the communicative explanation the fundamental assumption is that phenomena occur through a certain order or pattern, but the fundamental focus of attention is on certain moments of such an order, relative to the inquiring community. Thus these modes of explanation are exclusive and contradictory only to the extent that one assumes that laws of nature are prehensible independent of an inquiring community, whereas that the other assumes that no nomic patterns are discernible in the world.

Both demands are, of course, inessential to preserve the power of each alternative view of a recorded phenomenon. If we can provide a nomic basis to a phenomenon, an operational description, then a teleological explanation only consists of putting in parentheses the intermediate steps of a chain of causal events and concentrating on those patterns that are particularly interesting to the inquiring community. Accordingly, Pittendrich introduced the term teleonomic to designate those teleological explanations which assume a nomic structure in the phenomena, but chose to ignore intermediate steps in order to concentrate on certain events (Ayala 1970). Such explanations introduce finalistic terms in an explanation while assuming their dependency in some nomic network, hence the name teleonomic.

3.3.2 Symbolic Explanations

As we discussed in AS, the connection between an operational description such as autopoiesis and a finalistic description lies in the observer who establishes the nexus. Thus, we concluded that purpose plays no causal role in autopoiesis and therefore no role in the description of the system's organization. The same conclusion was valid for the notion of message, information, and code. What is significant in both of these classes of notions—purpose and information—is that the observer chooses to ignore the causal connection between classes of events and to concentrate on the ensuing relationships. This is an important idea, and it is insufficient to discuss it as merely a pedagogical maneuver. This possibility of choosing to ignore intervening nomic links is at the base of all symbolic descriptions. What is characteristic of a symbol is that there is a distance, a somewhat arbitrary relationship, between signifier and signified. This is very immediate in human discourse: words and their contextual meaning have such a remote and involved historical and structural mode of coupling that any effort to follow such nomic connecting is hopeless. Thus in order to understand language, we do not trace the sequence of causes from the waveform in the air to the history of the brain operations, but simply take it as a fact that we *can* understand. And precisely because we cannot make everything reducible to causal explanation, since we live and grow inside language, human life has the openness it has.

Thus we come to the conclusion that purpose and symbolic understanding are interrelated in a manner parallel to that of the favored pair of operational explanation and prediction. Under "symbol" we here subsume all its forms, such as code, message, and information.

So far we have argued that operational and symbolic descriptions do not contradict each other, since they belong to different levels of descriptions among a community of observers. Unless we keep clear in our minds that by changing modes of explanation we also change the frame of reference in which we operate, the whole issue becomes muddled. If not, teleonomic–symbolic terms can be reduced to an operative component, as, say, in specifying the components of a system through only one component (DNA), a typically useless form of reductionism. The two modes of explanation are distinct, yet they can be related without reducing one to the other.

The question we now want to ask is whether we *need* both forms of explanation? Can we not be satisfied with only operational explanations of a phenomenon? In the case at hand, these questions would amount to asking whether the autopoietic characterization is enough to explain the entire phenomenology of living systems. In a sense, in AS we have already shown that, indeed, the autopoiesis of each individual suffices to generate all of the phenomenology of the autonomy of living systems, and that, through their coupling and complexification, we can see in them the foundation for evolutionary and historical phenomena. Thus, in principle, all biological phenomena can be reduced to autopoietic mechanisms.

This, however, is reminiscent of the statement that all of the history of the universe could be determined if only we knew the positions and momenta of all the particles of the universe so that their future trajectories could be calculated. These kinds of assertions are, above all, epistemological. What we are saying in the case of autopoiesis is that if we could follow all the appropriate contingencies, the biological phenomenology would unfold from the autopoietic mechanism. What is obvious, however, is that this assertion, although it points to the sufficiency of autopoiesis as an operational explanation, says nothing about whether it is *cognitively possible or satisfactory*. Let us examine this in more detail.

We must raise, at this point, the question whether dissipative chemical structures, such as the Belusov–Zhabotinsky reaction (Tyson 1976), should be included among autopoietic systems. At the least, they are serious candidates. However, from the little we know of studies in the origin of living systems and protocellular systems, the mere production of a boundary through a chemical dynamics can hardly be a sufficient condition to be a precursor to cellular systems. A fundamental issue here, as pointed out by Pattee (1972), is the reliability of component specification versus the variability available for selection. Selection and evolution cannot exist without reproduction. Autopoietic systems *can* become reproductive systems, as we discussed in AS. However, their reproduction can become evolutionarily *interesting* only if (1) the process of specification of components is reliable so that there is continuity of structures through time, and (2) they are flexible enough to generate variety of components for selection to operate.

Living systems actually evolved through an appropriate combination of processes of specification and constitution, paradigmatically seen in the coupling between nucleic acids and proteins. Nucleic acids fulfill an essential role in specifying the protein components of cells, which are mostly responsible for processes of constitution and order. This is neatly seen in Eigen's (1971) work on the early evolution of living systems, where the minimum structure capable of generating a sequence of cell-like units takes the form of "hypercycle" (i.e., organizational closure), where there are "informational" components (i.e., nucleic acid) and "structural components" (i.e., proteins). Of course, the "informational" molecule is in *no* way different from any other molecule, its process of interaction from that of chemical species. The reason the name "informational" comes up at all is that we can change the time scale of our observation, consider the realization of these units through several generations, and observe the continuity and reliability of their process of specification of components in an evolutionary process. In other words, we abstract or parenthesize in our descriptions a number of causal or nomic steps in the actual process of specification, and thus reduce our description to a skeleton that associates a certain part of the nucleic acid with a certain protein segment. Next we observe that this kind of simplified description of an actual dynamic process is useful in following the sequences of reproductive steps from one generation to the other, to the extent that the dynamic process (i.e., the kind responsible for bonding, folding, and so on) stays stable. This seems to be the origin of the idea of genetic material as the central element of study for evolution and historical processes in biology. A symbolic explanation, such as the description of some cellular components as genes, betrays the emergence of certain coherent patterns of behavior to which we chose to pay attention. Pattee has discussed the physical basis of these reliable dynamic patterns in terms of constraints (1972).

Note that in switching from one mode of description (the processes determining the autopoiesis of an individual) to a symbolic description (dealing with the evolutionary sequences of autopoietic structures), we perform a leap in time that betrays our radical change in perspective. What we rediscover is the classical duality between physiological time and evolutionary time. Both seem necessary if we are to have a satisfactory explanation of the phenomenology of living systems. If we do not accept the change from a causal description, the actual handling of evolutionary phenomena, which depends on questions of reliability and reproduction, becomes literally impossible to comprehend. How are we to conceive and think at all about sequences of autopoietic units in purely operational terms, where all the components participate in the specification of the unit, if we can hardly do so for a single individual? We must reduce our explanation to a explanation in symbolic terms embodied in the idea of genome and proceed from that form of noncausal, symbolic description to cover the evolutionary phenomena by adding perhaps new causal notions such as natural selection. In other words, it is true that all historical and evolutionary phenomena are ultimately reducible to the coupling of autopoietic units to their ambients. However, this is true on purely logical grounds. For the cognitive capabilities of the observer–community, purely causal explanations are in no position to satisfy the degree of detail we need for ontogenetic and phylogenetic explanations, and a change in explanatory mode is mandatory. Thus, autopoiesis is, on logical grounds, necessary and sufficient to characterize living *systems,* as claimed in AS. What is incomplete here is that autopoiesis is necessary but *not* sufficient to give a satisfactory explanation of the living *phenomena* on both logical *and* cognitive grounds.

3.3.3 Complementary Explanations

When we say that teleonomic–symbolic explanations are not really necessary, it seems to me that we are succumbing to a prejudice of our historical tradition that it is time to revise, because in actual practice we cannot do without both operational and symbolic explanations. Our preference for causal explanations seems to be rooted in the understanding that "causes" are "out there" and reflects a state of affairs independent of the describer. This is, by the very argument used in AS, untenable. Operational causes and "laws of nature" are modes of descriptions adopted by inquiring communities for some intentional purpose (such as manipulation and prediction), and they specify modes of agreement and thus of coupling with the environment. However, ultimately, an operational description results only from one mode of agreement between an inquiring community, and in no way has an intrinsically superior status to a symbolic explanation. They just have different consequences: a symbolic explanation generates a form of agreement in the inquiring community, and thus a coupling with the environment that is not so dramatically visible in manipulation, but is more visible in more diffuse modes of relationships. A good example is the form in which Darwinian thought has modified our entire view of human affairs—not through manipulation but through an agreement about issues that are central to man's image,

such as origin and descent. It is very unfortunate that operational explanations are normally indentified with explanations. Both modes of explanation are, ultimately, in the domain of discourse of the
observer–community, and their only difference lies in the mode in which
they generate agreement.

It seems to me that there are tremendous advantages to maintaining
this duality of explanations in full view. By staying with purely operational
descriptions, we are forced to use other descriptive modes in a rather
sloppy and careless way, as is typical in molecular biology. This kind of
attitude is a remnant of a strong epoch of logical positivism, with its
insistence on *methodological monism*.

At the other extreme, the vitalist attitude, or, more importantly, the
system-theoreticians attitude, which takes information as "stuff," is
equally misguided. This attitude is interesting, for it has taken the same
kind of methodology implicit in operational descriptions and applied it to
a domain where it simply does not work. This is typical in computer
science and systems engineering, where information and information processing are in the same category as matter and energy. This attitude has
its roots in the fact that systems ideas and cybernetics grew in a technological atmosphere that acknowledged the insufficiency of the purely
causal paradigm (who would think of handling a computer through the
field equations of thousands of integrated circuits?) but had no awareness
of the need to make explicit the change in perspective taken by the inquiring community. To the extent that the engineering field is prescriptive
by design, this kind of epistemological blunder is still workable. However,
it becomes unbearable and useless when exported from the domain of
prescription to that of description of natural systems, in living systems
and human affairs. To assume in these fields that information is some
thing that is transmitted, that symbols are *things* that can be taken at face
value, or that purposes and goals are made clear by the systems themselves is all, it seems to me, nonsense. The fact is that information does
not exist independent of a context of organization that generates a cognitive domain, from which *an observer community* can describe certain
elements as informational and symbolic. Information, *sensu strictu,* does
not exist. Nor do, by the way, the laws of nature.

Thus, by putting these two modes of explanation, historically antagonistic, into a dualistic perspective, we gain in power of explanation. Moreover both modes of explanation are significantly modified. On the one
hand, teleonomic–symbolic explanations cannot be adduced without
embedding them in a nomic substrate that can, in principle, account for
them, that is, a network of processes that is abstracted in the process of
defining a symbol. This is clearly seen in the transition from the name
teleological to the name teleonomic: No goal or purpose is assumed without a frame of abstracted chain of events from which we are abstracting.

On the other hand, the causal explanation is also modified, for it no longer holds its position of methodological king and must make way for nonoperational explanations as equally valid. This amounts to no more and no less than a change in the authority images of our inquiring community; it has nothing to do with standards of science or a romantic revolution. To neglect this shift in authority implies, to say the least, a sloppy use of symbolic explanations in the natural sciences and a split between natural and human sciences in which the role of communication and understanding gains a central importance in disfavor of causal mechanisms for which we cannot possibly hope. In brief, then, the dual interplay between these two modes of explanation is the fertile road, which can be taken when and only when both are related to each other in a generative form by making explicit where the change of frame of reference occurs.

An elementary case of such dualistic operation is apparent in our understanding of the origin of life and in the use of genetic material as an explanatory device in evolution and development. By way of another example, consider the interaction of hormone molecules with the receptor surface of a cell. This kind of interaction is best described by abstracting the actual process of interaction and the detailed description of the autopoietic dynamics and phrasing them in terms of a symbol (or signal) with a regulatory effect, a description that emerges through a contracted account of the autopoietic dynamics of the individual cell. At the risk of being repetitive, I ask that you notice that there is *nothing* in the hormone molecule that is informational; its symbolic content is given first by the kind of dynamics determined by the autopoietic unity and its domain of interactions and second by the observer who wishes to follow a certain coherence in the individual dynamic and thus chooses to contract a long and complex sequence of nomic chains.

To regard this cell–hormone interaction in any sense as "intrinsically" informational, or to assume that the organism is "picking up information" from the environment, would be fundamentally wrong. But it seems equally wrong not to see in these kinds of events the beginning of symbolic[3] interactions so prevalent in higher organisms and man or the importance of their continuity with operational explanations.

3.4   Conclusion

This chapter is intended as a review of the idea of autopoiesis. It briefly pointed at those aspects of autopoiesis that still seem valid and useful, some of the subsequent work sparked by the idea, and some of its most

---

[3] I am eschewing the obvious question raised by the foregoing discussion: When are we entitled to say that a symbolic description is admissible for a process? I shall not attempt to go into this important point here (see Varela 1979).

important limitations. Most of the discussion centered on the question of the role and validity of explanatory terms based on the notions of information and purpose, or symbolic explanations. The basic conclusion is that autopoiesis, as an operational explanation, is not quite sufficient for a full understanding of the phenomenology of the living, and that it needs a carefully constructed complementary symbolic explanation.

I have put strong emphasis on duality and complementarity and on the need to make explicit the interdependency of alternative perspectives. I am convinced that this is a central theme that needs substantially more work in systems science and, conversely, that clarifying this basic issue has consequences for many fields in science at the empirical and theoretical level (Goguen and Varela 1979). Not to make explicit the ways in which the observer–community participates in the constitution of a phenomenon is not only to forsake a complete understanding, but to be stuck with bad science.

ACKNOWLEDGMENT
This work has been partially supported by the Alfred P. Sloan Foundation, through a fellowship to the author.

References

Ayala, F. J. (1970), Teleological explanations in evolutionary biology, *Phil. Sci.* 37, 32.

Beer, S. (1975), *Autopoietic Systems* (preface), in Maturana and Varela (1980).

Eigen, M. (1971), Self-organization and the evolution of biological macromolecules, *Naturwissenschaften* 58, 465–523.

Goguen, J. A., and Varela, F. (1979), Systems and distinctions; duality and complementarity, *Int. J. Gen. Systems* 5, 31–43.

Maturana, H., (1974), Cognitive strategies, in *L'Unite de L'Homme* (F. Morin and M. Piatelli, eds.), Seuil, Paris.

Maturana, H. (1975), The organization of the living: A theory of the living organization, *Int. J. Man–Machine Studies* 7, 313–332.

Maturana, H. (1979), The biology of language, in *The Biology and Psychology of Language* (G. Miller and E. Lenneberg, eds.), Plenum, New York.

Maturana, H., and Varela, F. (1973), *De Máquinas y Seres Vivos,* Ed. Universitaria, Santiago, Chile. English version: *Autopoiesis and Cognition,* Boston Studies in the Philosophy of Science, D. Reidel, Boston (1980).

Miller, J. (1965), Living systems, *Beh. Sci.*

Pask, G. (1977), Organizational closure of potentially conscious systems (this volume).

Pattee, H. (1977), Dynamic and liguistic modes of complex systems, *Int. J. Gen. Systems* 3, 259–266.

Pattee, H. (1972), The nature of hierarchical controls in living matter, in *Foundations of Mathematical Biology,* Vol. I (R. Rosen, ed.), Academic Press, New York.

Radnitsky, G. (1973), *Contemporary Schools of Metascience,* Gateway Books, Chicago.

Tyson, A. (1976), *The Zhabotinsky Reaction,* Springer-Verlag, New York.

Varela, F. (1975), A calculus for self-reference, *Int. J. Gen. Systems* 2, 5–24.

Varela, F. (1977), On being autonomous: The lessons of natural history for systems theory, in *Applied General Systems Research* (G. Klir, ed.).

Varela, F. (1979), *Principles of Biological Autonomy,* Elsevier-North Holland, New York.

Varela, F., and Goguen, J. (1976), The arithmetic of closure, in *Progress in Cybernetics and Systems Research,* Vol. III (R. Trappl, G, Klir, and L. Ricciardi, eds.), Hemisphere Publishing Co., Washington, D.C.

Varela, F., Maturana, H. and Uribe, R. (1974), Autopoiesis: The organization of living systems, its characterization and a model, *Biosystems* 5, 187–196.

Wright, G. von (1971), *Explanation and Understanding,* Cornell Univ. Press, New York.

Zeleny, M., and Pierre, N. (1976), Simulation of self-renewing systems, in *Evolution and Consciousness* (E. Jantsch and C. Waddington, eds.), Addison-Wesley, Reading, Mass.

Uribe was instrumental in constructing the first computer model of autopoiesis, and he presents a series of its outputs as a pictorial adornment at the end of his paper. However, Uribe does not deal rigorously with the technical issues of modeling autopoiesis but rather takes a broader view of the relationship between observers and their created models.

In the first part of his paper, Uribe introduces some basic notions of models as observer-dependent categories. The ideas of dividing systems into open and closed unities with respect to the observer, or, respectively, into deterministic or probabilistic models with respect to the observer, are not intended as absolute categories. A living system is seldom either completely open or completely closed in relation to *any* observer. The relationship is more of an intermediate nature, whether or not the observer is coupled with the observed system as a constituent part of some larger autopoietic unity.

Uribe is pointing out that modeling of autopoiesis can be only inadequately performed through the traditional input–output approach (although the approach is not necessarily impossible in simple cases). That is, closed unities, with no apparent inputs and outputs, must be mimicked through similarly closed models. Uribe presents a series of simple examples indicating how initially a random movement of components could cease to be random with respect to the larger unities they integrate. Coupling of components restricts or "deforms" the space in which the original unities (uncoupled components) move. If such couplings, resulting from random encounters, are sufficiently rich in variety and hierarchically interconnected, a random, unpredictable, and closed system can transform itself into a deterministic, predictable, and open system.

Most of the examples are very simple (almost naive), clear, and easy to understand. It would be a pity if the unusual clarity of this paper should detract some readers' attention from the ABC of modeling autopoiesis. Uribe is building up, slowly and didactically, all the necessary elements on which the autopoiesis models must be based. Uribe is a great teacher; he lets the reader mutter, "Hey, I knew that."

But what about statements like, autopoietic systems are necessarily allopoietic (i.e., with apparent inputs and outputs) with respect to the larger systems they integrate. Or, there is no hierarchy within living systems. That's what Uribe is contemplating. An important step has been taken. In addition to the customary observer–system relationship, Uribe contemplates a triad: observer–system–other system. You see, the "other" system could be an observer as well.

Uribe is the only one of the "autopoietic troika" who does not shy

away from "social autopoiesis." His "ants (autopoietic) become allopoietic with respect to the anthill (autopoietic)" leaves no doubt about
the direction of his "metaphor."

Uribe's concluding paragraphs are strong and challenging, not devoid
of compassion, and go beyond the original outlines of autopoiesis. His
assessment of "training" versus "learning" and their effects on the
"allopoietization" of our societies are insights of great value. It *is* possible
to become autopoietic, organizationally closed, and free (autonomous)
again, for people and their societies.

Ricardo B. Uribe was born in Santiago, Chile in 1935. He received both B.S.
(1957) and M.S. (1960) from the University of Chile. During 1960–1977 he held
various appointments in the Department of Electrical Engineering in the School
of Engineering of the University of Chile. He also served as an Associate Professor
at the Catholic University of Chile (1971–1972). Currently, he is associated with
the Department of Electrical Engineering at the University of Illinois and with
the Department of Cybernetics at Brunel University in London. His interests are
centered around the study and development of organizationally closed systems
(and the organizations that may evolve from their interactions) as well as the
tools, like networks of microprocessors, that may lead to their realization. *Address:* University of Illinois, 155 Electrical Engineering Building, Urbana, Illinois
61801.

Chapter 4
Modeling Autopoiesis

Ricardo B. Uribe

Dieu donne le feu pour faire l'enfer,
le Diable, le miel pour faire le ciel.

4.1 Knowledge

Observers know and create their environment through interactions with it. This interaction involves an explicit or implicit prediction about the environment (Uribe 1977). In order to predict, observers construct models (abstract or otherwise) of the unities (organizations or systems) (Uribe 1977; Maturana 1980) that constitute their environment. This construction allows deduction (computation) in the model, and thus confident prediction of the environment. Usually the model is a deterministic unity with input and output (Ashby 1956; Löfgren 1962), which assumes that the modeled unity is a deterministic unity with input and output. It is "known" that unities can be intrinsically nondeterministic and still be treated as deterministic so long as the observer assumes the proper perspective. For example, a table is composed of electrons whose positions and momenta cannot be ascertained simultaneously (Heisenberg); however, the observer usually assumes with great confidence that breakfast can be had on the table.

If an observer assumes that perspective from which the table is made of electrons, the table will have neither inputs nor outputs and hence cannot be modeled by a unity with inputs and outputs. It can only be "modeled" by a unity without inputs or outputs. The table (made of electrons) and the "model" for it are closed unities for that observer.

The observer who assumes the perspective from which the table is a simple (not composite) unity (on which one can have breakfast) can construct a model of it with inputs and outputs that (in this special case) coincide, as do the inputs and outputs in the table itself. The behavior of the table is also a special case, since it changes into itself at each instant of time (Pask 1961).

Consequently, models can be either deterministic or probabilistic with respect to the observer. The former are open unities; the latter are closed unities. They have neither inputs nor outputs and can never strictly represent the modeled unity, no matter how closely the "model" represents the modeled.

4.2   Relativity

Everything said is said by one observer to another, who could be that observer (Maturana 1980).

Time is a concept relative to the observer (Spencer Brown 1969). Hence behavior, which develops in time, is also relative to the observer. Modeling, which mimics behavior, will also be relative to the observer. And since knowledge comes from modeling, knowledge is itself relative to the observer (i.e., knowledge is subjective—von Foerster 1976).

While it is in principle possible to make infallible deductive inferences (computing in the model), it is in principle impossible to make infallible inductive inferences (predicting outside the model).

4.2.1 An Example

Suppose that an observer is looking at a screen on which a series of numerals appears one at a time in the sequence shown in Figure 1. The observer who belongs to the species homo sapiens and is still curious about the environment may be intrigued and may try to understand this phenomenon. The way to attain this understanding comes naturally to the observer, who will try to construct a model, abstract or otherwise, that mimics the observed phenomenon. The construction of the model can be either a conscious or an unconscious process. The model itself will depend on the observer's previous experiences and will probably range from the naive "another homo sapiens is presenting the numbers behind the screen" to the most elaborate construct of a digital counter or accumulator with the associated circuitry for displaying the numerals on the screen. However, common to all the possible models will be the notions

Figure 1

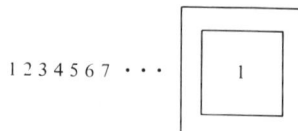

1 2 3 4 5 6 7 · · ·   1

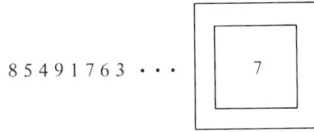

8 5 4 9 1 7 6 3 · · ·   7

Figure 2

of input and output, which are (apparently) indispensable for comparing the behavior of the model with the modeled. In the model and in the modeled: "If this goes in, that must go out."

If the screen's behavior can be modeled, it will be modeled by the behavior of a deterministic unity, and the observer will be able to predict the behavior of the screen, that is, predict the next numeral. The observer will understand the screen's behavior.

A different situation arises when the screen shows, for example, the sequence in Figure 2. Again, the observer may try to model the observed behavior, but now it is difficult or even impossible to construct a deterministic unity that will mimic this behavior. The (strict) connection between input and output is lost for the observer. Although the existence of an "input" that produces the "output" can be assumed, this "input" actually has no influence on the "output"—at least from the point of view of the observer, who is finally resigned to calling the output random. The sequence will not appear random to an observer who knows that the order of the numerals is alphabetical with respect to their names. Randomness is relative to the observer.

Up to now, the observer has had no influence on the screen's behavior; that is, the observer cannot perform experiments that could aid in constructing and testing the model. The observer who is not satisfied with this situation, or whose curiosity is not exhausted, will look behind the screen, where we shall suppose there is a storage element that can be accessed. The observer can improve the model by performing the following experiment: a number is stored (input I) differing from the one displayed, after which it is observed that the numbers displayed (output) and stored are the same (see Figure 3). The observer may venture the hy-

Figure 3

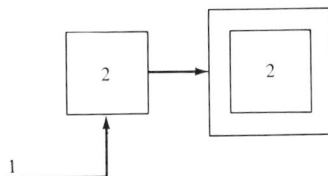

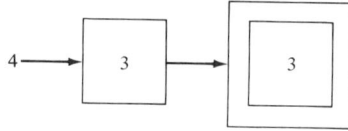

Figure 4

pothesis that the number displayed is the number stored and check this theory by repeating for several numbers the experiment of storing a number and observing the number displayed. In this example, the experiments confirm this hypothesis. The observer now ventures that the original series of numbers (1,2,3,4,5,6, ...) that was displayed on the screen will not be affected by storing new numbers; that is, a model is assumed (constructed) in which each new number of the series replaces the previous one in the storage element (see Figure 4). Suppose it happens that when the observer performs the corresponding experiments to check the theory, his or her input (the act of storing a number) does indeed turn out to affect the series of numbers displayed (the output). After this input, then, the series of numbers is observed to increase starting from the stored number and does not continue the series observed prior to the input. Some degree of control over the output (numbers displayed) exists; for example, the observer can restart the series at any desired number. The input is not merely stored, and the observer may guess that it is incremented to form the display: that is, a 1 is added each time to the number stored to form the series. The model is then reconstructed as shown in Figure 5.

The new model suggests to the observer what new elements to look for behind the screen, such as the input +1 or the (stored) addend, which are indeed there. We now suppose that the observer has more control over the output than before, since, for example, the input +1 can be set to 0 and the series displayed can be stopped at any number, or, of course, the series can be restarted. Now, all the experiments are performed in the behavior space of the model and/or the modeled (i.e., numbers—

Figure 5

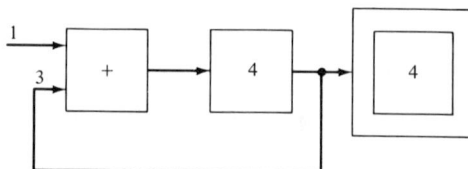

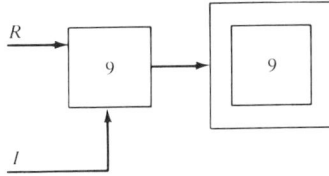

Figure 6

integers). The observer can apply neither heat to the storage nor blows to the screen.

On the other hand, when the series is random from the observers viewpoint, the situation changes abruptly, because his or her curiosity and imagination can only result in the construction shown in Figure 6 (in which R is a random number). This is no longer a model. No "input" I from the observer will affect the series of numbers displayed, either in the "model" or in the modeled, since the series will continue to be random with respect to the observer. Hence, the observer cannot achieve any degree of control over the "output." Consequently, the input–output approach loses all sense: It is no longer possible to test the behavior of the "model" using this approach. In fact, that which is to be modeled has neither input nor output from the observer's point of view, neither in the behavior space of the "model" nor in that of the modeled. The "model" will no longer mimic the behavior of the modeled; that is, while the observer's construction is another unity (organization or system) (Varela et al. 1974; Maturana and Varela 1976), it is not, as before, a model (Ashby 1956; Klir and Valach 1967). The observer could, of course, assume that the stimulus (the new number) is an input to the modeled (or the model) and that the next numbers on the screen are an output (response) of the modeled, but this would obviously be a delusion.

The observer may then conclude that the screen (and associated elements) displaying a random number series is a closed unity, and that a model of it cannot be constructed.

In general, then, a closed unity cannot be modeled [this is not the closure to which Ashby (1956) refers]. In order to study these closed unities it is necessary to expand the concept of model. As has already been shown, the observer can only construct another closed unity. This new unity will be called the closed model of the original unity.

4.3   Closed and Open Unities

The concept of randomness (relative to the observer) is akin to that of closed unities (relative to the observer).

A unity, then, can be closed or open with respect to an observer. If it is open, the observer can write a protocol of inputs and outputs (Ashby 1956) and derive a model (abstract or otherwise) from it. In this way the observer comes to know about these unities. However, until a model can be constructed for the unity, it remains closed for the observer, and its behavior will appear random. Only if the observer can construct a model of a unity can its behavior be predicted (deduced, in the model).

Example 1. Two unities *A* and *B* move toward each other in a unidimensional space (Figure 7). *A* moves to the right a random number of steps, and *B* moves to the left a random number of steps. Both are closed unities with respect to an external observer and with respect to each other since their behavior is random with respect to all concerned. Eventually *A* and *B* will meet and will not separate afterwards, since *A* will continue to try to move to the right and *B* will try to move to the left. This new unity *AB* will move randomly either to the left or right depending on the result of the attempted movements of *A* and *B*. For example, if *A* tries to move 7 spaces (to the right) and *B* tries to move 5 spaces (to the left), the movement of *AB* will be 2 spaces to the right, and so on. Thus *AB* moves randomly with respect to the external observer, but *A* and *B* no longer move randomly with respect to *AB*; that is, *AB* is now a closed unity with respect to the observer and *A* and *B* are open unities with respect to *AB*. Of course, *A* and *B* remain closed with respect to the observer.

A similar situation arises when unities "move" randomly with respect to each other in spaces of more dimensions and fewer restrictions. Couplings more elaborate than simple contact will then be necessary, as in the examples that follow and in the closed model for autopoiesis (Varela et al. 1974), but once the couplings have been made, the closed unities will integrate and open with respect to a new and larger closed unity, which in turn can become part of an even larger closed unity, and so on and on. Of course, the process can also be reversed by the spontaneous decay of the component unities or by the spontaneous breaking of the coupling. The components would become closed unities again. The different spaces and couplings may be not only physical but, for example, also chemical, electric, biological, magnetic, mechanical, gravitational, psychological, geographical, or social. In general, the couplings will restrict the movement of the original unities, but if the unities are complex

Figure 7

Figure 8

they can generate a movement into a different space, thereby disengaging their couplings and recovering their original, or even a new freedom.

Example 2. Unities $A,B,C,D,...$ move a random number of steps to the left or to the right in their unidimensional space (Figure 8). They are closed unities with respect to each other and with respect to an external observer as in Example 1. Eventually, two or more unities can meet and we shall assume that they move together afterwards forming, and opening with respect to, a larger unity. We shall also assume that the component unities can spontaneously separate from the larger unity, for example, through an intended movement above a certain threshold (e.g., $\tau = 10$), thereby becoming closed again.

An observer could take one of the component unities from a larger unity and let it go "free." He could consider this action as an "input" into the larger unity and the consecutive behavior of the larger unity as an "output" of it. He would then be deluding himself, because what he would be doing is no different from the spontaneous escape of a component unity from the larger unity. Therefore the larger unity has neither input nor output and consequently is a closed unity from the point of view of the observer, it cannot be modeled with open models, and its behavior cannot be predicted. However, the component unities become open with respect to each other and with respect to the larger unity that they integrate.

Example 3. If two substrate unities $(a,b,c,d,...)$ become simultaneously in contact with the catalyst K (see Figure 9), a new unity, an unbonded link $(A,B,C,D,...)$, is created and moves randomly in the unidimensional universe. The following are the possible interactions:

$$K + 2 \text{ substrates} \Rightarrow K + \text{unbonded link} \qquad (1)$$
$$\text{unbonded link} + \text{unbonded link} \Rightarrow 2 \text{ bonded links} \qquad (2)$$
$$\text{Bonded link} \Rightarrow 2 \text{ substrates} \qquad (3)$$

Figure 9

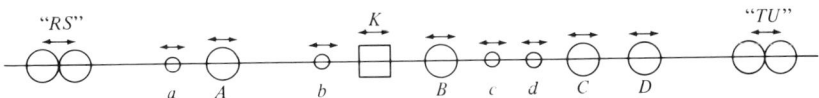

If the unbonded link encounters another unbonded link, they become bonded links and form a new unity that moves very little with respect to the other unities or not at all. As in Example 2, the substrate unities that compose a bonded (or unbonded) link can spontaneously escape from the larger unity (the link), disintegrating it, and becoming closed again, they can move freely in the unidimensional space until they again become attached. If one of two bonded links, for example, T of TU, disintegrates as a result of interaction (3), the other bonded link, U in this case, becomes unbonded (closed again) and moves randomly in its universe.

K, $A,B,C,D,...$, and $RS,TU,...$ are permeable to the substrate $(a,b,c,d,...)$; that is, if a substrate unity collides with one of these unities, for example, from the right, a substrate emerges from the left and similarly from left to right, much as when a marble strikes the end of a string of marbles. The catalyst K, which also moves randomly in the unidimensional space, can push the unbonded links $(A,B,C,D,...)$ but cannot push or go through the bonded links.

Therefore, the following circularity is established when links bond on each side of K: K becomes restricted to the region in between and keeps generating new unbonded links from the substrate, thereby replacing the decaying bonded links and securing its own enclosure.

Notice that the closed unities that form the substrate $(a,b,c,d,...)$ open with respect to each other and with respect to the unbonded link that they integrate as a result of interaction (1). This unbonded link is closed with respect to other unbonded links and with respect to an external observer, since it moves randomly with respect to all concerned. The unbonded links become open with respect to each other and with respect to the larger unity that they integrate as a result of interaction (2). Substrate and links can become closed again when bonded links disintegrate as a result of interaction (3). K, the unbonded links $(A,B,C,D,...)$, and the bonded link pairs $(RS, TU,...)$ are closed unities that open totally or partially with respect to each other and with respect to the larger unity that they integrate (defined by the circularity mentioned above). The term "partially" has been used to mean that these unities are confined to a region that can be ascertained by other unities or external observers, but their exact location remains unpredictable.

4.4 Autopoietic Unities

Autopoietic unities are closed unities (neither inputs nor outputs) formed by originally closed unities that become part of and open with respect to the autopoietic unity which they integrate.

The closed model for autopoiesis (Varela et al. 1974) is an abstract autopoietic unity whose building blocks (catalyst, substrate, and unbonded links) are closed unities (with respect to each other and an external observer) that move randomly in the habitat (restricted two-dimensional space); unbonded links meet, become bonded links and move no more (at least with respect to the chain of bonded links), and consequently they become part of and open with respect to the autopoietic unity which they integrate. They can still be closed with respect to an external observer.

The progression of events defining an autopoietic unity may now be conveniently summarized as in Figure 10, taken from Uribe (1975), in which

* represents a catalyst,

· a substrate,

○ an unbonded link,

+ a bonded link at the end of a chain, and

● a bonded link.

Figure 10

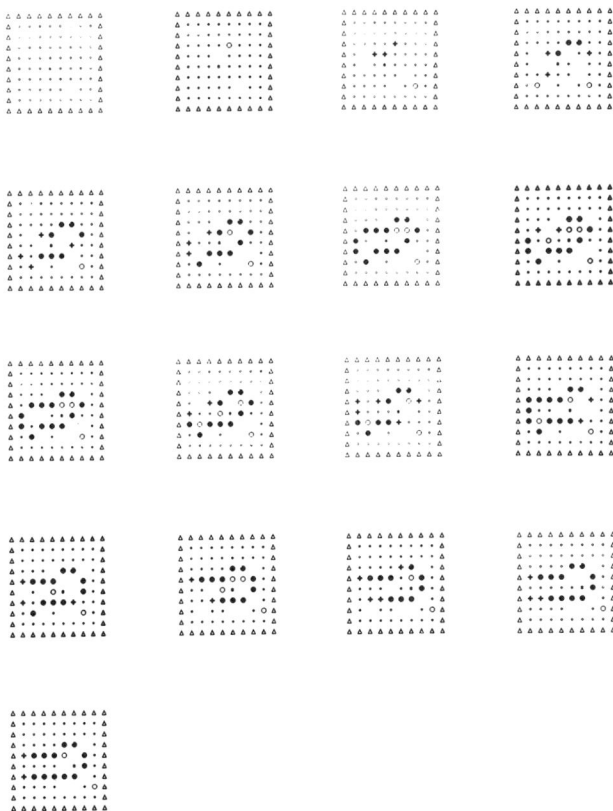

In Figure 10, the possible interactions are as follows:

Composition: $* + 2 \cdot \Rightarrow * + \circ$ (4)

Concatenation: $\circ + \circ \Rightarrow + +$ (5)

$+ + + \circ \Rightarrow + \bullet +$

$+ \bullet — — — — \bullet + \ + \circ \Rightarrow + \bullet\bullet — — — — \bullet +$

Disintegration: $\bullet \Rightarrow 2 \cdot$ (6)

Autopoiesis and self-organization are concepts relative to an observer (Pask 1961, 1962): If an autopoietic unity opens with respect to an observer, it becomes allopoietic with respect to the observer. Thus, if an observer knows, for example, the seed of the pseudorandom number that generates the closed model for autopoiesis (Varela et al. 1974), and hence can construct a model of the closed model and deduce with certainty its behavior, the closed model becomes open and therefore no longer an autopoietic unity, for the observer.

Autopoietic unities are autonomous unities and hence can relate to other autopoietic unities only through a reciprocal opening that will tend to create a new, larger unity (which may be either autopoietic or not). For example, cells, which are autopoietic unities, become open (allopoietic) with respect to the multicellular living organism (autopoietic unity) that they integrate, neurons (autopoietic) become open with respect to the nervous system (allopoietic) but not necessarily with respect to other neurons, and ants (autopoietic) become allopoietic with respect to the anthill (autopoietic). Cells, neurons, and ants remain autopoietic with respect to an external observer.

In the above examples (1–3) and in the closed model for autopoiesis, the interacting unities were relatively simple and their behavior was restricted to few dimensions. Autopoietic unities such as living organisms are more complex and behave in spaces of many more dimensions (physical or otherwise, as stated before). This richness of behavior is restricted, however, by the behavior (including the special cases mentioned at the outset) of the unities that constitute its environment through the couplings, more or less rigid and more or less permanent, that will develop among unities. As a result of these couplings, autopoietic unities can become allopoietic with respect to the larger unity (autopoietic or not) that they integrate; examples are the cell in the multicellular organism, the ant in the anthill, the human being in society. When this is the case, the decay (death) or generation (birth) of the component unities are simply aspects of the behavior of the larger composite unity.

4.4.1 The Observer

Autopoietic unities define the environment in which they can exist in relation to their autopoiesis, and some (the observers) interact recursively with this environment (acting and modeling) through their descriptions, it being impossible for them to step out of this relative descriptive domain through descriptions (Maturana and Varela 1976).

Human observers are autopoietic unities who learn and create their environments through modeling (intention of predicting the environment's behavior). In this environment dwell other autopoietic unities (including the observer), and they are subject to the same intention of the observer to predict (model) their behavior. Eventually two or more observers (autopoietic unities) become coupled through one or more domains of interaction; they integrate as a larger unity (couple, family, tribe, community, or society); they lose their autonomy and eventually their own autopoiesis (i.e., they become allopoietic, at least with respect to the larger unity).

The complexity (and richness) of human behavior allows the breaking of the couplings and the restoration of autonomy and autopoiesis in certain cases. But excessive training (as opposed to learning), of whatever sort, reduces the dimensions of behavior of an autopoietic unity and makes probable its coupling with other autopoietic unities subject to a similar training. New, larger unities are formed that render the component unities allopoietic. Examples of these are the military (and similar) institutions. Sometimes the training is forced (as in oppressive governments or institutions), sometimes it is more subtle, in such forms as education, cultural legacies, economic pressures, and the like. Either way, the results are the same: closed autopoietic unities open with respect to the larger unities that they integrate. It is possible that some may spontaneously, creatively, become closed again.

ACKNOWLEDGMENTS
The author wishes to express his gratitude to Richard Herbert Howe and Paul Weston for their constructive suggestions and encouragement.

References

Ashby, W. R. (1956), *An Introduction to Cybernetics,* Chapman and Hall, London.

Foerster, H. von (1976), An epistemology for living things, in *Collected Works of the Biological Computer Laboratory* (K. Wilson, ed.), Illinois Blueprint Corp., Peoria, Ill.

Klir, J., and Valach, M. (1967), *Cybernetic Modeling,* Illife Books, London.

Löfgren, L. (1962), Self-repair as the limit for automatic error correction, in *Principles of Self-Organization* (H. von Foerster and G. W. Zopf, eds.), Pergamon, Oxford.

Maturana, H. (1980), "Representation and Communication Functions," in *Encyclopedia Pleiade* (J. Piaget, ed.) (in press).

Maturana, H., and Varela, F. (1976), Autopoietic systems, in *The Collected Works of the Biological Computer Laboratory* (K. Wilson, ed.), Illinois Blueprint Corp., Peoria, Ill.

Pask, G. (1961), *An Approach to Cybernetics,* Hutchinson, London.

Pask, G. (1962), A proposed evolutionary model, in *Principles of Self-Organization* (H. von Foerster and G. W. Sopf, eds.), Pergamon, Oxford.

Spencer Brown, G. (1969), *Laws of Form,* Allen and Unwin, London.

Uribe, R. (1975), Autop, in the *Univ. of Illinois Computer Based Educational System* (PLATO),

Uribe, R. (1977), Organizations and uncertainty, in *Proc. Soc. for Gen. Systems Res., Denver* (coord. J. D. White).

Varela, F., Maturana, H., and Uribe, R. (1974), Autopoiesis: The organization of living systems, its characterization and a model, *Biosystems 5,* 187–196.

Wittgenstein, L. (1961), *Tractatus Logico-Philosophicus,* Routledge & Kegan Paul, London.

Jantsch starts rather boldly by asserting that this volume mirrors "the unusual degree of parochialism, defensiveness and quasi-theological dogmatism that has arisen around autopoiesis." This "observation" is based on a little chat he had with Maturana and Varela, as he dutifully acknowledges in the footnote. Such a gossipy introduction should not distract the reader from Jantsch's major observation: "Autopoiesis is a central aspect of dissipative self-organization."

Jantsch attempts to explore the relationship of autopoiesis to dissipative structures of Prigogine and the hypercycles of Eigen. The reader thus receives a useful discussion of these two concepts as an additional benefit; Jantsch displays a confident grasp of both.

There is rather free talk about production and dissipation of *entropy,* almost as if it were some sort of "substance." Readers should remind themselves that entropy is an abstraction enabling us (observers) to measure chaos. References to its "production" or "dissipation" should be viewed in their proper metaphorical framework.

Jantsch introduces a new term: dissipative self-organization. The autopoietic distinction between structure and organization is thus complemented by the notion of "function" (an observer-dependent notion, teleonomic description), and their mutual interdependency is emphasized. Although the organization might be primary and the structure only emergent, the structure (not "function") does affect the organization itself, leading to what is being perceived as self-organizing phenomena. The word "dissipative" is given a prominent role, although it is more the order through fluctuation than dissipation of entropy that Jantsch is trying to convey.

This paper is very interesting in its efforts to subordinate autopoiesis to the notions of self-organization (see Morin on the same issue, p. 128). Yet the concept of autopoiesis permeates the paper on a much more profound level than the notions of dissipation or hypercycles. Jantsch intuitively embraces autopoiesis, although rationally he still feels obliged to frame it within the older paradigms. This propensity to preservation is further underlined by his inexplicable optimism about the usefulness of the purely structural catastrophe theory and its "graphical appeal." But he does confirm his intuitive-theological grasp of autopoiesis by concluding, "I am sure some [new insight] must exist."

The concept of the hypercycle might not become entirely clear from contemplating Figure 1. In simpler terms, a hypercycle refers to a cyclical interdependency of several nucleic-acid cycles; that is, nucleic acid may "code" for the making of another protein molecule while the protein molecule promotes further nucleic-acid reproduction. In order to establish a hypercycle, one of the nucleic-acid cycles has to produce a protein that would help a second nucleic-acid cycle to work more efficiently. That cycle in turn makes a protein assisting a third nucleic-acid cycle, and so on, until the nth protein becomes instrumental in enhancing the *first* nucleic-acid cycle. This is a simple example of organizational closure, an underlying concept of autopoiesis.

The reader will find a simple description of the Belousov–Zhabotinsky reaction, the one mentioned also in Varela's paper. Its closed organization of component-producing processes is readily apparent. One may wonder why there is so much effort expended to keep so many concepts separated.

Jantsch then presents a series of examples of organizationally closed systems, including fission-breeding cycle, ecosystem, the Gaia system, and so on. He concludes his paper with the notion of ultracycle, forming a basis for coevolution, learning, and communication.

Erich Jantsch was born in Vienna in 1929. He received a doctorate in astrophysics from the University of Vienna in 1951. Dr. Jantsch is a well-known international lecturer and writer. His major visiting appointments included M.I.T., Technical University of Hanover, Portland State University, University of Bielefeld, University of California at Berkeley, Technical University of Denmark, University of Paris, Institute of Advanced Studies in Vienna, University of Lund, Graduate School of Economic and Social Sciences in St. Gall, and the University of Kassel. Since 1971 he is living mostly in Berkeley, California. Dr. Jantsch has published eight books, including *Technological Forecasting in Perspective, Technological Planning and Social Futures, Design for Evolution, Evolution and Consciousness* (co-editor with C.H. Waddington), and *The Self-Organizing Universe. Address:* 1962 University Avenue, Apt. 4, Berkeley, California 94704.

Chapter 5
Autopoiesis: A Central Aspect
of Dissipative Self-Organization

Erich Jantsch

5.1 Introduction: A Challenge to System Theory

Autopoiesis has been called a kind of system dynamics through which the system regenerates its own components and thus itself (Maturana and Varela 1975). It implies organizational closure, in other words, cyclical or recursive arrangement of the processes directly involved in this self-reproduction. As realistic examples, biological cells and organisms are usually cited. However, as will be exemplified in this paper, metabolism and self-regeneration (or, more generally, self-reproduction)—which are the familiar dynamic features addressed by the new word—are by no means restricted to the realm of the living, nor are the capabilities of evolving through mutations and of competing for selection in a scarce environment. Although life is often defined by these dynamic features, they are, at varying degrees, characteristic of a wider class of matter systems. If this appears confusing on the one hand, it implies a great opportunity on the other, that is, the formulation of a truly general dynamic system theory embracing the nonliving as well as the living. The unifying perspective, as will be shown in this paper, is provided by the paradigm of dissipative self-organization.

As Roland Fischer (1976) remarks, the recent surge of interest in self-referential systems, exemplified by the discussions triggered by the concept of autopoiesis, may be interpreted as a manifestation of the renewed and pressing search for our own self. This aspect may help to explain the unusual degree of parochialism, defensiveness, and quasi-theological dogmatism that has arisen around autopoiesis—and which is mirrored in this volume.

The self-referential aspect of autopoiesis apparently has led the authors of this concept to insist on purely self-referential descriptions, not just of the system in question, but also of its dynamics.[1] This has led to considerable confusion and tends to obscure the issues. Instead of becoming a fruitful concept in the development of a process-oriented system theory—going far beyond the old structure-oriented theory—autopoiesis is in danger of becoming a sterile tautology. Varela's chapter in this volume includes a partial and intuitive attempt to overcome this danger, but it fails to point out a systematic perspective. Such a systematic perspective is provided by modern theories of self-organization that can no longer be ignored in serious discussions of autopoiesis. This paper attempts to briefly sketch this perspective. As a phenomenological concept, autopoiesis is most useful where it provides a focus for studying the holistic behavior of a self-organizing system. The underlying dynamics of self-organization, however, have been the subject of intense theoretical and empirical study over the past decade and are now basically understood (Eigen 1971; Glansdorff and Prigogine 1971; Nicolis and Prigogine 1977; Eigen and Schuster 1977). Building on this foundation, autopoiesis may now be recognized as one of the ways in which the self-organization of nonequilibrium systems manifests itself.

5.2   Two Kinds of Self-Organization

Self-organization is possible in two entirely different ways. One way arises from the interaction of *conservative* attractive and repelling forces and leads to stable *equilibrium structures,* which may either be static (e.g., crystals) or dynamic (e.g., planetary rotation in stable orbits). There is no energy flow involved in the maintenance of an equilibrium. In contrast, *dissipative structures* represent a kind of self-organizing dynamic order that maintains itself through continuous exchange of energy with the environment, in other words, through maintaining a metabolism. They arise spontaneously under far from equilibrium conditions and establish for the system a certain *autonomy* from the environment. Whereas equilibrium structures, such as crystals, may grow indefinitely, dissipative structures find and maintain their optimal size, independent of the environment.

One-way reactions or reaction systems, such as chains or treelike pat-

---

[1] I promised this footnote to Humberto Maturana and Francisco Varela in order to make sure that the reader is not mistaken about their unhappiness with my understanding of autopoiesis. Here are characteristic samples of personal communications, faithfully recorded over a sumptuous dinner in San Francisco on 11 January 1978: "Autopoiesis is autopoiesis and nothing else" (Maturana). "Autopoiesis is *not* self-organization; however, it may be linked to the latter" (Varela). "A dissipative structure is not an autopoietic system, because the reference to dissipation is already a reference to something else but autopoiesis; autopoiesis refers only to itself" (Maturana). "Social systems are *not* autopoietic systems" (Varela).

terns, move towards equilibrium. They represent *allopoiesis*. Only a cyclical organization of processes is capable of maintaining nonequilibrium indefinitely. Continuous autopoiesis, in turn, depends on such a maintenance of nonequilibrium.

In autopoiesis, the system components are continuously generated and may degenerate at the same global rate (self-maintenance or self-regeneration) or at a lower rate in periods of system *growth,* if the latter is included in the concept. This implies first that autopoietic systems are *open* towards the exchange of matter and/or energy with the environment, second that since the generation of components involves work, entropy is produced and has to be *dissipated,* and third that if reactions are supposed to run irreversibly in the same direction—or at least have a net vector in that direction—the system has to be *far from equilibrium.* At equilibrium, the reactions may reverse themselves and fluctuate randomly in both directions. The three conditions are interdependent and are basic to self-organization in nonequilibrium systems, or dissipative self-organization.

We may distinguish between the organization, function, and structure of a system. *Function* refers to the nature of the essential processes, the type of reactions and reaction kinetics they represent, and the reaction participants they involve. Thus, function also includes, at a more coarsegrained level of perception, the logical *organization* of these processes, in particular their cyclical organization in dissipative self-organization. *Structure* refers to the space–time structure, or dynamic regime, that is assumed by these processes and their participants. Function is not directly visible; structure is. Structure represents macroscopic order; its spatial aspect is form, its temporal aspect the changes this form undergoes. In dissipative structures, function and structure are complementary:

$$\text{Structure} \longleftrightarrow \text{Function}$$

This complementarity is one of the principles underlying self-organization at a profound level.

Dissipative structures mediate *cooperative behavior.* In other words, processes do not depend on the random encounter of reaction participants. In particular, a dissipative structure may regulate the function of the system in such a way that degeneration and regeneration of its components are balanced and the structure maintains itself in the same dynamic regime and in the same dynamic (usually pulsating) shape. In this case, we may even speak of autopoiesis of the structure—with the understanding, though, that Maturana and Varela applied this term only to organization, not structure, or strictly speaking, not even to the processes in the organization, only to the components (which, in principle, may be regenerated along more than one pathway).

The same organization may underlie different configurations of function and structure. Function and structure can evolve within the same organizational logic. This is the case, for example, if the nonequilibrium changes significantly. The same basic reaction cycle may operate in different dynamic regimes and in structures that look different. This is the root of *morphogenesis*, the emergence and evolution of form, which is of particular importance in biology. The same basic biochemical organization of cells may underlie vastly different cell functions, depending to a large extent on the spatial positioning of the cell in the development of the embryo. Typically, cells may undergo an evolving sequence of different functions (histogenesis). Positioning, in turn, depends on the space–time structure of the processes. The buildup of solid or regenerative elements permits the reinforcement of processes (for example, by enzyme fixation on membranes) and thereby the "miniaturization" of pathways.

The "aloofness" of structure and function from the basic organization of the dynamic system is responsible for the tremendous morphological and functional variety found in nature. A recent mathematical development based on differential topology, *catastrophe theory* (Thom 1972; Zeeman 1977), has become a valuable tool for the study of morphogenesis. However, it cannot express self-organization, but only the movement of a system through a "morphogenetic landscape," with the dynamics imposed from outside. In contrast, the theory of dissipative structures deals with the nonlinear self-reinforcement of fluctuations within the system, which may drive the system into a new dynamic regime (function). Around the instability region between the old and the new regime, the "law of large numbers" becomes invalid and it is not the system state (measured in any terms) that is decisive, but the way in which the fluctuation propagates itself and breaks through. Therefore, the order principle at work beyond the thermodynamic (equilibrium) branch has been appropriately named *order through fluctuation* (Glansdorff and Prigogine 1971). The microscopic view, focusing on the individual fluctuation, has led to a stochastic description (Nicolis and Prigogine 1977) that extends the original complementarity of levels of description to a triangle:

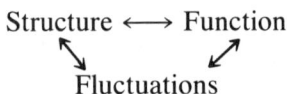

$$\text{Structure} \longleftrightarrow \text{Function}$$
$$\searrow \qquad \nearrow$$
$$\text{Fluctuations}$$

In other words, the dissipative structure may itself be understood as a giant fluctuation.

While structure and function evolve, the basic organization—in particular, the basic autopoietic cycle—may remain the same. Ultimately, of course, the basic organization may itself evolve, through mutations in the reaction participants, changes in the cyclical pathways, or both. Apart from external impact (such as molecular changes due to cosmic radiation), the evolution of the structure will usually lead to, or at least be closely associated with, such a system evolution. Generally, such an evolution will be open, leading irreversibly away from the original system. However, an example will be outlined in Section 5.4.2, in which the evolutionary sequence itself forms a closed cycle, so that we may speak of an autopoietic system that is cyclical with respect to its own mutants.

I hope that this sketchy outline helps to make it clear why the dynamics of dissipative self-organization inherently defy any attempt to deal with them in terms of an equilibrium-oriented system theory in the spirit of, say, Ross Ashby (1956, 1960). But the question may also be asked at this juncture, what may be gained by employing the nonequilibrium concept of autopoiesis, if the latter disregards structure as well as its modes of evolution. This question is the more justified, since a sound theoretical basis already exists in the wider framework of self-organization, including the aspect of autopoiesis. Catastrophe theory, while dealing only with certain aspects of morphogenesis, has made a successful claim for recognition because of a certain graphical appeal and simplicity in a range of applications. What new insight, what kind of appeal, what extra simplicity and flexibility will the concept of autopoiesis have to offer? I am sure some must exist. One way to capture more of what is going on in autopoietic dynamics would be to recognize both autopoietic system organization and autopoietic dissipative structure in the same system. After all, a dissipative structure also regenerates itself continuously; the emphasis here, however, is on cooperative process patterns, or dynamic regimes.

5.3   A Hierarchical Typology of Reaction Cycles Underlying Self-Organizing Systems

In the introduction to their seminal trilogy on hypercycles and precellular evolution, Manfred Eigen and Peter Schuster (1977) sketch a simple hierarchy of cyclical reaction systems that are of importance in natural dynamics: A transformatory reaction cycle acts globally as a catalyst, a cycle of catalysts acts globally as an autocatalyst, and a catalytic cycle of autocatalysts acts globally as a hypercycle. The concept of the hypercycle, introduced by Eigen (1971), becomes clear from Figure 1. In this figure, the simple scheme by Eigen and Schuster is extended and modified so as to provide a unifying perspective to a wider range of cyclical phenomena.

Whether all levels except that of equilibrating systems should be called autopoietic, or just the level of self-regenerative or self-maintaining systems, is a matter of definition. Maturana and Varela (1975) seem to focus mainly on self-maintaining systems, but they do emphasize occasionally that self-reproduction qua growth follows from autopoiesis, or is even

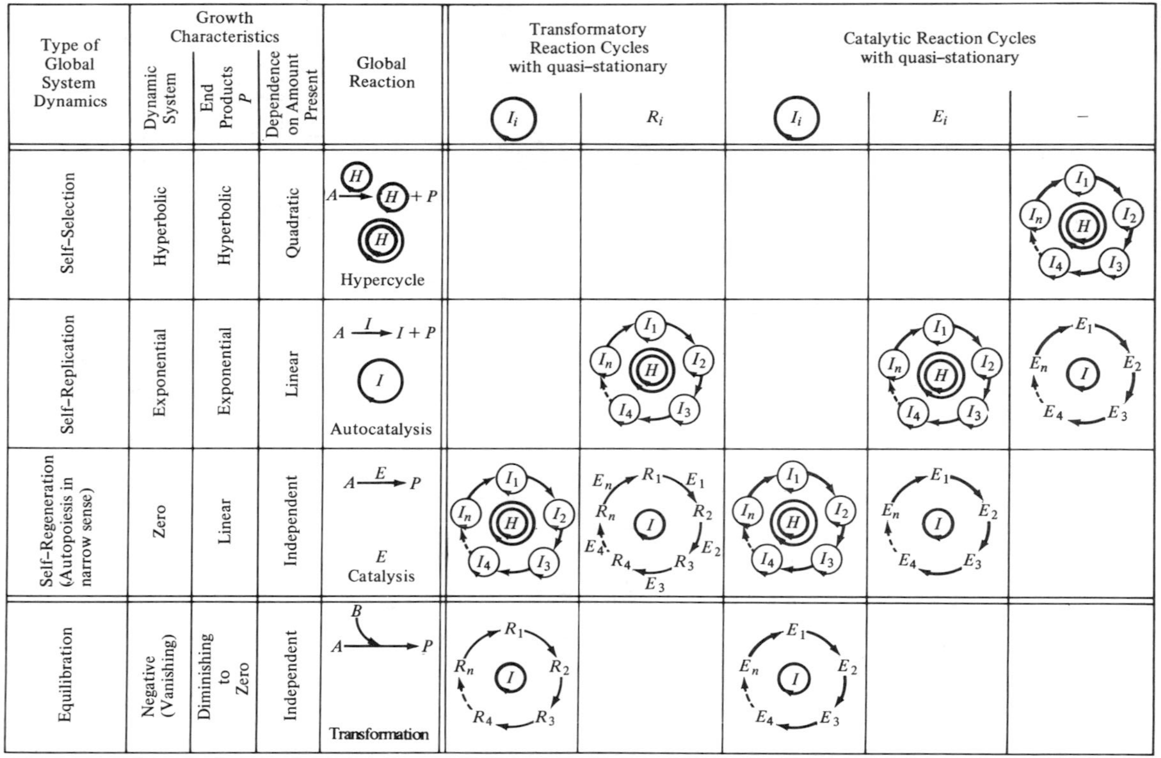

Figure 1. A generalized hierarchical scheme of reaction cycles with varying degeneration characteristics: $\longrightarrow$ transformation, $\longrightarrow$ catalytic action.

included in the concept. In either case, autopoiesis in the meaning of the system's self-making is involved; the difference lies in the balance or imbalance of the rates of generation and degeneration. Self-maintaining dynamic systems are of particular importance in nature, especially where their balance is mediated by dissipative structures.

Both transformatory and catalytic reaction cycles degrade and dissipate energy, either by converting free energy directly into heat (as in photosynthesis), or by converting energy-rich materials into energy-deficient ones. Transformatory cycles may be aided in this task by catalysts that are not themselves part of the cycle; catalytic cycles may be added by auxiliary transformation cycles. Biological systems usually combine many such cycles or parts of cycles in an intricate web, which forms a global cycle itself and which makes frequent use of widely applicable intermediary products, such as ATP (adenosine triphosphate), the "energy coin" of the cell.

The reactions in a transformatory cycle may be of any kind, in particular nuclear or chemical. Catalytic action may be either chemical (adding a transformatory reaction) or template action, in which a positive form instructs the synthesis of a negative form, from which a positive form may be obtained in turn. In the former case, chemical catalysis is usually of the heterogeneous type, which means that it constitutes the global effect of a small transformatory cycle in which the initial reactant is adsorbed onto the surface of the catalyst and transformed, the end product separated, and the catalyst recycled. In the latter case, complex information may be passed on by means of template action, usually involving chains of nucleotides whose molecular form is very conspicuous and easily recognizable. Autocatalysis, in this latter case, is then self-instruction for reproduction and always involves a catalytic cycle in which positive and negative forms alternate. Such a microscopic view reveals the realistic basis of the hierarchy noted by Eigen and Schuster (1977).

For hypercycles to attain their characteristic dynamics, it is not necessary that all participants in the cycle are autocatalytic. Usually, the minimum requirement is only one autocatalytic step.

The same type of cycle appears at different hierarchical levels in Figure 1. The type of *growth* associated with the cycle is determined not only by the generation mechanisms but also by the built-in degeneration or diffusion mechanisms. The latter may be characterized by the level of cycle participants that remain quasi-stationary (their concentration may fluctuate a little), the participants R_i in the transformatory cycle, the catalysts E_i, the autocatalytic units I_i, or the hypercycle H itself. There are rarely natural dynamics leading to higher than hyperbolic growth. Therefore, whereas it would be possible to continue indefinitely with higher hierarchies, there is little practical interest in it.

Since all cycles are dissipative and maintain exchange with the environment, they may be thought of as mediators of global reactions, ultimately turning starting products that enter the cycle into end products that leave the cycle. There is always net entropy production associated with self-organization. There is always metabolism or, in terms of the new words, *autopoiesis always goes together with allopoiesis.*

The frequently cited tesselation models and computer simulations of autopoietic systems (Varela, Maturana, and Uribe 1974; Zeleny and Pierre 1976; Zeleny 1977) constitute the only serious work done so far with the focal concept of autopoiesis in mind. They seem to deal mainly with transformatory reaction cycles evolving to a dynamic state with quasi-stationary R_i, aided by a static catalyst, and address thus more or less the type exemplified in Section 5.4.2. It is interesting to note that these simulation studies do *not* deal with the basic organization of the system whose cyclical nature is nowhere evident in the models. Rather, they permit tracing of the evolution and global stabilization of function and structure and include explicitly the open (allopoietic) metabolic reaction chain while neglecting the latter's significance from a point of view of energy degradation: No distinction is made between the quality of starting and end products. Thus, the only cycle that is graphically visible is the wrong one; products rejected by the metabolism may be readily picked up again to feed it—which, of course, would make the system dynamics stop or run down quickly. I conclude that the models present superficial analogies rather than homologies (dynamics related in kind). Once homologous models will have been achieved, the real task remaining is the homologous translation of the model rules into realistic (especially biological) principles, and vice versa. Only then will it become possible to learn something new from these models and simulation runs.

Eigen and Winkler (1975) have systematically explored the simulation of the quantitative evolution of most of the types appearing in Figure 1. Again, the logical link between stipulated rules of the tesselation game and realistic principles has yet to be established in an explicit way.

5.4   Realistic Examples of Reaction Cycles in Self-Organizing Systems

"Pure" examples of cyclical systems are not so easy to find. However, this should not lead to the conclusion that cyclical organization plays a relatively minor role in nature. It is present in the global link-up of many full and partial cycles that feed into each other. Their microscopic representation may surpass by far the complexity of the basic hypercycle. The representation chosen depends on the level of resolution sought. With an appropriate sacrifice in resolution, each self-organizing system may be represented by the types of cycles appearing in Figure 1 and reaching up to the hypercycle.

5.4.1 Equilibrating Systems: Vanishing Dynamics

Transformatory or catalytic cycles continue to run down until the system reaches its equilibrium. Around the equilibrium, the reactions become reversible and the system fluctuates around its equilibrium. For any practical purposes, its dynamics vanish. Equilibrium is equivalent to death.

5.4.2 Self-Regenerative (Self-Renewing) Systems: Zero Overall Growth

This level represents autopoiesis in the usually employed, narrow sense of the term. The global effect of these systems is that of a catalyst for the open metabolic reaction chain.

Transformatory Hypercycles with Quasi-Stationary I_i. Chemical reaction systems of this type give rise to truly and spontaneously self-organizing process structures under conditions far from equilibrium. These chemical reaction systems are usually referred to as the simplest examples of dissipative structures beyond the thermodynamic branch of equilibrium systems. The precondition for the onset of cooperative behavior is at least one autocatalytic step, to be precise, a trimolecular step (Nicolis and Prigogine 1977, p. 90 ff.). The best-studied realistic reaction cycle of that type is the *Belousov–Zhabotinsky reaction,* which may be written in the simpler form of the theoretical study model "Oregonator" as follows:

$$A + Y \rightarrow X$$
$$X + Y \rightarrow P$$
$$B + X \rightarrow 2X + Z$$
$$2X \rightarrow Q$$
$$Z \rightarrow fY$$

Starting products	$A = B = [BrO_3{}^-]$
Intermediary products	$X = [HBrO_2]$
	$Y = [Br^-]$
	$Z = 2[Ce^{4+}]$

End products P, Q

(stoichiometric factor f)

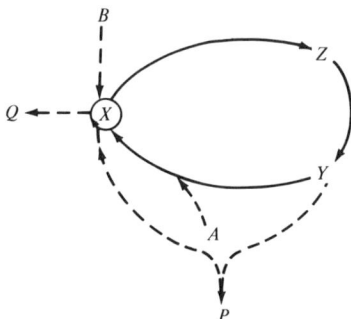

Figure 2. The Belousov–Zhabotinsky reaction is organized as a transformatory hypercycle with one autocatalytic step. (For the meaning of symbols see the system of equations in the text.)

The intermediary products X, Y, Z are organized in a transformatory hypercycle, with X representing the only autocatalytic unit, as shown in Figure 2. The reaction system is maintained by the dissipative structure as long as the environment is supportive with respect to free energy and metabolic reaction participants A, B.

A controlled *fission breeding cycle* also constitutes a hypercycle of this type, as shown in Figure 3. A straight fission reaction, without breeding material, would constitute a single autocatalytic unit, the neutron. In such a nuclear exogenous cycle, the "metabolism" consists of dissipation of energy liberated by fission reactions within the system itself; no energy is needed from the environment. In critical assemblies of high energy flow density, the dissipative structure would blow the system apart were it not contained or fixed in solid form (although dilute liquids may conceivably

Figure 3. A fission breeding cycle with uranium U^{238} organized as a transformatory hypercycle in which the autocatalytic unit is the neutron n. There are two steps of radioactive decay. The same cycle may use thorium Th^{232} instead of U^{238} as breeding material, in which case the participants in the cycle are Th^{233}, Pa^{233}, and U^{233} (U—uranium, Np—neptunium, Pu—plutonium, Th—thorium, Pa—protactinium).

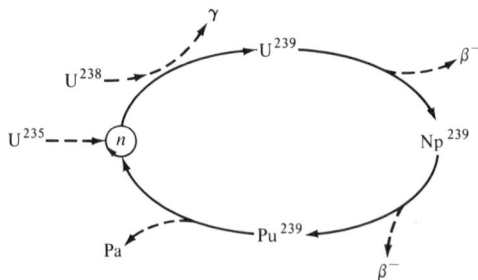

give rise to pulsating dissipative structures). In high-energy-density assemblies, such as reactors, the "degeneration" of neutrons has to be mediated by an auxiliary control function regulating escape and capture of neutrons outside the cycle. It is only because of delayed neutrons that this is generally an easy task.

A mature (globally stable) *ecosystem* is an example of a hypercycle in which practically all of the participating matter—even the metabolic end products—is recycled (Figure 4). Herbivores eat plants, carnivores eat herbivores (and possibly plants, too), carnivorous predators eat carnivorous prey, and so forth, until the last predator that is nobody's prey dies a natural death and its matter is broken down by all sorts of animals and bacteria to the original molecules and atoms and reenters the cycle through the plants, as do metabolic waste products. The global reaction, the overall metabolism, in this case is the degradation of energy-rich photons entering through photosynthesis and leaving as energy-deficient photons or heat radiation.

A particularly interesting hypercycle of this class, linking the smallest living units with the macrosystem of the planetary atmosphere, is the *Gaia system* (Margulis and Lovelock 1974). Among other things, it provides a cycle between oxidizing and reducing reactions, turning oxygen into carbon dioxide and vice versa. The autocatalytic units managing these processes are almost exclusively procaryotes (nucleus-free single-cell organisms, such as soil bacteria) or their descendants in the eucaryotic cell, the mitochondria (oxidizing) and the chloroplasts (reducing), which—

Figure 4. A mature ecosystem is organized as a transformatory hypercycle in which all materials are recycled (dotted arrows inside the cycle indicate that also metabolic end products are recycled by plants. *P*—plants, *H*—herbivores, C_i—carnivores, at various steps in the cycle. The cycle catalyzes the degradation of energy-rich photons in the visible light range (γv) to energy-deficient photons in the infrared region (γ_{IR}).

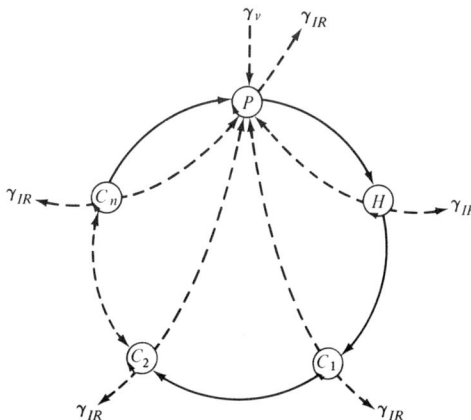

according to the endosymbiotic theory of the same Lynn Margulis (1970)—are former procaryotes joining in the more complex eucaryotic cell. The fascinating aspect of the Gaia hypothesis lies in the postulated autopoietic and self-regulating characteristics of the hypercyclic macrosystem, which apparently stabilized itself dynamically about 1.5×10^9 years ago, when free oxygen reached its present concentration. This globally stable state is far from equilibrium. Whereas oxygen and nitrogen dominate in a static view, other gases, such as hydrogen, methane, ammonia, carbon monoxide, and carbon dioxide, all show approximately the same through-flow in a dynamical view. Only carbon dioxide increases markedly due to nonbiological (industrial) activity. The Gaia hypothesis provides the most sweeping illustration for one of the most basic evolutionary principles—namely, that life itself creates the conditions for its further evolution. Life and its environment are linked in a hypercycle.

Transformatory Reaction Cycles with Quasi-Stationary R_i. The basic biochemical reaction cycles, involved in the energy management of the cell, are of this type. Eigen and Schuster (1977) cite the tricarboxylic or citric acid cycle for the oxidation of fuel molecules, also called the Krebs cycle after its principal discoverer. Another example is the glycolytic cycle, which extracts energy from glucose and stores it in ATP; it is linked to the Krebs cycle. All these biochemical cycles cannot run by themselves, since they do not include an autocatalytic step. Therefore, practically every step needs a catalyst provided from outside the cycle.

A nuclear reaction cycle of this class, however, can well run by itself if one or more steps involve spontaneous radioactive decay. The effect is the same as if the decay product were acting as an autocatalyst, "pulling" its own kind from the precursor nucleus. The most interesting example of such a cycle, also cited by Eigen and Schuster (1977), is the *carbon cycle,* proposed in the 1930s by Hans Bethe and Carl Friedrich von Weizsäcker (Figure 5). It is assumed to play an important role in the energy production of younger stars, such as our sun, in which the fusion of hydrogen to helium is the basic energy process. A precondition for the onset of this cycle is the "pollution" with some heavier elements (about 2% in the sun), which have been formed in older stars and distributed by supernova explosions. Evidence is mounting that such a supernova explosion triggered the formation of the sun and the solar system from a protostellar cloud. In the carbon cycle, the reaction participants are reconstituted in each cycle almost identically, except for a few nucleons that they may have exchanged. The important point is that radioactive decay slows down the cycle tremendously. Direct fusion, involving the strong nuclear force, would imply a vast fusion explosion instead of controlled burning. However, the direct fusion of hydrogen nuclei (protons) to helium is not possible; just a few percent are lacking in the strong forces

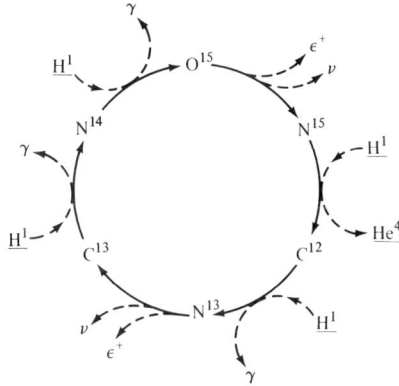

Figure 5. The carbon cycle according to Bethe and Weizsäcker. It catalyzes the fusion of four hydrogen nuclei (protons) H^1 into one helium nucleus He^4. The isotopes of carbon (C), nitrogen (N), and oxygen (O) are reconstituted in the cycle. Energy is dissipated in form of γ radiation, positrons (ϵ^+), and neutrinos (ν).

to overcome a threshold. The weak nuclear force governing radioactive decay acts slower by a factor of about 10^{18}—a fact of tremendous importance for the unfolding of complexity in evolution.

Catalytic Hypercycles with Quasi-Stationary I_i. A particularly interesting hypercycle of this type may be recognized in the evolution of the *influenza virus A*, which has been clarified through the work of Stephen Fazekas de St. Groth (see Staehelin 1976). This virus evolves within a subtype through a number of years, characteristically by means of replacing one amino acid with a bigger one. The "forward" evolution to escape immunization by coevolving antibodies leads sooner or later into a cul-de-sac. Thus, every decade or so, the basic form of the subtype (which must have survived in small numbers) mutates in another amino acid position to form a new subtype. When the first strain of the new subtype appears, it usually causes a so-called worldwide pandemic with the most severe cases of illness. Only in such areas where an intermediate strain—a necessary step in the evolution of a new subtype—has caused local epidemics has the human immune system halfway caught up and can provide some protection. Because of the limited availability of positions in which amino acid mutations can occur in a virus, the evolution of subtypes form a closed cycle (Figure 6). It seems hardly accidental that the period of the cycle is approximately 70 years, the average lifetime of the human "prey." Were this time shorter, the antibodies formed in a young person, and becoming part of a lifelong personal library of antibodies, would leave the influenza virus little chance. Of course, the autocatalytic units in this

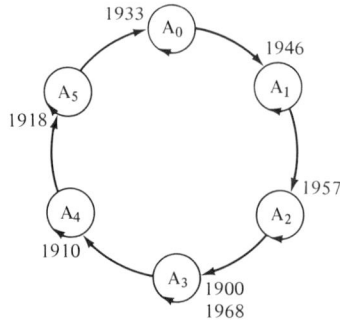

Figure 6. The cyclical evolution of subtypes of the influenza virus A forms a catalytic hypercycle that stretches over approximately 70 years, the average human lifespan. In this way, the subtypes escape immunization by antibodies remaining in the human organism from earlier attacks of a subtype.

cycle (the individual subtypes) are not stationary in a strict sense, but peak with the epidemic they cause. The cycle as a whole, however, is autopoietic and self-regenerative, strung out over a long period of time.

In the socioeconomic domain, a cycle of this type may be found in the way in which a number of service sectors of the economy stimulate each other in a balanced way (Figure 7).

Catalytic Reaction Cycles with Quasi-Stationary E_i. In the cell metabolism, there are many self-regulating cycles that mediate the production and activity of enzymes (biological catalysts). The production of an enzyme is catalyzed by nucleic acids, but the compound that the enzyme synthesizes in turn may interact with this enzyme to inhibit further activity. Inhibition and activation may be considered equivalent to catalytic

Figure 7. An example of a catalytic hypercycle linking several service sectors of the economy.

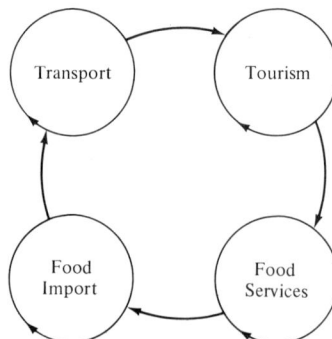

action. Any such negative- or positive-feedback coupling to a precursor forms an autopoietic cycle and may give rise to a dissipative structure that usually manifests itself through oscillations.

Intricate cross-regulation occurs in intercellular communication, for example in the synthesis of cyclic adenosine monophosphate (cAMP; cyclic AMP), which is the basis of chemotaxis, or mutual chemical attraction, in the aggregation of single-cell amoebae to form the slime mold *Dictyostelium discoideum* (Figure 8). ATP pyrophosphohydrolase (E_1) is activated by cAMP and catalyzes the production of 5'AMP from ATP. Adenyl cyclase (E_2) is activated by 5'AMP and catalyzes the production of cAMP from ATP. A third enzyme, phosphodiesterase (E_3) regulates the transformation of cAMP into 5'AMP, and a fourth enzyme, 5'nucleotidase, mediates the degradation and exit from the system of 5'AMP. In slime mold aggregation, this cycle transcends the single cell because acrasine, which contains cAMP as its active element, is secreted by the individual amoebae. When food (bacteria) is scarce, acrasine secretion assumes a marked pulsatory mode that is felt as a frequency signal in the respective biochemical cycles of other amoebae, which, in turn, start secreting in pulses. It is this signal that triggers chemotaxis and aggregation. The global effect may again be depicted as a catalytic cycle in which the individual amoebae appear as catalysts, each one inducing a pulsating mode in others and receiving the same signal, each one attracting and being attracted at the same time. The result is a truly self-organizing system of the slime mold. This system is also a good example for biological information in which each receiver is also a sender.

5.4.3 Self-Replicating Systems: Exponential Growth

The overall system acts as an autocatalyst. Growth is exponential, which means that the increase per unit time is proportional to the amount in existence. Doubling time remains constant.

Figure 8. The catalytic cycle that mediates intercellular communication and chemotaxis in the amebae aggregating to form the slime mold. E_1—ATP pyrophosphohydrolase, E_2—adenyl cyclase. The scheme is highly simplified.

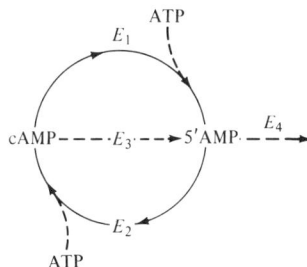

Transformatory Hypercycles with Quasi-Stationary R_i. This type of hypercycle applies, for example, to an uncontrolled, supercritical fission breeding cycle that is "running away." It may also apply to pioneering ecosystems, such as may be encountered on small islands or in areas that have burnt down. The niches of the new ecosystem are first filled by those species that are fastest in two respects: reproduction and the building of relations within the ecosystems. Exponential growth is soon damped as the system approaches limits and matures. The overall growth characteristics, therefore, resemble an S-shaped logistic curve.

In the social domain, full or partial recirculation of the materials involved in economic sectors may be depicted as a hypercycle that is either balanced or is growing exponentially; the latter case is, of course, at present still dominant. Even if recirculation is not an avowed policy, it can be found in many classical relationships among economic sectors that exchange materials, such as coal and steel production. Materials are also recycled to make up part of the material requirements for the same product; for example, old cars are partially recycled back to the steel industry in order to make up part of the material flow, which is again turned into cars. Recycling of nonrenewable resources is, of course, a major theme of a future-oriented economy, although it may never become as complete as the recycling of materials in mature ecosystems.

Less visible, but important in some cases, is the recycling of energy. The power generated in power plants is partly used for mining and processing fuel, for building new power plants, and for operating them. Full recycling would mean that a huge machinery maintains itself without any spin-off to other activities in society. It has only become clear in the recent past that, in some cases, reality is no longer very far off from such an idling, self-serving technological energy cycle. Whereas materials are most useful when they are conservatively recycled, energy can only be useful in economic and other activities if it is degraded—that is, if maximum entropy is produced at every step. Therefore, ideal recycling means something different for energy than for materials. For materials it means that the material flow is closed; for energy it means that energy flows through many steps and is degraded to the maximum extent possible between the point where it enters the system and the point where it leaves it. Natural ecosystems operate nearly optimally, both with respect to material recycling and to energy utilization.

Catalytic Hypercycles with Quasi-Stationary E_i. An example of this is a cycle of service sectors such as depicted in Figure 7, but growing exponentially by mutual stimulation.

Catalytic Reaction Cycles with No Degeneration Effect. As an example for this class, Eigen and Schuster (1977) cite the enzyme-free rep-

lication cycle of single-stranded RNA, which involves the mutual replication of plus and minus strands through template action. Metabolism is
built into this system by the use of energy-rich nucleotide triphosphates
as building materials and the rejection of pyrophosphate as waste. The
overall cycle acts like an autocatalyst of both strands as long as the degeneration of RNA may be neglected.

Another example is the growth of activity and knowledge in new scientific or technological fields. A certain level of activity generates publications that, in turn, generate more activity and thus, again, more publications. The nearly exponential growth of scientific and technological
literature in the 1950s and 1960s may be ascribed to the compounded effect
of many such hypercycles. In maturing fields, degeneration effects (e.g.,
due to redundancy) slow down the growth of both activities and publications.

5.4.4 Self-Selective Systems: Hyperbolic Growth

The system acts as an autocatalyst of an already autocatalytic system,
which leads to faster than exponential growth. Hyperbolic growth implies
quadratic increase in relation to the amount present and a reduction of
the doubling time by half with each doubling. As Eigen (1971) has shown,
hyperbolic growth leads to "once-forever" selection, or, in other words,
to selection as the sole survivor in competition. Therefore, this hierarchical level of system dynamics may be called self-selective.

Catalytic Hypercycles with No Degeneration Effect. A simple realistic
example, cited by Eigen and Schuster (1977), is the RNA-phage (virus)
infection of a bacterial cell. The phage lacks both the metabolic and the
translation systems necessary for self-replication. Therefore, in order to
reproduce itself, it enters a host cell and uses the latter's translation mechanism to synthesize a protein which, together with other proteins from
the host cell, forms an RNA-replicase, which becomes the basis for the
production of both plus and minus strands of the phage. For this task,
the "tricked" host cell even lends its metabolic system. The result of this
simple hypercycle formed by the autocatalytic phage and the catalytic
replicase, is hyperbolic growth of the phage until the host cell runs out
of its metabolic supply when a number of the order of 100 phages has
been synthesized.

A most imaginative type of hypercycle involves two kinds of molecules,
polynucleotides and proteins (Figure 9). It has been proposed by Eigen
(1971) as an important stage in precellular evolution. The polynucleotides
are excellent information carriers and the proteins good catalysts (enzymes). Their "symbiosis" results in a hypercycle capable of a high degree of error correction in multiplication, and thus of transmitting infor-

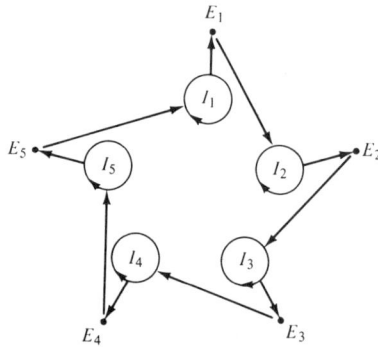

Figure 9. A catalytic hypercycle of second degree that, according to Eigen and Schuster (1977), probably played an important role in precellular evolution. Each information carrier (I_i) instructs its own self-reproduction as well as the production of an enzyme E_i. The latter provides catalytic aid for the formation of the next information carrier I_{i+1}. A hypercycle of this type is capable of a high degree of error correction in its self-reproduction, and thus of the transfer of complex information. In competition, it forces "once-forever" selection.

mation of a complexity needed in biological evolution. Each information carrier I_i is capable of instructing its own self-reproduction as well as the production of an enzyme E_i. The enzyme E_i, in turn, provides catalytic aid for the self-reproduction of the next information carrier I_{i+1} in the cycle. If the cycle is closed, so that the last enzyme acts as a catalyst for the first information carrier, the cycle represents a hypercycle of second degree (because it fulfills two catalytic functions, one autocatalytic in the I_i and one catalytic through the E_i). In general, such hypercycles will be of higher degree, since there has to be at least a built-in or auxiliary metabolic function to provide and dissipate energy for the process. If

Figure 10. Hyperbolic growth of the world's population is made possible through a hypercycle that involves increase in land use and increase in agricultural productivity, or either of these two factors.

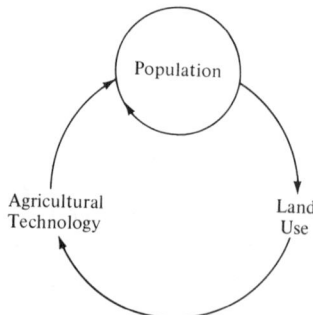

copying errors are not repaired but passed on, the hypercycle may evolve through coevolution of information carriers and enzymes until it reaches a new level of optimal stability.

The most conspicuous hyperbolic growth in the world occurs in human population dynamics. For the past 300 years or so, the world's population has increased following more or less a hyperbolic law, and the population of many of the less developed regions still does. To a large extent, this is due to advances in food and health technology, removing traditional limits to population growth. The hypercycle involving land use and food production efficiency is of the general form shown in Figure 10.

5.5 Dissipative Structures: Cooperative Behavior

The metabolism of self-organizing matter systems would soon die down were the exchange processes with the environment left to chance encounters of the participants involved in the reactions. Entropy would accumulate in the reaction volume and move the system towards its equilibrium, at which all processes come to an end, except for thermal (Brownian) motion. This is the way equilibrating systems run themselves down.

Under conditions far from equilibrium, the processes within the system as well as its exchange processes with the environment assume a distinct order in space and time, called a *dissipative structure*. It constitutes the dynamic regime through which the system gains autonomy from the environment, maintains itself, and evolves. In particular, it is the dynamic regime that keeps the system self-regenerative, that is autopoietic in the narrow sense of the word. Where random behavior is characteristic for equilibrium systems, this kind of *cooperative behavior* is a correlate of nonequilibrium system dynamics.

Dissipative structures have a life of their own. In general, they do not mediate a steady-state flow between reaction steps. In autopoiesis, they become typically manifest in pulsations of the limit-cycle type, in concentric waves and interference patterns among such waves and in chemical vectors and polarization phenomena. The high internal nonequilibrium that becomes visible in such behavior is continuously maintained by exchange with the environment. No equilibration, but maintenance of nonequilibrium, is the core notion of this type of dynamics. The dissipative structure stabilizes itself globally at a level at which high energy penetration is assured. High nonlinearity, due to the autocatalytic step in the reaction cycle, gives rise to self-reinforcement of fluctuations in the mechanisms (the function) of the system, which may ultimately drive the structure over an instability threshold into a new dynamic regime: This is the meaning of "order through fluctuation."

Autopoietic systems of the self-regenerative kind are characterized by such dissipative structures. The accompanying phenomena, in particular oscillations, are now being studied in a wide variety of systems, ranging from chemical, electrochemical, and biochemical reaction systems to neuron populations, aggregation and social behavior of organisms, eco-systems, urban and interurban systems. Rapidly growing empirical research shows almost dramatic agreement with theoretical predictions. [For overviews, see the monograph by Nicolis and Prigogine (1977) as well as the concise chapter by Prigogine (1976).]

Less is known about cooperative behavior in the structures of self-reproductive *growing* systems. In the case of the RNA-phage multiplication mentioned in Section 5.4.4 there is apparently a corresponding multiplication of dissipative structures mediating the metabolic process for the spreading units of self-reproduction. It is hardly conceivable (to me, at least) that in a hypercycle of the type sketched in Figure 9 the hyperbolically multiplying self-reproductive units can form new hyper-cycles without some sort of cooperative behavior, mediated by dissipative structures. It is perhaps the dissipative structures that limit and control growth, not the multiplication of the basic organization. It is interesting that the theoretical study of evolving structures in self-reproductive systems has yielded some of the first results in the area of urban and regional development (Allen et al. 1977).

5.6   Evolution of System Organization

Dissipative structures may evolve to new dynamic regimes while the basic reaction cycle remains the same. If the organization of the cycle changes, for example through mutation of one of the reaction participants or through the introduction of a new participant, we may speak of evolution of the organization of the system. In addition to metabolism and self-reproduction, the matter systems in question exhibit now the capability to mutate, evolve, and compete for selection of the mutants. The accumulation of these capabilities, often mistakenly assumed to define life, is a general characteristic of self-organization in nonequilibrium matter systems that are not restricted to the domain of what is usually called the living.

In the protogenetic and genetic mechanisms dominating prebiotic and early biotic steps in the evolution of life, a system's organization evolves if errors in the information transfer (or mutations due to cosmic radiation and other factors) lead to a selective advantage. But even then it is not a single, sharply defined molecular species that is selected, but a statistical distribution—a "quasi-species" (Eigen and Schuster 1977). Coupling among species, for example through horizontal gene transfer among bacteria, contributes to this effect.

Cyclical organization, characteristic of autopoiesis, is a prerequisite for the evolution of complexity. Eigen and Schuster (1977) have shown that the type of cycle and the functions it incorporates determine the maximum complexity of information that may be passed on through template action: For enzyme-free RNA replication, based on a catalytic cycle as discussed in Section 5.4.3, the maximum digit content (the number of nucleotides) is limited to about 10^2; single-stranded RNA replication by means of specific replicases, the simple hypercycle discussed in Section 5.4.4 for RNA-phage infection, is limited to about 10^4 digits; double-stranded DNA replication by means of polymerases including proofreading by exonuclease—a hypercycle of third degree—is limited to about 5×10^6 digits (close to the actual maximum number of nucleotides found in big bacteria); and sexual DNA recombination in eucaryotic cells—representing the coupling of hypercycles—extends this limit to something like 5×10^9 digits (with the 3×10^9 nucleotides in human DNA coming close to that limit).

Starting with the eucaryotic cell, *epigenetic* evolution increasingly complements genetic evolution. Epigenetics, a term introduced by Conrad Waddington in 1947 (see Waddington 1975), is the process by which the phenotype uses available genetic information (the genotype) selectively for optimal alignment of its own ontogeny with environmental conditions. We may speak of internal self-selection of subsets of genetic information. This means that structure and function of the metabolic system mediating the development of the organism become complementary in the same way as in simple dissipative structures. The chromosome becomes the core element of a dissipative, autopoietic unit in which the genetic material is continuously broken down and rebuilt (Lima-de-Faria 1976).

In particular, epigenetics gives rise to the coevolution of two or more systems (Figure 11). Coevolution is usually open, not cyclical as discussed in Section 5.4.2 for the influenza virus A above. It is easily demonstrated in the morphological and behavioral features of predator–prey cycles. More generally, complex cyclical systems, such as ecosystems, may evolve along a path that turns the cycle into a helix. For the description of this type of systemic coevolution, and emphasizing the connection with Eigen's hypercycles, Ballmer and Weizsäcker (1974) have introduced the term *ultracycle*. Each autocatalytic unit in a hypercycle (or each unit in a cycle of hypercycles, depending on the level of resolution chosen for the description) represents a niche. This term is used by Ballmer and Weizsäcker in a broad sense and denotes itself a smaller ecosystem with several participating species that relate to each other in specific ways. Typically, species participate in more than one niche. Each mutation in a niche—in terms of genetic mutants or new species entering the niche, as well as in terms of new relationships—stimulates changes in other niches. In particular, an increase in complexity stimulates a corresponding increase in complexity in neighboring niches, with multiniche species providing the simplest mechanism of stimulation. In other words, we have here a coevolution of niches with positive feedback. From such a coe-

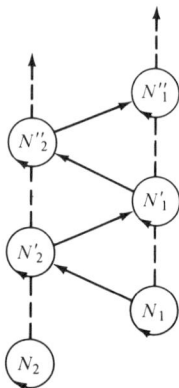

Figure 11. Coevolution of two niches in an ecosystem through an ultracycle according to Ballmer and Weizsäcker (1974).

volution at the level of subsystems emerges the evolution of the overall system. I should like to emphasize that ultracycles have nothing to do with the law of requisite variety in equilibrium theory (Ashby 1956); coevolution is not about control.

Complex economic systems and supersystems may evolve through similar ultracycles. Certain economic sectors and clusters of sectors may coevolve in the overall system of the national economy, which, in turn, may coevolve with other national economies within economic blocks or regional economies.

The ultracycle is a model for the basic learning process in general. In line with Maturana's (1970) very useful model of cognition, communication involves no transfer, but rather the reorientation of the set of processes linking the other system to its environment and defining its cognitive domain. Pointing to a beautiful sunset leads to communication at a deep level; describing it in words does not. Learning may be viewed as the coevolution of experiential systems—whether the subject of learning is relationships within real systems or within concepts, theories, paradigms, and other mental structures. Information is generated by such a learning ultracycle, not just processed and transferred as it is in classical information theory of the Shannon–Weaver type.

5.7   Conclusions

Autopoiesis is a central aspect of dissipative self-organization and is fully included in the theory of the latter. There is no real need for a focal concept of autopoiesis, except for its graphical—and even poetic—quality, which may induce system theoreticians to turn away from the machine syndromes of the "old" general system theory and to focus their research on the self-organization dynamics leading to life and maintaining it. However, it would be naive and backward to proceed with intuitive arguments while ignoring the theoretical foundation that, *grosso modo,* has already been established in considerable detail. It would be even more naive and backward to attempt descriptions in terms of the old equilibrium concepts, which are concepts pertaining to death. One might as well attempt the design of a mechanical *perpetuum mobile.*

These issues are not just academic. There is an important incentive to clear up the mess in which system theory finds itself due to the sterility of equilibrium concepts. Just as reality splits into equilibrium and dissipative structures, so does system theory with regard to its basic concepts. Equilibrium theory has become most fruitful in the design and operation of technological systems but dangerous or even detrimental in its applications to biological and human systems. The theory of dissipative self-organization in nonequilibrium systems embraces the core of the basic "logic of the living" and its autogenesis. It may also underlie cosmic autopoiesis and evolution in many more ways than have hitherto been explored. It certainly governs the chemical and prebiotic stages preceding the evolution of life (Eigen 1971; Prigogine, Nicolis and Babloyantz 1972; Eigen and Schuster 1978). It is also the principle active in manifestations of life beyond biology, in sociobiological and sociocultural systems as well as in the living systems of the mind, in paradigms, vision, ideologies, and religions (Jantsch 1980).

The new paradigm of self-organization, and with it the focal concept of autopoiesis, ends the alienation of science from life. It forms the backbone of an emergent science of life that *includes a science of our own lives,* the biological as well as the mental and the spiritual aspects, the physical as well as the social and the cultural. It is only in our days that the dreams of Ludwig von Bertalanffy and Norbert Wiener may come to fruition.

References

Allen, P. M., Deneubourg, J. L., Sanglier, M., Boon, F., and de Palma, A. (1977), *Dynamic Urban Growth Models,* Report TSC-1185-3, prepared for Dept. of Transportation, Transportation Systems Cr., Cambridge, Mass.

Ashby, R. W. (1960), *Design for a Brain,* 2nd Ed., John Wiley, New York.

Ashby, R. W. (1956), *An Introduction to Cybernetics,* John Wiley, New York.

Ballmer, T. B., and Weizsäcker, E. von (1974), Biogenese und Selbstorganisation, in *Offene Systeme I: Beiträge zur Zeitstruktur von Information, Entropie und Evolution* (E. von Weizsäcker, ed.), Klett, Stuttgart.

Eigen, M. (1971), Self-organization of matter and the evolution of biological macromolecules, *Naturwissenschaften* 58, 465–523.

Eigen, M., and Schuster, P. (1977), The hypercycle: A principle of natural self-organization; Part A: Emergence of the hypercycle, *Naturwissenschaften* 64, 541–565.

Eigen, M., and Schuster, P. (1978), The hypercycle: A principle of natural self-organization; Part C: The realistic hypercycle, *Naturwissenschaften* 65, 341–369.

Eigen, M., and Winkler, R. (1975), *Das Spiel: Naturgesetze steuern den Zufall,* Piper, Munich.

Fischer, R. (1976), Transformations of consciousness: A cartography; Part II: The perception–meditation continuum, *Confinia psychiatrica* 19, 1–23.

Glansdorff, P., and Prigogine, I. (1971), *Thermodynamic Theory of Structure, Stability, and Fluctuations,* Wiley-Interscience, New York.

Jantsch, E. (1980). *The Self-Organizing Universe: Scientific and Human Implications of the Emerging Paradigm of Evolution,* Pergamon, New York.

Lima-de-Faria, A. (1976), The chromosome field (in five parts), *Hereditas* 83, 1–22, 23–34, 139–152, 175–190; 84, 19–34.

Margulis, L. (1970), *Origin of Eukaryotic Cells,* Yale Univ. Press, New Haven, Conn.

Margulis, L., and Lovelock, J. E. (1974), Biological modulation of the earth's atmosphere, *Icarus* 21, 471–489.

Maturana, H. R. (1970), *The Biology of Cognition,* Biol. Computer Lab. Res. Report 9.0, Univ. of Illinois, Urbana.

Maturana, H. R., and Varela, F. (1975), *Autopoietic Systems,* Biol. Computer Lab. Res. Report 9.4, Univ. of Illinois, Urbana.

Nicolis, G., and Prigogine, I. (1977), *Self-Organization in Nonequilibrium Systems: From Dissipative Structures to Order through Fluctuations,* Wiley-Interscience, New York.

Prigogine, I. (1976), Order through fluctuation: Self-organization and social system, in *Evolution and Consciousness: Human Systems in Transition* (E. Jantsch and C. H. Waddington, eds.), Addision-Wesley, Reading, Mass.

Prigogine, I., Nicolis, G., and Babloyantz, A. (1972), Thermodynamics of evolution, *Physics Today* 25, 23–38, 38–44.

Staehelin, T. (1976), Hoffnung an der Grippefront, *Neue Zürcher Zeitung* (foreign edition No. 14), 20 January, 27–28.

Thom, R. (1972), *Stabilité Structurelle et Morphogénèse: Essai d'une Théorie Générale des Modèles,* Benjamin, Reading, Mass. English version: *Structural Stability and Morphogenesis* (D. H. Fowler, transl.), Benjamin, Reading, Mass. (1975).

Varela, F., Maturana, H. R., and Uribe, R. (1974), Autopoiesis: The organization of living systems, its characterization and a model, *Biosystems* 5, 187–196.

Waddington, C. H. (1975), *The Evolution of an Evolutionist,* Edinburgh Univ. Press; Cornell Univ. Press, Ithaca, N.Y.

Zeeman, E. C. (1977), *Catastrophe Theory: Selected Papers 1972–1977,* Addison-Wesley, Reading, Mass.

Zeleny, M. (1977), Self-organization of living systems, *Int. J. Gen. Systems* 4, 13–28.

Zeleny, M., and Pierre, N. A. (1976), Simulation of self-renewing systems, in *Evolution and Consciousness: Human Systems in Transition,* (E. Jantsch and C. H. Waddington, eds.), Addison-Wesley, Reading, Mass.

INTRODUCTORY REMARKS

It is rather awkward to comment on one's own contribution, even in this age of enhanced self-reference. It should be noted, however, that the following paper is an attempt for an *intuitive* synthesis of what the author has learned from studying autopoiesis, hypercycles, and dissipative structures, as they bear on the theme of the origins of life. Nothing more is being claimed. A short abstract follows.

An alternative view of the emergence of "living" systems and a special concept of "life" itself are being advanced. In this conception of life, uniformitarian in its premises and holistic by its design, the cell is considered to be the primary organizational entity capable of "life." This orientation includes both procaryotes and eucaryotes, as well as organelles. By "cell" we mean a protocell (or a protobiont), the first autopoietic unity.

It is the autopoiesis—the cyclical and unity-maintaining organization of even the simplest components—that may lead to the initial auto-organization of life, or autogenesis. Earlier experiments of synthetic biologists, coupled with Lamarckian respect for the environment and a proper dose of modern biochemistry, provide the necessary empirical framework for postulating the relatively rapid emergence of living phenomena and forms.

The emergence of a protocell cannot be separated from the variety of igneous rocks evolved and crystallized some 4 billion years ago. These primal crystalline formations, still uneroded and highly compartmentalized, exposed to the unceasing tidal rhythms of primeval oceans, became the cradles of life. In their innermost compartments, the biomatrices, the archaic nuclei and membranes developed their simple and elegant catalyzing interaction, fully protected in their womblike milieu from the harsh and tempestuous environment of the prebiotic seas.

By means of absorption and adsorption of primordial monomers, under the favorable thermodynamic conditions of the vast tidal zones, their structures were gradually transformed from predominantly inorganic-molecular to mostly organic-polymeric.

Subsequent erosive, volcanic, and mechanical events released the already complex protocells from their biomatrices in rich varieties of local forms and functional distinctions. Continually they spilled out into the waters and started on their parallel, predifferentiated evolutionary paths.

Milan Zeleny was born in Prague in 1942. He received an Ing. degree from Prague School of Economics, and M.S. and Ph.D. degrees from the University of Rochester. Dr. Zeleny has published over 100 papers and articles on topics of multiple criteria decision making, mathematical programming, computer simulation, autopoietic systems, game theory, economics, the theory of risk, critical path analysis, and judgmental psychology. Among his many books are, for example, *Linear Multiobjective Programming, Multiple Criteria Decision Making, Uncertain Prospects Ranking and Portfolio Analysis, Towards a Self-Service Society,* and *Autopoiesis, Dissipative Structures and Spontaneous Social Orders.* He serves on editorial boards of *Computers and Operations Research, Fuzzy Sets and Systems, General Systems Yearbook,* and *Human Systems Management.* He is currently on leave from Columbia University on a von Humboldt Award. *Address:* European Institute for Advanced Studies in Management (EIASM), Place Stéphanie 20, B-1050 Brussels Bte 15–16, Belgium.

Chapter 6
Autogenesis:
On the Self-Organization of Life

Milan Zeleny

6.1 Introduction

The question of the origin of life, possibly the greatest biological question, cannot be discussed yet outside the realm of conjecture and speculation. Not many biological scientists are willing to trod such doubtful and unrespected ground. Yet, it seems obvious that humans should keep asking questions about the origins of life and updating their conjectures. The essential incompleteness of scientific theories will always demand that their closure, however imperfect, be attempted.

This paper, written by a nonbiologist, is a simple presentation of several conjectures, possibly novel in character, that could point to fruitful directions for further research. They are meant as nothing more and the author will not take offense at being accused of simplicity.

Before discussing an origin of any kind, one must address some philosophical arguments *against* the possibility of spontaneous generation (i.e., life arising "by itself") (Farley 1977). These are exemplified by such Virchowian statements as "life is produced by life," "*omnis cellula e cellula*," "*omnis organisatio ex organisatione*," and other derived varieties of this theme. These considerations invoke the need for a preceding creator, a program, or at least an inner "language" to trigger the initial description–interpretation process.

One simple argument against *generatio de novo* could go as follows. For anything to produce itself is to act; if it acted before it was, it was then something and nothing at the same time. It acted before it brought itself into being. How could it act (process) without a being (state), unless it was? Nothing can act before it is. If something were the cause of itself, it must be before itself; it was before it was: *omne systema e systema*.

The precursor system must provide a "program" for the succeeding system.

The problem with this "reasoning" is that its presupposition, that the origin of a system is conceivable only through another system, and not as an outcome of certain *favorable conditions* occuring in the space of potential components, amounts to *denying* that the organization of a system must have "nonorganization" as its precursor; it then leads inevitably to the *belief* in special creation.[1]

In a solution of molecular components certain conditions (density, temperature, mixture, etc.) can arise such that the molecules concatenate themselves into particular spatial relations, forming a crystal. It would be difficult to argue that such a crystalline concatenation is due to random encounters: it emerges *whenever* the conditions are satisfied. A system, in this case allopoietic, arises from a "nonsystem"—an unordered solution of molecules. (Anyone who has experienced the stunning phenomenon of a sea "freezing over," the tremendous crystallization of waters within minutes of the occurrence of proper temperature conditions, can confirm the compelling unavoidability of such a phenomenon.)

The question arises, can an autopoietic system, which is not defined simply as a spatial organization of components but as a circular concatenation of component-producing processes, similarly emerge under certain favorable conditions characterizing the environment of its components? Can life be compared with a particular type of "crystallization" of matter in the sense of being bound to appear when conditions are right?

Several comments should be made with respect to this question. First, some components must exist before the autopoietic unity does; that is, their production processes are not yet a privy to the autopoietic system but are inherent in the molecular interactions in the physical space. The component-producing processes must exist by virtue of the existence of their products; what does not exist is the particular organization necessary for their autopoiesis. Accordingly, autopoiesis may arise in a molecular system if the relations of production become concatenated in such a way that their products specify the system existing only while it is actively being produced by such a concatenation of processes (Maturana and Varela 1973).

Second, the actual (favorable) conditions existing on preautopoietic Earth are not necessarily deducible from the *presently* observable modes of existence of autopoietic systems. That is, organic macromolecules, nucleic acids, protein systems, and the like are the components of already *evolved* autopoietic systems, not necessarily of protoautopoietic systems. Such conditions might not exist anymore and even the possibility of their laboratory reconstruction could be debatable.

---

[1] George Wald (1954) cautioned, "It is a symptom of the philosophical poverty of our time that this necessity [of spontaneous generation] is no longer appreciated."

Third, the phenomenon of life cannot be fully predicted or explained from the quantum mechanics or chemistry of the molecular components. One can use the analogy with waves or waterfalls (i.e., dissipative structures). If life can be compared to a standing wave, a whirlpool, or a waterfall, then a detailed knowledge of the molecules taking part in its autopoiesis is not sufficient to explain it as a phenomenon. It is the *relations* among molecule-producing processes that are of primary interest.

Finally, there is the question of graduality, or continuity of nonlife and life. It is often alleged that a system either is or is not autopoietic (Maturana and Varela 1973). Intermediate steps are excluded and the natural extension of this idea is that anything could be either living or nonliving. Some sort of "vital principle," an autopoietic organization of life "snapping" itself into existence, is thus implied. One way of resolving this dilemma is to assume that autopoietic and yet nonliving systems are at least conceivable. Stéphane Leduc (1971) insists that there is no sharp demarcation line separating life from the other phenomena of nature—the passage from the inanimate to the animate is gradual and fuzzy.

6.2   Circular Organization

Autopoietic organization is characterized by the circularity and interdependence of component-producing processes. Although such organization either is or is not (circular), and the establishment of an autopoietic system cannot be gradual in a strict sense, there could be a variety of intermediate types of systems in which the autopoietic organization is only a short-lived, transient, or alternating phenomenon. An essentially allopoietic system can attain autopoietic organization for a short time, but being unable to sustain it, because of prevailing unfavorable conditions, it reverts to its allopoietic existence. With "improving" conditions, such autopoietic organization might be able to gradually extend its longevity of existence and become observable and identifiable as an autonomous system. Consequently, transitional systems *are* conceivable in the sense described above.

Consider the minimal model of an autopoietic system as presented in this volume's introductory tutorial, or in many other related expositions (Varela, Maturana, and Uribe 1974). We assume three types of processes: production (P), bonding (B), and disintegration (D). (The reader is advised to consult the summary of the model on page 8.)

Look at each of the processes as acting independently. Production *alone* would continue to form more complex macromolecules from simpler elements. As the environmental substrate becomes exhausted, production would either cease or, under certain conditions, resume on the higher level of "elements" (macromolecules). Bonding *alone* would ultimately concatenate the products into chains (closed or open) and subchains; then it would cease. For example, stable crystalline patterns might set in. Disintegration *alone* would simply transform all "organized matter" into a uniform and inert environment of substrate (a state of maximum entropy).

The preceding *hypothetical* examples of singular processes rarely appear in nature. Let us consider multiple processes, say, pairs: Production + bonding transforms the substrate into "crystals." Bonding + disintegration transforms the macromolecules into the uniform environment of substrate. Production + disintegration results in a mixture of substrate elements and higher macromolecules, possibly oscillating around some "equilibrium" proportion. In neither of these cases would we observe an autopoietic system, only allopoietic structures, inert environment, or chaotic mixture.

All three processes are necessary for autopoiesis. But they are not yet sufficient. The processes, P, B, and D, must be conceived of as being *interrelated* if a system is to be identified—they must be organized. We have analyzed P, B, and D in isolation. Suppose that we identify an interdependence between two such processes (P,B). We then test the third process (D), whose properties in isolation we already know, in its dependency on the state of the two others. Can we gain a higher synthetic insight into the behavior of (P,B,D) by such stepwise addition? We could only if the original processes were actually relatively independent.

In autopoietic systems the three processes must be both present *and* interdependent. They must be organized in a circular fashion (their organization must be closed). The products of one process are necessary for another process to take place. That is, disintegration "produces" the substrate necessary for production; production "produces" a catalyst necessary for the production itself and a higher macromolecule necessary for bonding. Bonding "produces" components that assure that their disintegration will "produce" the substrate in a particular locality rather than in a dispersed fashion.

Thus, in this case, we cannot achieve a stepwise assembly of (P,B,D) by first joining P with B and then adding D, for in the absence of D, neither P nor B could take place. The products of D (substrate) are necessary for P and through P for B as well. The coexistence and cooperation of all three processes, P, B, and D, is indispensable for the existence and operation of *any* of them. They must be studied as a whole. This necessity illustrates the dilemma of reductionism and the inevitability of holism.[2]

---

[2] It is thus inadequate to concentrate only on the *production* of proteins. The breakdown of proteins, protein *degradation,* is equally important and indispensable. All proteins are degraded and resynthesized many times within the life of a cell. Protein synthesis and degradation are interdependent processes and their understanding cannot be achieved through their separation.

Yet, the circular organization of processes is not sufficient in itself: it does not necessarily determine the topology of an autopoietic system. A cleavage, or a distinction, from the environment is needed. A membrane, boundary, or any other kind of separation (or protection) from the environment is a common characteristic of living systems.

6.3 Autopoiesis and Allopoiesis

Many autocatalytic processes as well as some physicochemical dissipative structures are characterized by the circularity of their organization. Although their organization does not determine their topology, the organization itself is continually being renewed and maintained. As long as the ambient of their autopoiesis is favorable and supportive, there is *no need* for their topological separation and definition. It is only when, gradually, the environment becomes less conducive to autopoiesis that a distinct cleavage and separation of the autopoietic unity becomes necessary for its further maintenance and survival. To define autopoietic systems only through that particular protective structure (membrane) would be overly mechanistic and unnecessary. We raise the following question: Is it possible that, under certain conditions, a circular organization of production processes *and* an independently emerging local enclosure or membrane could be coupled into a symbiotic and mutually enhancing coexistence, ultimately achieving a unity of organization and structure?

What do we mean by this notion of "symbiosis" of organization and structure? An autopoietic organization can be conceived of either as operating on an essentially unordered environment of components *or* as acting upon an already ordered, structured milieu, favorable to its enhancement and maintenance. This kind of reasoning is pointing toward the inseparable unity of allopoietic and autopoietic systems.[3]

Allopoiesis is defined as a production of "something else" than itself (Maturana and Varela 1973). This is intuitively clear with respect to man-made machines and artifacts. But what about "natural" allopoietic systems, such as crystals? Are they not capable of producing themselves, even though through a linear, unidirectional organization? The point seems to be in the organization (closed versus open) and not in the production of "self" or "something else." Can any system produce itself before being able to produce something else? Can we draw a strict line

---

[3] The usefulness of the organization–structure duality appears in thinking about such problems as morphogenesis and polymorphism. Both arms and legs are composed of the same types of differentiated cells. The key to their structural differences is not in the cells as such but in the organization of cells within a higher-level unity. Similarly, certain kinds of aphids, although organizationally and genetically identical, exhibit a structural polymorphism of wingless and winged forms.

of independence between allopoiesis and autopoiesis? Allopoietic systems are undoubtedly the precursors of autopoiesis—allopoiesis is the framework, a condition, within which autopoiesis can take place.

Autopoiesis is inseparable from allopoiesis as much as organization is inseparable from structure. The two emerge and evolve in mutual symbiosis.

6.4 Conditions of Biopoiesis

It now appears almost certain that life first emerged nearly 4 billion years ago. The conditions favorable to life on the Earth, characterized by a solid crust, cooler temperatures, and liquid water, could not have been realized, according to our present knowledge, earlier than 4.4 billion years ago. It seems that simple eukaryotic organisms, *Isuasphaera,* might have been extant at least 3.8 billion years ago (Pflug and Jaeschke-Boyer 1979). This would indicate that the emergence of life must have been closely connected with the cooling-off conditions and the accompanying phenomena occuring on the newly formed Earth. In a stricter sense, the emergence of life must have been concurrent with, and actually brought about by, the conditions accompanying formation of the Earth's crust. This short time span of roughly half a billion years was all that was available for the emergence (or an outburst) of life.

It was demonstrated that the molecules necessary for life could have been generated by abiotic processes (Miller and Orgel 1974). We should also include meteorites as other possible sources of highly concentrated and localized organic macromolecules. Earth's primordial atmosphere was probably strongly reducing, hydrogen-rich, devoid of oxygen, and saturated by volatiles of carbonaceous chondrites. A liquid, shallow hydrosphere condensed over relatively uniform and yet uneroded crust formations.

The conditions of prebiotic Earth and its short time span indicate that life emerged very switfly and that the early protobionts lived interspersed within their still favorable surroundings. Any self-replicating genetic system of reasonable complexity could not have assembled itself within such a short "snap" of time; its current uniqueness and complexity must be a result of a long evolutionary process. Similarly, topological boundaries and membranes, appearing as early as 3.8 billion years ago, would not be producible through such primitive "programs"—they would have to be self-organized with the help of other, obviously abiotic, processes. Such topological *biomatrices* then became the most favorable environment for protection, maintenance, and evolution of so far ephemeral, circular, or even autopoietic organizations of component-producing processes. The currently pervasive "organic plasticity" of the topological matrices was acquired gradually through symbiosis and evolution of their environmental coupling.

It is therefore submitted that the attempts to explain the origin of life (and autopoiesis) on the basis of exclusively organic components are going to be increasingly frustrated by largely inadmissible retrograde extrapolation of modern life forms and conditions to some 4 billion years ago. It is in the inorganic chemistry of prebiotic Earth (and other planets) where the origins of living organizations must be sought.[4]

Under the assumptions sketched above, life arose *polyphyletically;* that is, many parallel, different, and relatively independent protopopulations occurred at different times and at different locations on Earth. Each such protopopulation gave rise to its own phylogeny, which shared the same basic principles of biochemical organization, reflecting the essential unity of Earth's environment; yet they exhibited often remarkable differences according to their individual localities of origin. Ultimately, they interlocked their evolutionary paths and assumed the dominant isotropy and symbiosis as observed today.

In this sense, all contemporary living systems have already evolved their respective organizations and optimal couplings with their environment. They are fully endowed, completely formed, and relatively perfect end products of these diverse evolutionary pathways. (This is not to deny their continuing evolution, even if proceeding at a much slower rate and in an increasingly ordered and more predictable environment). Bacteria, protists, and unicellular and multicellular organisms are not linearly interdependent "milestones" of a unitary evolutionary process, but are parallel products of intricate symbiotic interactions.

The often noticed morphological homologs between the successive forms characterizing the development of an embryo and the deceptive "succession" of forms acquired by individuals in the "evolution" of species—the "ontogeny recapitulates phylogeny" notion—thus attains a new and potentially powerful paradigmatic quality by contemplating the implications of its reverse: "Phylogeny recapitulates ontogeny."

6.5 Inorganic Imbedding of Life

We can characterize life as a phase in the incessant flux of matter, from the mineral to the living and back again from the living to the mineral world. Organisms are closely connected and imbedded in the mineral world. For example, the salt-loving bacteria, halobacteria, can live in dry crystals of salt. They actually disintegrate at low salt concentrations, and

---

[4] The 1976 Viking mission to Mars also did not identify any organic molecules in the Martian soil. Yet, some metaboliclike chemical processes were undoubtedly detected. The possibility of ice-eating crystophages or rock-dwelling "bacteria" (see also note 5) cannot be excluded. Searching for organic debris stems from our defining life as a substance, not as a particular (autopoietic) type of organization of substance-producing processes. The failure to interpret the Viking landings phenomena satisfactorily is a sad monument to our naive conception of life; witness "Organic molecules not only make life, they define it and characterize it. We don't find them where there is not life" (Cudmore 1977).

it takes a temperature of 800°C to burn them off.[5] Generally hostile conditions of the purely inorganic world could thus conceivably provide a generating environment for a large variety of protoorganisms during the early stages of Earth's crust formation.

The purely *inorganic* osmotic growths of Leduc display life *analogies* of form, organization, and function (Leduc 1911). An osmotic growth suggests the idea of a living thing at first sight. Osmotic growths are organized of cells or vesicles separated by osmotic membranes. An osmotic stem, formed by a row of cellular cavities, resembles the knotted stems of bamboo.

The analogies of function are also notable. Nutrition consists of the absorption of alimentary substances from the surrounding medium, the chemical transformation of such substances, their fixation by intussusception in every part of the organism, and the ejection of the products of "combustion" into the surrounding medium. Osmotic growths similarly absorb materials from the medium in which they grow, submit them to chemical metamorphosis, and eject the waste products of the reaction into the surrounding medium. They "select" among the substances offered for consumption, absorbing some greedily and entirely rejecting others.

Phenomena of growth and development follow the absorption and fixation of nutrients by osmotic production. It grows, its form develops and becomes more complicated, and its weight increases by hydration. Osmotic growths undergo ontogenic development. In their early youth the phenomena of exchange, growth, and organization are very intense. As they grow older, their exchanges gradually slow down and their growth is arrested. With age the exchanges still continue, but at a much slower rate; even that gradually fails and an osmotic growth dies: it decays, losing both its structure and its organization.

The phenomena of osmotic growth demonstrate how ordinary mineral matter—carbonates, phosphates, silicates, nitrates, and chlorides—may induce the forms of living matter without the intervention of living organisms. Ordinary physical forces are quite sufficient to produce forms like those of living systems, closed compartments containing liquids separated by osmotic membranes, with tissues similar to those of the vital organs in form, color, function, and development.

The chemical composition of the first protobionts and their membranes did not have to be identical with those of modern biological membranes.[6]

---

[5] In addition to halobacteria, living organisms are also found at the other extreme: "ice worms," *Mesenchytraeus solifugus,* living in solid blocks of ice of the Artic, exhibit autolysis (i.e., disintegration by its own enzymes at higher than 0°C temperatures).

[6] The self-organizing processes in chemistry, apart from synthetic biology, have a long tradition in the literature. Periodic precipitative rings of Liesegang (1898) [see Hedges and Meyers (1926)] or spatial and temporal patterns of A. J. Lotka (1924) are well known. Recently, the oscillating Belousov–Zhabotinsky reaction was invoked by Nicolis and Prigogine (1977) as one of the examples of their dissipative structures.

More than 3 billion years of evolution of complex protein–lipid interaction suggest that it was mostly inorganic substances that participated in forming the first membranes. These protocells served as templates or functional matrixes for subsequent and gradual "proteinization" of their structures.

Leduc's report on the status of *synthetic biology* contains a wealth of experiments with inorganic osmotic growths (Leduc 1911). These growths exhibit the phenomena of circulation and respiration; they reproduce by budding; they traverse the cycle of growth, plateau, death, and decay. Through their semipermeable osmotic membranes the processes of nutrition are carried on. An injury to such an osmotic system is repaired by coagulation of its internal plasma. They are capable of performing periodic movements; they even float freely in their medium or grow out of the solution into the air.

Leduc offers a simple and basic exercise in osmosis: a fragment of $CaCl_2$ (calcium chloride) is covered by a liquid solution of potassium carbonate, sodium sulphate, and tribasic potassium phosphate. The calcium chloride surrounds itself with an osmotic membrane. An osmotic production, half aquatic and half aerial, emerges. It absorbs water and salts by its base and loses water and volatile products by evaporation from its crown, while at the same time it absorbs and dissolves the gases of the atmosphere.

It is stimulating to contemplate the contrast between the hard crystalline forms of ordinary chalk and these soft, transparent, elastic membranes that have the same chemical composition.

We can induce osmotic growths to produce terminal organs resembling flowers and capsules. When the growth is considerably advanced we can diminish the solution concentration a hundredfold by adding a large quantity of water. Spherical terminal organs will then grow out from the ends of the stems, acquiring conical or other shapes during their further growth.

Corallike forms may be grown from a semisaturated solution of silicate, carbonate, and dibasic phosphate by adding a concentrated solution of sodium phosphate or potassium nitrate. "Corals" are formed by a central nucleus from which large leaves radiate like flower petals. The presence of a nitrate produces pointed leaves with thornlike structures resembling the aloe or the agave.

Funguslike forms may also be obtained, as well as mushrooms, shells, amoeboids, medusas, and many other forms and shapes. Many of these can be allowed to thicken and be taken out of the liquid without breaking. Some growths first rest immobile at the bottom of the vessel. As they grow, they absorb water, their specific gravity diminishes, and they rise up and acquire a considerable degree of mobility.

It should be emphasized that in an osmotic growth the active growing portion is the gelatinous contents of the *interior*, the external visible growth being only a skeleton or a shell. One can break the calcareous sheath and draw out a translucid gelatinous cylinder, separated from the liquid by a fine colloidal membrane.

Several hundred other scientists have engaged in synthetic biology, and the literature they left behind is voluminous. For example, Martin Kuckuck of St. Petersburg, Russia, published his *Archigonia, Generatio Spontanea* in 1907. He used a mixture of gelatine, glycerol, and common salt as a substrate, and a crystal of radium as a catalyst. A peculiar culture appeared on the gelatine after 24 hours: a population of cells that grow, divide, and manifest other *external* signs of life.

Soviet academic A. I. Oparin, the proponent of the theory that highly unstable coacervate bags of molecules have mysteriously acquired a living organization, dismissed Kuckuck's experiments as follows: "This was obviously the work of a dilettante who was not sufficiently familiar with colloidal systems, and is, of course, devoid of any real significance" (Oparin 1953).

The point is that synthetic biologists *did* obtain the results they described. They never asserted that their growths were living but were struck by the close resemblance of the growths to the forms of nature. They left hundreds of photographs and detailed descriptions of their experiments. Whether they sufficiently understood the colloids or not, their productions were closer *analogs* of life than anything produced in Soviet biological laboratories so far.

6.6 Hypercycles

We have argued that complex organic molecules were conceivably produced by abiotic processes and that inorganic matter is capable of self-organization through basic physical forces, such as for example by osmosis. Compartments, membranes, and other enclosures could emerge from the inorganic milieu of prebiotic Earth. We now have to postulate a self-enhancing symbiosis of such compartments and membranes with the circular organization of component-producing processes.

Eigen and Schuster have proposed the *hypercycle* of nucleic-acid cycles as the minimal system that could bridge the gap between nonlife and life (Eigen and Schuster 1977).

Among the early organic macromolecules were nucleic acids (RNA, for example) and proteins (e.g., polymerases). Proteins can reproduce themselves only by a very roundabout process, and their evolutionary appearance and existence in the absence of RNA is difficult to imagine. On the other hand, RNA molecules replicate simply but rather imprecisely in the absence of proteins. How do we explain the appearance of proteins from the replication of RNA molecules? A minimal amount of specific proteins is needed to assure the accuracy of RNA replication—but the proteins themselves are coded for by that RNA! As Smith summed up: one cannot have accurate replication without a length of RNA of 2000 or more base pairs, and one cannot have that much RNA without accurate replication (Smith 1979).

What is needed to resolve the dilemma described above is a highly potent event: one of the nucleic-acid cycles must join with the existing proteins to produce a protein that would assist a second nucleic-acid cycle to replicate more precisely and faster. That cycle would have to produce a protein assisting a third nucleic-acid cycle, and so on, until the nth protein involves itself in the replication of the *first* nucleic-acid cycle, forming a hypercycle.

It is easier to propose the "hypercycle" as a necessary intermediate stage between nonliving and living than to explain its emergence. Some sort of "natural selection" process must be involved. The circular organization of production and replication processes must be stable, precise, and *protected* from the turbulent environment. A sort of free-flowing molecular soup or broth is certainly excluded.

Amino acids seemed to have been quite abundant in the primitive oceans. In a free ocean the natural selection would favor the growth and more precise replication of each constituent nucleic-acid cycle, but not necessarily the hypercycle as a whole. Given the short time span available for the emergence of the first protobionts, the hypercycles must have evolved relatively fast and in parallel fashion. Compartmentalization of nucleic-acid cycles and membranous enclosures of individual hypercycles would provide the necessary favorable conditions for their further evolution.[7] Enclosed and protected hypercycles could then grow and divide, possibly by budding or fission, and thus establish the first phylogenetic lineages of individuals.[8]

Such compartmentalization could take place within the cavities of the rocky crusts of vast primordial tidal zones. The inorganic osmotic growths of Leduc's type could also prepare the necessary enclosures for the macromolecular concentrates.[9] Their symbiotic self-enhancement would lead to the *autogenesis* of life.

---

[7] Evolution by natural selection thus must take place on the molecular level of hypercycles. The DNA $\rightarrow$ RNA $\rightarrow$ Proteins $\rightarrow$ DNA dynamic systems are selected as wholes even before expressing themselves fully in a particular organism.

[8] There is now evidence that $3\frac{1}{2}$ billion-year-old primitive spheroids, although not necessarily biogenic, exhibit phenomena of cell division (Knoll and Barghoorn 1977).

[9] Compartmentalization, the formation of vesicles, could also occur through self-organizing forces of the hydrophobic effect. See Tanford (1978), based on the early work of J. W. Gibbs on chemical potentials. Compare also with Leduc (1911) and note 6. Tanford emphasized that hydrophobic forces lead to plastic, deformable structures, uniquely suited for the first critical steps in the organization of living matter. Multiwalled vesicles form spontaneously whenever phospholipids are dispersed in an aqueous medium. Some hydrophobic spherules are formed with the help of electrical discharges.

6.7 Autogenesis

After reviewing some necessary building blocks for our main proposition, we shall attempt to summarize the basic time progression of autogenesis of autopoietic and living systems. The emphasis is on the conditions rather than the mechanisms of such occurrences.

6.7.1 Prebiotic Earth

The formation of Earth's crust and the crystallization of prebiotic rocks stabilized some 4 billion years ago. The interaction of the cooler primitive atmosphere with high-temperature magmatic materials of the crust induced the simultaneous evolution of the primitive hydrosphere as well as masses of igneous rocks. Gradually, standing bodies of liquid water and shallow tidal zones formed. Subsequent water erosion, weathering, and mechanical friction led to the accumulation of sediments and deposits, forming sedimentary rocks, clays, and silts.

The most ancient sedimentary rocks, around 3.8 billion years old, contain traces of cellular structures. For example *Isuasphaera* were detected within the quartz grains or in the silica cement (Pflug and Jaeschke-Boyer 1979). It is assumed that *Isuasphaera* may represent a halfway line between a microspherelike protobiont and the subsequent evolution toward eucaryotes.

The primitive atmosphere consisted largely of water vapor, ammonia, methane, molecular hydrogen and nitrogen, traces of carbon dioxide, and virtually no free oxygen. Earth's crust consisted of rocks made up of a number of different kinds of minerals organized in specific crystalline arrays. These were mostly silicates, sulphates, phosphates, carbonates, and oxides (i.e., inorganic compounds of oxygen, silicon, aluminum, iron, magnesium, calcium, sodium, and potassium).

Note that the macronutrients for most cells are calcium, phosphorus, chlorine, sulfur, potassium, sodium, magnesium, iodine, and iron. Micronutrients or trace elements are manganese, copper, zinc, fluorine, cobalt, vanadium, and selenium—most of these are indispensable for catalytic functions in many enzyme systems. In short, all the necessary components used in experimenting with osmotic growths of synthetic biology were present.

Gaseous bubbles caused countless inner compartments and cavities to solidify within the individual rock chunks. This initial compartmentalization was probably of a higher degree than that of younger metamorphic or sedimentary rocks.

High-energy sources of ultraviolet radiation, electrical discharges, and volcanic and solar thermal energy were abundant. Massive rains and accelerated condensation of waters led to substantial increases in the shoreline surface. The regular ebb and flow of the tides, alternately covering and exposing the littoral zone, became pronounced and relatively stable. Ocean waters cooled to the mild temperatures conducive to biological phenomena but the igneous rocks of the littoral zone were rhythmically exposed to intense solar heat and radiation easily penetrating the ozone-free atmosphere. Daytime and nighttime variations of temperatures on the shoreline were substantial.

6.7.2 Biomonomers

Simple reagents in the atmosphere, high-energy sources, and condensation of water into clouds and rains—these were favorable conditions for simple chemical reactions leading to the reactive compounds of nitriles and aldehydes. Volcanic activity brought simple carbides to the surface and their reaction with superheated water led to the formation of hydrocarbons. Reactive compounds and hydrocarbons are prone to react by themselves and with other reagents at a much faster rate than simpler compounds.

A number of pansyntheses and syntheses occurred and the major classes of biomonomers were produced, namely, alcohols, amines, amides, organic acids, sugars, nucleic acids, porphyrines, fatty acids, carbohydrates, purines, pyramidines, and the like. These monomers were washed down into oceans and water basins, forming dilute aqueous solutions.

The degradation of these compounds was prevented by their dissolution in water; most of the aromatic acids were relatively stable anyway. The concentration of biomonomers was quite low although their variety was significant. Many laboratories have succeeded in random synthesis of nearly all the biologically important monomeric substances, for example, α-amino acids (Miller and Orgel 1974 or Miller 1953).

The more concentrated the reactants are in a solution, the greater is their chance of colliding. Higher pressure effectively concentrates solutions, increases collision frequency, and speeds up reactions. An increase in temperature speeds up the movement of molecules. Adsorption of reactants onto a surface where they are brought closer together leads to a surface catalysis.

6.7.3 Polymers

The next stage is the polycondensation and polymerization of the accumulated simple monomers. New, less energetic reactions, involving dehydrations and phosphorylations, are needed to obtain polypeptides, nucleosides, nucleotides, polysaccharides, lipids, and polynucleotides. A very dilute aqueous solution of the primeval ocean is a quite improbable environment for dehydration.[10]

One of the principal ways in which more complex materials of life are compounded is dehydration synthesis through condensation, involving subsequent elimination of water. This could happen through evaporation in tidal pools, rock cavities, and temporary basins. The process can be carried out at temperatures above the boiling point of water.

We need a mechanism for concentrating solutes and some catalytic process that would drive dehydration reactions at appreciable rates. The low temperature and low concentration of solutes in primitive oceans virtually precludes any significant polymerization taking place in the seas.

The alternating tidal wetting and drying of littoral zones of the oceanic shores is a massive process that allows temperatures close to 100°C at low tide. Igneous rocks and their compartments absorb and adsorb water, and high evaporation rate leads to dehydration and concentration of solutes, as well as to increased catalytic properties of minerals present. Crystals of igneous rocks adsorb biological monomers concentrated in their inner cavities and catalyze dehydrations and phosphorylations. During the day the temperatures of concentrations inside metallic rocks could reach 70–140°C and attain highly pressurized conditions. Excess water is evaporated or even boiled away (low atmospheric pressure of the prebiotic Earth), and resulting polymers are screened from ultraviolet radiation and protected from being washed away by tides and rains.

Traces of microcrystalline clays, salts, and other sediments were also concentrating in the nooks and crannies of the rocks and rock surfaces. These clays enhanced the concentration and provided further catalytic capabilities.

It has also been established that the silicates, apatites, phosphates, and other inorganic minerals are not only capable of adsorbing polyphosphates, nucleosides, purines, and pyramidines from the concentrates of internal cavities, but are also acting as catalyzers if the concentration and temperatures are high enough. During low tides a massive evaporation of water left huge quantities of amino acids and some salts concentrated in the internal cavities in a relatively anhydrous condition. Adsorption of some of these acids by crystals of the inner walls of these compartments further facilitated their selective concentration.

For example, glycine and glutamic and aspartic acids are highly absorptive; less adsorptive acids stayed mostly in the solute. High temperatures, high concentration, catalytic action, high pressure—these were the

---

[10] An alternative mechanism is described in Matthews et al. (1977). Helical macromolecules of heteropolypeptides could have synthesized spontaneously in the stratosphere, from hydrogen cyanide and water, without intervening α-amino acids. Their descent on Earth would then contribute to the formation of proteinaceous matrix.

proper conditions for advanced polymerization of polypeptides, polysac-charides, polynucleotides, and other biologically important polymers.

The amino-acid polymerization is a nonrandom process, as are the in-itial syntheses of biomonomers. In the same fashion in which the differ-ence in the nature of reactivity of the units of a growing crystal determines the final constitution, so differences in reactivities of the various amino acids serve to promote a definite ordering in a growing peptide chain.

This nonrandomness is now further emphasized by the fact that po-lymerizations occurred in connection with different minerals, and con-sequently different adsorptive selections and rates, from a variety of levels of concentrations in cavities of infinitely many sizes and shapes, under different thermodynamic conditions of metallic and nonmetallic rocks, on different beaches, and at different geographical locations.

Still, the same basic monomers were being washed on all beaches and rocks, and the same basic minerals were present, as were the same rhythm of tidal waves and the same alternating solar heat and radiation. These conditions were responsible for the emergence of an only finite, and ac-tually quite limited, number of basic polymers. The enormous diversity in detail but overwhelming unity in essence were the main characteristics later imprinted on the emerging life.

A functional concentrating catalyst must not only provide a unique environment, distinct from the general medium, but must also restrain the reactants at its surface for a sufficient amount of time in order to permit the reactions of interest to occur.

Thus the protective environment of igneous crystalline compartments is absolutely essential because the evolution of biologically important polymers was influenced by some types of rather weak interactions; for example, hydrophobic and hydrogen bonds, both relatively weak and noncovalent linkages, are fundamental to the maintenance of proteins.

6.7.4 Interlude

So, here we are: myriads of igneous cavities on the primeval beaches, each containing a particular concentration of particular polymers, grad-ually increasing in density with the incessant tidal wetting and subsequent evaporation, engaging in higher levels of chemical reactions, forming more and more complex polymers at increasing rates, with the accom-panying increase of catalytic activity. But even the most complex polymer is not a living system. Morphological complexity and dynamic circular organization of the most primitive organisms are absolutely necessary for metabolism, assimilation, reproduction, and excretion.

What mechanisms organized biological monomers and polymers into autonomous, reproducing units? We do have protected inorganic "wombs," full of highly "pregnant" building materials, exposed to basic biological rhythms of tides, days and nights, influencing rhythmically the chemical processes inside. We know that the catalytic abilities of mineral crystals steadily increase as a result of the intimate interaction of organic and inorganic matter. The conditions approach the point when the emergence of membranous protocells becomes possible. These originally inorganic membranes continually adsorb organic molecules, and become more elastic and capable of survival, at least for short periods, even in the open waters of their immediate neighborhoods. They contain a variety of interdependent processes, amino-acid cycles, and perhaps some primitive hypercycles.

Membranes of comtemporary cells contain proteins and lipids. Simple lipids contain only carbon, hydrogen, and oxygen—lipids bridge the gap from water-soluble to water-insoluble organic substances. Hydrolysis of simple lipids then yields glycerol and fatty acids. In any case, these compounds could have arisen only through intensified catalytic and even enzymatic functions. These were exactly the processes appearing inside our igneous cavities.

The ability to act as catalysts is characteristic of the majority of substances, especially metals and their oxides, hydrogen and hydroxyl ions, halogens, salts, and the like. Many of these were present in the mineral cavities; others were brought in by tidal waves. Rhythmical changes in temperature led to the formation of unstable intermediate compounds, the substrate–catalysts, decomposing into more complex end-products and restoring the catalysts.

But the catalysts themselves were also changing. Adsorption of the organic material profoundly increased their potentials. Very powerful organic and inorganic catalysts exist, but it is the close interaction of both that leads to the formation of enzymes. Enzymes could be even 10 million times more powerful than the pure crystalline or organic catalysts. Nearly every enzyme requires some inorganic component that in combination with protein dramatically increases its catalytic power.

Enzymes originally evolved inside the igneous inorganic "wombs" at a very early stage. The conditions were perfect. Only basic proteins were needed.

Protein macromolecules are mechanochemical aggregates of common amino acids. Again, they evolved nonrandomly and selectively, at an accelerating rate, inside the cavities. The cavities were increasingly protected by lipids—high temperatures would inactivate most enzymes by denaturing their tertiary structures. As membranes form, dialysis and osmosis cause separation of monosaccharides, amino acids, and nucleotides (small molecules) from polysaccharides, proteins, and nucleic acids (large molecules). However, the simple polymerization, as we have described it, although sufficient for the formation of amino acids, might be too haphazard and undirected for efficient protein formation. A selective mechanism, governing amino acid sequences in proteins, must have been present.

The evolution of proteins is thus dependent on the existence of a membranous protocell, an ever more efficient hypercycle, and individualization and reproduction of protocells.[11] The "wombs" must ultimately break up and release their contents into the ocean. Forces of natural selection and competition complement the earlier forces of symbiosis.

6.7.5 Protocell

We said that compounding of proteins and nucleic acids could not be separated from a highly selective functioning of a primitive membranous enclosure of first hypercycles. Crystalline structures of inner walls of igneous cavities provided important surface catalytic effect, and with the help of the adsorption, simple and more complex lipids formed and accumulated internally.

The resulting membrane is however still too passive, does not allow dialysis, and cannot facilitate the formation of proteins.

Another kind of catalysis also took place *within* these spherical enclosures. In addition to the surface catalysis, a "nuclear" catalysis was taking place as well. Catalytic fragments, usually phosphates, often dropped in the middle of the substrate, functioning from within rather than from without the protocells.

When a small piece of crystal drops into the substrate solution, given the favorable conditions existing within the cavities, a catalytic process is started. In a spherical neighborhood of the catalyst, intermediate substrate–catalyst compounds are being formed and a spherical membrane around the nucleus is produced. Outside this inner membrane the substrate is still unchanged and freely interacting with the outer osmotic membrane of the cavity.

The membrane-building polymers are not yet fully bonded and the membrane is initially very fluid and fuzzily delineated. The original "protomembrane" is thus defined dynamically; it is highly volatile, alternately exhibiting disintegration and repair stages. The continuing tidal supply of substrate leads to its gradual strengthening and definition. Depending on the conditions, it might still totally disintegrate by night or at high tide, when the catalytic processes are dampened or the concentration lowered by excess water. But it is bound to reassemble again as conditions become more favorable during the day.

All the vital processes of concentration, dehydration, polymerization,

---

[11] Other types of "protocells" or Jeewanu—coacervates, liposomes, and proteinoid microspheres—have also been described in the literature (Oparin 1962; Calvin 1969).

and the like are continuing. The substrate is being enriched by larger and larger molecules that are being retained within the matrix. Even the catalytic crystal adsorbs additional organic compounds and grows larger and more powerful. The volatile membrane starts to acquire some opaqueness and plasticity and adsorb free porphyrines—the conditions for self-production processes are qualitatively changing and external energy receptors acquired.

There is still no reproduction, no evolution, just myriads of individual entities struggling for their independence from the environmental fluctuations. Their protected rocky "wombs" are indispensable. There are still many million years to go before the first membranes stay intact through the full cycles of outside temperatures and the tides.

In the manner thus described, most likely billions of membranes with or without well-formed nuclei have formed *simultaneously,* at different places on Earth. Fully protected from the harshness of the outside environment, they gradually built their complex structures of chemical relationships.[12,13]

The rhythms of temperatures and the tides, reliable and never failing, provided for the stability of their inner environments. Continuous restoration of identical, or almost identical, conditions allowed repeated restoration and duplication of identical molecular chains. Igneous rocks of primeval beaches became teeming with life about 3 billion years ago, 1.5 billion years after the emergence of first protobionts.

6.7.6 Breaking of "Wombs"

The continuing cooling of Earth's crust led into the period of volcanic eruptions, other geological disturbances, and formation of the first mountainous accumulations. The biomatrices were breaking up, releasing their contents into their immediate neighborhoods and ultimately into the forming oceans.

Most of the early protocells obviously did not survive their new environments. Low concentrations, unreliable and more equalized temperatures, ultraviolet radiation, more turbulent waters—these conditions

---

[12] The separate origin of eucaryotes and procaryotes is now postulated against the naive linear procaryotes to eucaryotes evolution hypothesis. See for example Darnell (1978). It is likely that the basic patterns of genome organization between present-day procaryotes and eucaryotes are correspondingly different. Their independent origin does not preclude their ultimate endosymbiosis, as postulated by Margulis (1970).

[13] The early reducing atmosphere (see also note 10) was probably conducive to the emergence of *methanogens,* anaerobic, methane-producing microorganisms, only vaguely related to procaryotes or eucaryotes. This "third class of life" further supports the idea of polyphylogenesis. Again, structural similarities do not necessarily imply organizational and genetic identity of such autopoietic systems.

took their toll. Yet, the complex polymers and macromolecules did not disappear. They found their way back into the remaining matrices and washed over the beach rocks, possibly at more distant places They became factors of the new environment for succeeding generations of autogenesis.

Only the most cohesive unities could survive outside, at least for a moment. Especially in the immediate neighborhoods, in small and partially protected pools of substrate, their survival rate was much higher and they continued to grow.

Ultimately, some of these protocells acquired a critical size and weight and divided through a simple cleavage, budding, or mechanical separation. They were able to devour their less fortunate companions, or their remnants, and incorporate them within their structures.

Their membranous enclosures, simple reproduction, and therefore emerging phenomena of heredity allowed the processes of natural selection and evolution to take place. Ontogeny of the individuals gave rise to shorter- or longer-lived phylogenic chains.

Disruptions of individual ontogenies occurred at different places and at different stages of their development. For at least a billion years, the biomatrices were releasing their contents, starting different phylogenic chains. This process continued until the exhaustion of the matrices or until the outside environment became inhabited by species strong enough to destroy or consume any newly emerging entities. The process of autogenesis slowed down and perhaps even came to a halt. The processes of natural selection, evolution, and structural environmental coupling took over.

It is implied that the basic phylogenic chains were established as a result of differentially interrupted individual primeval ontogenies. The framework of initial speciation, establishment of parallel phylogenic lineages, cannot be separated from autogenesis. *Ontogeny precedes phylogeny*— both ontogeny and phylogeny are phenomenological outcomes of the continuing autopoiesis. That is what we meant by our earlier stating that phylogeny recapitulates ontogeny.

The preceding considerations lead to the hypothesis of "instantaneous" appearance of species, at different places and times, and they weaken the notion of phyletic gradualism of ancestor-descendant evolution. Pervasive gaps (both morphological and distributional) in the fossil record are currently labeled as "imperfections." But as stated in Cracraft and Eldredge, for example, there is a growing awareness among paleontologists that our ability to resolve time in the fossil record allows only for "instantaneous" species *formation* (Cracraft and Eldredge 1979).

The emergence of complex forms (excluding protocells) in the fossil record can be characterized as "instantaneous." The problem of interspecific "missing links" is infamous and the theory of convergent evolution appears a bit artificial and forced. There are also frequent instances of younger fossils exhibiting more primitive traits than older fossils.

This theory has nothing to do with denying evolution: intraspecific variations and geographically isolated interspecific transitions are undeniable. The question concerns species formation, not their evolution and adaptation.

One conjectural implication of autogenesis is the possibility of correlating major outbursts of speciation, properly lagged, with the periods of geological upheavals and disturbances of Earth's crust—the times of breaking of "wombs."

6.8 Lamarckian Evolution

The recent resurrection and appreciation of Lamarck's thought does not need much additional documentation. Lamarck's *Philosophie Zoologique* (1809), after being cleaned of the mud of misunderstandings and misinterpretations, is revealing its elegantly crafted beauty of truthfulness. His intuitive respect for the environment is now being experimentally vindicated.[14] We shall mention only a few of Lamarck's thoughts, based on abiogenesis, that seem to be directly relevant to autopoiesis and autogenesis.

Lamarck recognized that in nature there are no such things as classes or orders or families, only individuals. This is a profound observation and is not foreign to the proponents of autopoiesis.

He also rebelled against the then supposed immutability of species. His explanation is a beautiful expression of system-environment structural coupling and of the idea of structural adaptations as compensations for environmental perturbations: " . . . when an individual of a given species changes its locality, it is subjected to a number of influences which little by little alter, not only the consistency and proportions of its parts, but also its form, its faculty, and even its organization; so that in time every part will participate in the mutations which it has undergone."

Such slogans as "function creates the organ" or "acquired characteristics are inheritable" have little to do with Lamarck's theory. He sum-

---

[14] For example, it appears that certain species not only lack detectable sex chromosomes (amphibians), but in many species (alligators, turtles, lizards) environmental temperature is a determinant of sex differentiation. See Bull and Vogt (1979). Constant incubation temperatures of 31°C and above produce females, cooler temperatures of 24–27°C produce males, and, in *Chelydra*, even cooler temperatures (20°C) again produce females. *No* peculiar assortment of genes at conception or even during cleavage can account for these sex ratio biases. Chromosomes seem to be capable of degrading and rebuilding their own components in response to environmental perturbations.

marized his doctrine in six points (Leduc 1911):

1. All the organized bodies of our globe are veritable productions of nature, which she has successively formed during the lapse of ages.
2. Nature began, and still recommences day by day, with the production of the simplest organic forms. These so-called spontaneous generations are her direct work, the first sketches of organization.
3. The first signs of an animal or a plant growth being begun under favorable conditions, the faculties of commencing life and of organic movement thus established have gradually developed little by little the various parts and organs, which in process of time have become diversified.
4. The faculty of growth is inherent in every part of an organized body; it is the primary effect of life. This faculty of growth has given rise to the various modes of multiplication and regeneration of the individual, and by its means any progress that may have been acquired in the composition and forms of the organism has been preserved.
5. All living things that exist at the present day have been successively formed by this means, aided by a long lapse of time, by favorable conditions, and by changes on the surface of the globe—in a word, by the power that new situations and new habits have of modifying the organs of a body endowed with life.
6. Since all living things have undergone more or less change in their organization, the species that have been thus insensibly and successively produced can have but a relative constancy, and can be of no very great antiquity.

There is a grandeur in this view of life.

6.9   Lessons of Synthetic Biology

We have described several of the experiments of Leduc (1911) and other synthetic biologists,[15] and attempted to incorporate some of their implications for autogenesis. In summary, at least the following can be noted:

1. The morphogenic action of diffusion produces osmotic growths of extreme variety, resembling closely the forms of shells, fungi, corals,

---

[15] Self-organization experiments with *organic* substances are represented by the work of Guiloff appearing in this volume. Earlier results with *Jeewanu* should also be noted. These "particles of life" consist of peptides, ascorbic acid, ammonium molybdate, a variety of inorganic ions, and a suitable buffer. They exhibit phenomena of growth, reproduction, and metabolic activity. They grow from "within," not from the "outside" like crystals or osmotic growths, their components are not those directly present in the medium but synthesized within the Jeewanu.

and algae. In addition, the analogy of function is comparable to that of form. The organizing action of osmosis on organic material has hardly been attempted.[16]

2. Many osmotic growths appear to be of great complexity, often *structurally* much more complicated than the simpler forms of living organisms. Osmotic morphogenesis shows that the ordinary physical forces have a power of organization infinitely greater than hitherto supposed. Our ability to distinguish between the osmotic and the living forms in the early fossil record must be carefully reexamined.

3. The growth of an osmotic form structurally resembles the ontogenetic development of the ovum (Weiss 1970; Zeleny 1980). It is at least conceivable to entertain the idea that the beginning of life was not the production of a simple primitive form, from which all others have descended, but that a number of such primitive forms may have been produced, forms that by a rapid physical development attained a high degree of complexity.

4. The physical and chemical conditions of prebiotic Earth were significantly different from anything we can deduce from our immediate experience. The "mineral" world was pervasive—the oxygen, hydrogen, nitrogen, carbon, phosphorus, silica, and lime that now form the substance of living organisms. All living beings are formed of the same elements as those of the mineral world. The term "organic," it should be emphasized, refers to the specific "*organiza*tion" of these elements, not to a specific organic substance.

5. Every living organism consists of liquids, solutions of crystalloids, and colloids separated by osmotic membranes. A study of the origins of life must include the study of solutions and the study of the physical forces and conditions that can produce cavities surrounded by osmotic membranes, that can associate and group such cavities, and that can differentiate and specialize their functions.

6. The primeval beaches of prebiotic Earth presented the particular conditions favorable for the production of osmotic growths. An exuberant growth of osmotic vegetation must have been produced in these primeval seas. The soluble salts of calcium, carbonates, phosphates, silicates, and proteinoids became organized as osmotic productions. Billions of these ephemeral forms are long dissolved and without a trace. Many of them left an impression that we often mis-

---

[16] The nature of species, especially in plants, is a matter of conjecture. D. A. Levin goes even further and submits that "species" are not natural units of evolution—they are *mental constructs* of a community of observers (Levin 1979). This understanding is in direct opposition to Mayr's "Species are the real units of evolution." Such uncritical re-creations of Darwin's *On the Origin of Species* include those of Mayr (1942), Dobzhansky (1941), Huxley (1940), and many others.

take for that of a modern living organism. Enough of them became the matrices of life.

6.10   Concluding Remarks

It would be inappropriate to follow the autogenesis further—to the stages of reproduction and evolution. Reproduction, or more precisely self-reproduction, and consequently evolution, appeared only at the later stages of autopoiesis. They are secondary, derived phenomena. Originally, only a multitude of relatively independent ontogenies, nonreproducing autopoietic systems, populated Earth. A successful mechanical fragmentation of an autopoietic unity is a form of self-reproduction—its emergence and subsequent selection were probably quite simple. The reader is advised to consult the recent works of Maturana (Maturana 1980) and Varela (Maturana and Varela 1973) on these topics.

The concept of evolution, as it becomes modified from the vantage point of autopoiesis, is a fascinating but too lengthy topic to handle within a single paper. There are basically two ways of treating evolution within autopoiesis, both radically different from the currently prevailing notions. In both, evolution is a secondary (i.e., derived) phenomenon of autopoiesis. This is in contrast to the pivotal position accorded to it, for example, by Dobzhansky et al. (1977): "Nothing in biology makes sense except in the light of evolution."

1. Evolution takes place only when there are organizational changes occurring throughout the sequentially interdependent self-reproductions of unities. Only sequentially reproducing patterns of *organization* can evolve—if changes are occurring during the reproduction of that pattern. An invariant organization, manifesting itself through a pattern of *structural* transformations of a unity, is undergoing *ontogenesis,* not evolution. In this view, nonreproducing entities (e.g., embryo, Earth, universe) do *not* evolve but are undergoing ontogeny. Also, organizationally invariant unities (e.g., species) do not evolve but only have a history of structural adaptations (i.e., undergo ontogeny) (Maturana and Varela 1973).

2. Evolution takes place whenever there is a hereditary *structural* change in the sequentially self-reproducing, but organizationally invariant, unities. In this view, establishment of a phylogeny is necessary and sufficient for evolution. In this framework, the nonreproducing entities do not evolve, but the organizationally invariant classes, as for example species, do. Evolution consists of the history of structural changes that the members of a phylogeny undergo.

Both of these views are consistent with autopoiesis. The first refers to the structural changes within a given identity class (invariant organization) as ontogeny and would label as evolutionary only the changes pertaining to the organization. The second view allows for evolution within a given identity class, thus labeling as evolutionary the changes of structure.

The second view is less radical and closer to our common usage and understanding of evolution. The first view is potentially more significant, but possibly should not be referred to as "evolution." Perhaps *structural* and *organizational* evolution, in all of their interdependence, might fruitfully reflect this duality.

References

Bull, J. J., and Vogt, R. C. (1979), Temperature-Dependent Sex Determination in Turtles, *Science* 206, 1186–1188.

Calvin, M. (1969). *Chemical Evolution: Molecular Evolution Towards the Origin of Living Systems on the Earth and Elsewhere,* Oxford University Press, Oxford.

Cudmore, L. L. Larison (1977). *The Center of Life,* Quadrangle/The New York Times, New York, p. 29.

Cracraft, J., and Eldredge, N. eds. (1979), *Phylogenetic Analysis and Paleontology,* Columbia University Press, New York.

Darnell, J. E., Jr. (1978), Implications of RNA·RNA Splicing in Evolution of Eukaryotic Cells, *Science* 202, 1257–1260.

Dobzhansky, T. et al. (1977), *Evolution,* Freeman, San Francisco.

Eigen, M. and Schuster, P. (1977), The Hypercycle: A Principle of Natural Self-Organization, Part A: Emergence of the Hypercycle, *Naturwissenschaften* 64, 541–565; Also see Part B: The Abstract Hypercycle, 65 (1978), 7–41, 341–369.

Farley, J. (1977), *The Spontaneous Generation Controversy from Descartes to Oparin,* Johns Hopkins University Press, Baltimore.

Fox, S. W., and Dose, K. (1972). *Molecular Evolution and the Origin of Life,* Freeman, San Francisco.

Hedges, E. S., and Meyers, J. E. (1926), *The Problem of Physiochemical Periodicity,* Arnold, London.

Huxley, J. S. (1940), *The New Systematics,* Clarendon, Oxford.

Knoll, A. H., and Barghoorn, E. S. (1977), Archean Microfossils Showing Cell Division from the Swaziland System of South Africa, *Science* 198, 396–398.

Leduc, S. (1911). *The Mechanism of Life,* Rebman, London.

Levin, D. A. (1979), The Nature of Plant Species, *Science* 204, 381–384.

Lotka, A. J. (1924), *Elements of Physical Biology,* Williams and Wilkins, Baltimore.

Margulis, Lynn (1970), *The Origin of the Eukaryotic Cell,* Yale University Press, New Haven, Conn.

Matthews, C. S., et al. (1977), Deuterolysis of Amino Acid Precursors: Evidence for Hydrogen Cyanide Polymers as Protein Ancestors, *Science* 198, 622–624.

Maturana, H. R. (1980). Autopoiesis: Reproduction, Heredity and Evolution, in M. Zeleny, ed. *Autopoiesis, Dissipative Structures and Spontaneous Social Orders,* Westview, Boulder, Colo., pp. 45–79.

Maturana, H. R. and Varela, F. J. (1973). *De Máquinas y Seros Vivos,* Editorial Universitaria, Santiago. This book is now available as *Autopoiesis and Cognition: The Realization of the Living,* Boston Studies in the Philosophy of Science, Vol. 42, Reidel, Boston. The continuity hypothesis is also reiterated in F. J. Varela (1979), *Principles of Biological Autonomy,* Elsevier, New York, pp. 26–29.

Mayr, E. (1942), *Systematics and Origin of Species,* Columbia University Press, New York.

Miller, S. L. (1953), A Production of Amino Acids Under Possible Primitive Earth Conditions, *Science* 117, 528.

Miller, S. L. and Orgel, L. E. (1974), *The Origins of Life on the Earth,* Prentice-Hall, Englewood Cliffs, N.J.

Nicolis, G., and Prigogine, I. (1977). *Self-Organization in Nonequilibrium Systems,* Wiley-Interscience, New York.

Oparin, A. I. (1953), *The Origin of Life,* Dover, New York, pp. 57–58. (Originally published in the USSR in 1936.)

Oparin, A. I. (1962), *Life, Its Nature, Origin and Development,* Academic, New York.

Pflug, H. D. and Jaeschke-Boyer, H. (1979), Combined Structural and Chemical Analysis of 3,800-Myr-old Microfossils, *Nature* 280, 483–486.

Smith, J. M. (1979), Hypercycles and the Origin of Life, *Nature* 280, 445–446.

Tanford, C. (1978), The Hydrophobic Effect and the Organization of Living Matter, *Science* 198, 396–398.

Varela, F. J., Maturana, H. R., and Uribe, R. B. (1974), Autopoiesis: The Organization of Living Systems, Its Characterization and a Model, *Bio-Systems* 5, pp. 187–196. For further elaboration of the model see M. Zeleny (1977), "Self-Organization of Living Systems: A Formal Model of Autopoiesis" *General Systems* 4, 187–196. See also M. Zeleny, ed. (1980). *Autopoiesis, Dissipative Structures and Spontaneous Social Orders.* Boulder, Colo.: Westview.

Wald, G. (1954), "The Origin of Life" *Scientific American* (August), pp. 1–11.

Weiss, P. A. (1970), The Living System: Determinism Stratified, in A. Koestler and J. R. Smythies, *Beyond Reductionism,* Macmillan, New York, pp. 3–55.

Zeleny, M. (1980). Autopoiesis: A Paradigm Lost? in Zeleny, M., ed., *Autopoiesis, Dissipative Structures and Spontaneous Social Orders,* Westview, Boulder, Colo., pp. 3–43.

INTRODUCTORY REMARKS

Guiloff's paper is the most ambitious in the whole volume. She has attempted a direct laboratory production of a molecular protobion, a first autopoietic unity. Her paper is thus bound to be quite controversial and subject to sweeping critiques—especially since she did not succeed. I am sure that Dr. Guiloff would add "yet" to my last sentence, and affirm that controversy may lead to new insights and more creativity. As she says, "Who knows? . . . sometimes our judgment fails, and surprises may happen."

In spite of the recommendations of some referees, I decided to include this paper in the volume. It stands a witness to the fact that autopoiesis does not have to be perceived as an inflexible scholastic theory, but can acquire a rather lively and exciting experimental content—something that cannot be said about many of the "competing" paradigms. Also, this editor believes, the efforts of a lonely pioneer are often more important than the opinions of the dons running the well-financed superlabs of today.

How does one go about creating an autopoietic unity (similar to the one we have seen emerging in our computer simulations) in the physical space of atoms and molecules? Guiloff provides some tentative experimental ideas about how the making of a protobion should proceed. She concluded that constructing a lipoproteic permeable membrane (i.e., having it emerged in a spontaneous fashion) would go a long way toward validating some of the main precepts of autopoiesis.

Guiloff actually succeeded in obtaining rather beautifully shaped mem-

Gloria D. Guiloff was born in Santiago, Chile, in 1944. She received "Licenciada en Ciencias con mención en Biología" from the University of Chile in 1974. Thesis: "An electrophysiological study of the retinal ganglion cells of the toad, *Bufo spinulosus*." After being exposed to autopoiesis, she started preliminary experiments aimed at a synthesis of a molecular model of a minimal autopoietic system. After the Ph.D. qualifying examination (dissertation: "The origin of the eukaryotic cell") she wrote a formal thesis project on neobiogenesis and autopoiesis. The experiments with autopoiesis were postponed in 1978, after $3\frac{1}{2}$ years of efforts. At the advice of Professor Maturana and other colleagues she returned to the fields of neurobiology and vision in order to be able to complete her official Ph.D. requirements. Although working on the anatomy of the toad's visual system at present, Dr. Guiloff plans to resume the autopoiesis project in the near future. *Address:* Department of Biology, Faculty of Sciences, University of Chile, Casilla 653, Santiago, Chile.

branous structures, as can be seen from the reproduced photographs. She has not been able to establish whether these are images of dynamic and permeable membranes or just static fixtures.

It seems that Guiloff views the difference between a two-dimensional computer model of autopoiesis and its realization in physical space as being simply a matter of the added dimension. Of course, a three-dimensional *computer* model of autopoicsis is theoretically realizable, but the insights gained might still not be sufficient for the physical realization of autopoiesis. Thus it appears that the best strategy is to obtain a variety of hypothetical autopoietic unities, by whatever means (including those of synthetic biology), and to establish reliable validation procedures and tests to determine whether the proposed systems are autopoietic. (One should remark here that the "tests" used recently on Mars would not of course be satisfactory examples of such procedures.)

Chapter 7
Autopoiesis and Neobiogenesis

Gloria D. Guiloff

7.1 Introduction

The full understanding of the living organization requires a theoretical formulation that allows for an experimental approach to neobiogenesis. This is true not only because it is necessary to define some procedure for the experimental realization of an artificial living system, but also because it is necessary to be able to recognize when the intended aim has been obtained.

Neobiogenesis has been attempted on many occasions by many authors (Oparin 1962, 1965; Hayakawa et al. 1967; Keosian 1967; Calvin 1969; Fox and Dose 1972; Fox 1973. 1974a–c) but without the use of a precise theory of the living organization, because this theory had not been available (Varela et al. 1974; Maturana 1975; Maturana and Varela 1975); now, the theory of autopoiesis seems to fill this gap.

7.2 Present State of Affairs

The theory of autopoiesis is a precise characterization of the relations that define a living system as a unity (organization) but establishes no constraints about its structure; this can be any, so long as it satisfies the autopoietic organization. The authors of the theory have taken advantage of this precision by making a computer model of an autopoietic system that they call *protobion*.

My present aim is to take into consideration the possibility of making a molecular protobion, based on my rather modest experience in the attempt at neobiogenesis.

7.3   What the Theory Does

The theory of autopoiesis does essentially three things:

1. It defines, without reference to the whole, the relations that the components must satisfy in the integration of an autopoietic system.
2. It makes a clear distinction between the phenomenic domain of the components of an autopoietic system (as the domain of its states in autopoiesis) and the phenomenic domain in which the autopoietic system operates as a unity (as the domain of its relational states), showing that these two phenomenic domains do not intersect.
3. It shows that reproduction is not a definitory feature of the organization of living systems, but that it is necessary for evolution.

The computer model protobion (Varela et al, 1974) realizes a minimal autopoietic system in a bidimensional space. It has one basic limitation: It contains a permanent component that is not produced by the autopoietic network and therefore the system cannot undergo reproduction. However, the computer model shows that, in practice, it is sufficient to specify the properties of a set of components and of certain ambient conditions for an autopoietic system to arise spontaneously, as had already been accepted in principle (Maturana and Varela 1975).

7.4   Proposition

I propose an experimental approach to neobiogenesis directed by the theory of autopoiesis, an approach whose purpose would be to find a minimal structure that satisfies the autopoietic organization in physical space, in other words, to construct an autopoietic unity whose components are atoms and molecules. I shall refer to such a minimal case as a *molecular protobion*.

Any attempt to make a molecular protobion should

not include a permanent component. To do so would preclude the possibility of reproduction and hence, of evolution, and would consequently be an unsuccessful and insufficient model for neobiogenesis.

abandon the attempt to reproduce any particular function or set of functions that an observer sees as a distinctive feature of the operation of a known living system, or of their known components; to do this would result in a confusion of phenomenic domains.

define the conditions of production of the boundaries of the unity. If this were not done, it would not be possible to identify the unity, as seems obvious.

In these circumstances, I can define a molecular protobion's minimal structure as a unity in the physical space, limited by a "membrane" that is continually being formed and degraded as a component of the autopoietic network. It seems obvious that the system's border "membrane" must both enclose and participate in the network of processes of production that generates it.

Since the molecular protobion's "membrane" is a central element in this design, the following considerations seem pertinent:

1. According to the computer model (Varela et al, 1974), such a border should be permeable. The "membrane" could not be only a lipid bilayer, since it is known that artificial lipid bilayers are practically impermeable, even to small ions (Mueller et al. 1962; Alvarez et al. 1975); when proteins are added to the bilayer, it becomes permeable.

2. The membranes of cells of contemporary organisms are lipoproteic in their composition, but the way in which protein and lipid interact to form membrane is not known; that there is a turnover of the cell's membranes is apparent, but the processes involved in it are not yet elucidated. Though the biosynthesis of proteins and of lipids are fairly well-known processes, the mechanisms of formation and degradation of the cell's membranes are still unclear.

These considerations may be used to define the molecular protobion's border "membrane" as a lipoproteic permeable structure whose chemical composition would not necessarily be the same as that of known biological membranes.

7.5 Some Preliminary Observations

Whatever experiments one does, the main problem seems to be the recognition of a molecular protobion; points 1, 2, and 4 of the "key for the identification of autopoietic unities in any space" (Varela et al. 1974) do not offer great empirical difficulties, but this is not so with point 3, which is crucial because it represents the difference between a static and a dynamic system.

Discrete multimolecular "membranous" systems may be obtained by mixing

hinoquitol [$C_{10}H_{12}O_2$, cedar oil],

thermal proteinoids [Asp:Glu:(Ala:His:Leu) = 2:2:1(w/w)],

salts [NaCl, $Ca(NO_3)_2$, KH_2PO_4, $MgCl_2$, Na pyruvate, choline chloride, $NaNO_2$, NH_4AlSO_4, and $Fe_2(NO_3)_3$],

sugars [glucose, arabinose],

H_3PO_4, and

sand.

This mixture is incubated for 2 h at 70° C; water is then added and the mixture is incubated for 1 h more at 70° C; next, it is cooled to room temperature, and finally to 4–5° C.

These systems are stable, and under the compound microscope their forms are varied (nearly spherical, ellipsoid, and irregular shapes), and their size ranges from a few micrometers to around 30 μm in diameter (see Figure 1). They can be centrifuged to obtain a pellet for electron microscope. The border's appearance under the electron microscope, after glutaraldehyde–osmium tetroxide treatment, resembles that of biological membranes; that is, it is a trilaminar structure of 80–220 Å width. Biological membranes in similar conditions appear in the range of 65–120 Å. Under conditions of cryofracture using liquid nitrogen, and after freeze-etching, one can obtain a carbon-metal replica for electron microscopy observation of the border structure.

If crude or partially purified thermal proteinoids (Fox and Harada 1960; Fox and Dose 1972) are added to a lipid bilayer formed with lecithin–cholesterol in a KCl 0.1M Tris-Cl, pH 7.0 solution (Alvarez et al. 1975), the bilayer becomes permeable to ions, as observed by a rise of its conductance; the same phenomenon has recently been observed to occur with synthetic peptides (Kennedy et al. 1977). Single amino acids or sets of them, when added to the bilayer, do not affect its conductance (unpublished observations). A bilayer formed with hinoquitol in the conditions specified by Alvarez et al. (1975) interacts with thermal proteinoids also exhibiting a rise of conductance to ions.

For example, one can perform experiments using radioactive thermal proteinoids to form these "membranous" systems so as to be able to determine whether their border is permeable to proteinoids in the medium, whether there is a turnover of proteinoids, or where the proteinoids are structurally located. If these systems are succesively centrifuged (20,000 rpm, 10 min), the loss of radioactivity is smaller when they are in a proteinoid–salt solution than when they are in water alone or salt solutions; however, the kinetics of this loss does not yet allow me to conclude that there is a turnover of proteinoids in these systems.

If one uses one ^{14}C-proteinoid to make the systems and successively centrifuges them in a ^3H-proteinoid-salt solution, and if there is a turnover of proteinoids, the predicted curve will be that shown in Figure 2, provided that the number of systems remains constant throughout the experiment.

Very sensitive techniques are available today enabling the location of proteinoids or any other putative component in the systems (Feder et al. 1977; Sayre et al. 1977).

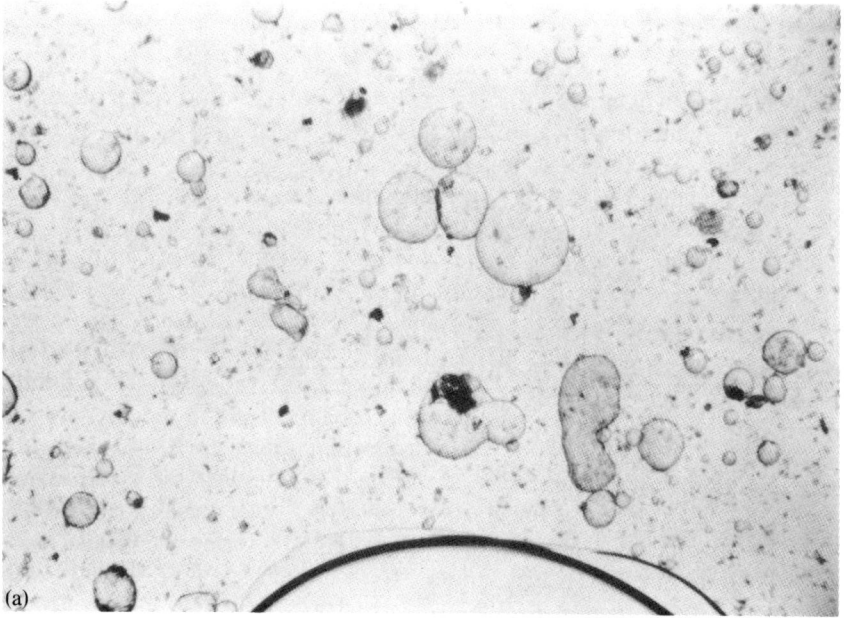

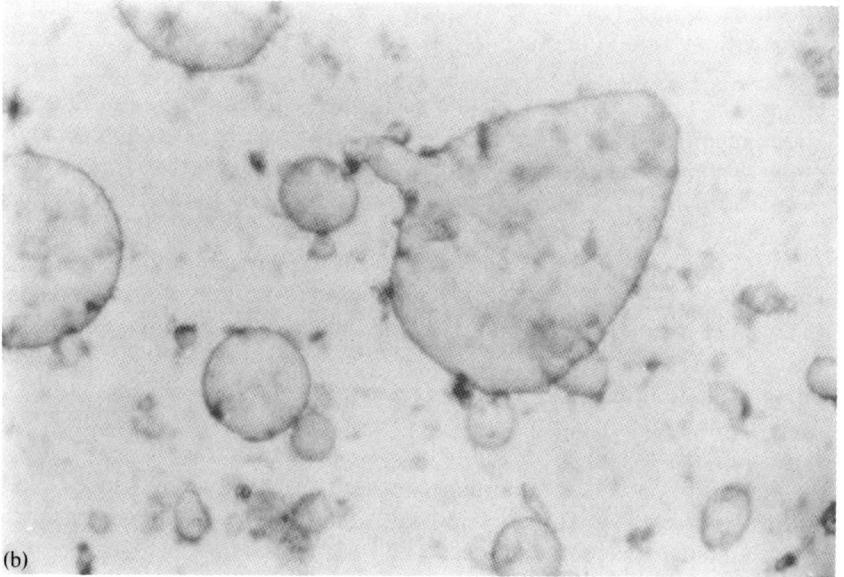

Figure 1. Scale: (a) 2.3 cm = 500 μ, (b) 3.5 cm = 200 μ.

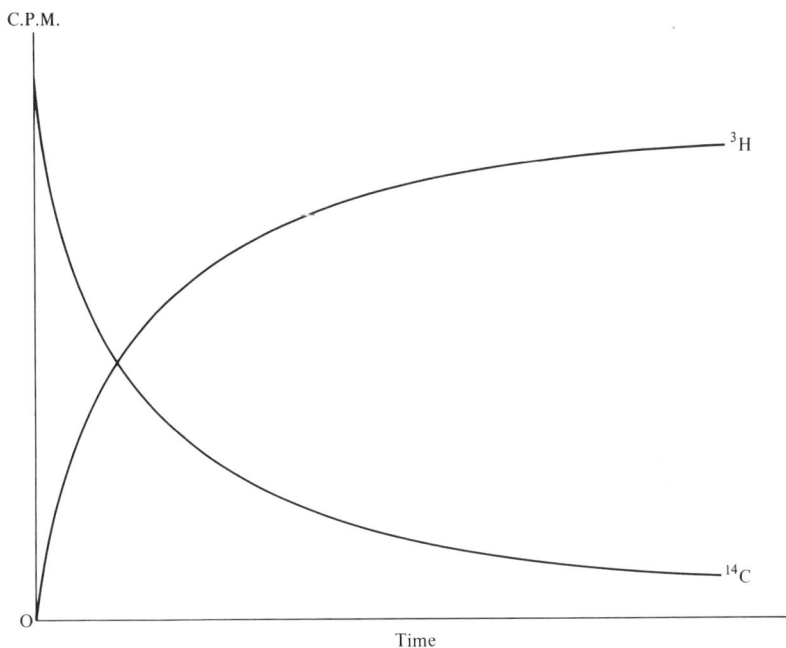

Figure 2

These few observations serve to illustrate the fact that one can design experiments to follow the "key for the identification of autopoietic unities in any space" if one is dealing with a multimolecular system of the type described briefly above; of course, this amounts to an indirect approach to neobiogenesis, which may well be a necessary stage before the step-by-step construction of a molecular protobion.

7.6   Additional Remarks

The difference between the computer protobion model and molecular protobion model, conceptually, is the addition of another dimension. In practice, this actually means qualitatively different components and processes; the peculiarity of the phenomena of physical space, as compared to those in a two-dimensional space, seems to be a consequence of the properties of its components, as revealed through their interactions. The main problem in dealing with these components in the concrete construction of a molecular protobion is our relative ignorance with respect to atomic and molecular interactions. Nevertheless, one possible approach is to obtain hypothetical minimal autopoietic molecular entities and to study them in the laboratory until there is evidence that the proposed systems are or are not autopoietic networks.

Another difficulty arises when one wants to restrict the term "dynamic" with respect to a putative molecular protobion. The criteria for the distinction between a static multimolecular system and a dynamic one (not necessarily autopoietic) would have to be the times of occurrence of the detected processes, that is to say, the kinetic constants. A molecular turnover of the type produced by mere thermal agitation would be considered proper to a static system and therefore it would not be a process within an autopoietic network.

It would be advantageous to have a three-dimensional computer model of a minimal autopoietic system without a permanent component. Such a model could provide some insight as to what relevant properties the components should have so as to generate the network; from this insight the possibility of step-by-step construction of a molecular protobion seems greater.

I believe that the use of the theory of autopoiesis as the conceptual framework for an experimental approach to neobiogenesis leads to the conception of a minimal molecular structure, which is necessarily a "membranous" unity that is continually building and degrading its border. In conclusion, then, I think that the possibility of obtaining a molecular protobion opens a new perspective on the experimental investigation of the origin of living systems and may lead to a novel insight upon biological phenomena.

ACKNOWLEDGMENTS

This work was partly supported by the Chilean Oficina de Desarrollo Científico y Creación Artística of the University of Chile and was presented in a modified version under the title "Autopoiesis and the Origin of Living Systems" at the 1st International Conference on Applied Systems Research: Recent Developments and Trends, sponsored by Scientific Affairs Division of NATO, SUNY, Binghamton, New York, 17 August 1977.

References

Alvarez, O., Díaz, E., and Latorre, R. (1975), Voltage-dependent conductance induced by hemocyanin in black lipid films, *Biochim. Biophys. Acta,* 389, 444–448.

Calvin, M. (1969), *Chemical Evolution: Molecular Evolution Towards the Origin of Living Systems on the Earth and Elsewhere,* Oxford Univ. Press.

Feder, R., Spiller, E., Topalian, J., Broers, A. N., Gudat, W., Panessa, B. J., Zadunaisky, Z. A., and Sedat, J. (1977), High-resolution soft x-ray microscopy, *Science* 197 (No. 4300), 15 July, 259–260.

Fox, S. W. (1973), Molecular evolution to the first cells, *Pure and Applied Chemistry* 34, 641–669.

Fox, S. W. (1974a), Origins of biological information and the genetic code, *Mol. Cell. Biochem.* 3 (No. 2), 129–142.

Fox, S. W. (1974b), The proteinoid theory of the origin of life and competing ideas, *American Biology Teacher* 36 (No. 3), 161–172.

Fox, S. W. ed. (1974c), Thermodynamic perspectives and the origin of life, in *Quantum Statistical Mechanics in the Natural Sciences*, Plenum, New York, 1974.

Fox, S. W., and Dose, K. (1972), *Molecular Evolution and the origin of life*, W. H. Freeman, San Francisco.

Fox, S. W., and Harada, K. The thermal copolymerization of amino-acids common to protein, *J. Am. Chem. Soc.* 82 (No. 14), July 20, 3745–3751.

Hayakawa, T., Windsor, C. R., and Fox, S. W. (1967) *Arch. Biochem. Biophys.* 118, 265.

Kennedy, S. J., Roeske, R. W., Freeman, A. R., Watanabe, A. M., and Besch, H. R., Jr. (1977), Synthetic peptides form ion channels in artificial lipid bilayer membranes, *Science* 196 (No. 4296), 17 June, 1341–1342.

Keosian, J. (1967), *The Origin Of Life*, 3rd Ed., Reinhold Publ. Corp., New York.

Maturana, H. R. (1975) The organization of the living: A theory of the living organization, *Int. J. Man–Machine Studies* 7, 313–332.

Maturana, H. R., and Varela, F. G. (1973), *De Máquinas y Seres Vivos*, Ed. Universitaria, Santiago de Chile. English version: *Autopoietic Systems*, Biol. Computer Lab. report 94, Dept. of Electrical Engineering, Univ. of Illinois, Urbana.

Mueller, P., Rudin, D. O., Ti Tien, H., and Wescott, W. C. (1962), Reconstitution of excitable cell membrane structure in vitro, *Circulation* XXVI, 1167–1170.

Oparin, A. I. (1962), *Life: Its Nature, Origin and Development*, Academic Press, New York.

Oparin, A. I. (1965), The pathways of primary development of metabolism and an artificual model of this development, in *The Origins of Prebiological Systems and of Their Molecular Matrices*, Academic Press, New York.

Sayre, D., Kirz, J., Feder, R., Kim, D. M., and Spiller, E. (1977), Potential operating region for ultrasoft x-ray microscopy of biological materials, *Science* 196, (No. 4296), 17 June, 1339–1340.

Varela, F. G., Maturana, H. R., and Uribe, R. (1974), Autopoiesis: The organization of living systems, its characterization and a model, *Biosystems* 5, 187–196.

Morin concludes the Proposition with a piece of writing worthy of one of the best-known representatives of the recent French wave of systems humanists. It certainly belongs in Part I of the volume as it is a good example of incorporating the concept of autopoiesis into one's own frame of thought, without necessarily asking its "fathers" or creators for permission. I wish that many readers could follow Morin's self-creative attitude.

Morin starts by explaining the origin and meaning of the prefix auto-, or self-, used in our attempts to describe the autonomy of a living individual. He tries to clarify the enormous confusion between self-organization and autopoiesis as it is also witnessed by some writers in this volume. To that end he introduces and distinguishes the notions of *self* and *autos*.

Morin's autos incorporates and even controls the self. Thus self-organization is a narrower concept than autoorganization: "in fact, biological autoorganization contains and controls self-organization, which is brought about thermodynamically in and by 'dissipative structures.'" Obviously the reader will find it useful to compare Morin and Jantsch on these issues. Basically Morin refers to self-organization only in a purely physical domain (stars, dissipative structures, whirlpools, etc.), while autoorganization, although remaining physical, acquires the dimension of biological "computation." He considers the idea of computation (or program transcription) to be essential for our understanding of the autos and thus prepares the ground for Part II.

Edgar Morin was born in Paris in 1921. He received graduate licenses in history and geography and later also in law (1942). E. Morin was a combat volunteer in the resistance movement and became a lieutenant in the French Army during 1942–1944. Later he served as a chief in the propaganda office of the French military administration in occupied Germany (1945–1948). He continued his journalistic activities and served as a journal editor in Paris until 1950. From 1950 until 1979 he worked as a researcher at CNRS (Conseil National de la Recherche Scientifique). Professor Morin is currently the Director for Research at CNRS and the Director of CETSAS (Centre d'Études Transdisciplinaires—Sociologie, Anthropologie, Sémiologie) associated with École des Hautes Études en Sciences Sociales. He also directs a journal called *Communications*. *Address:* CETSAS, 6 rue de Tournon, 75006 Paris, France.

He says, "Both the genetic code and subjectivity are common to all creatures, from the bacterium to the elephant." The duality and complementarity of subjectivity and computation within the *autos* make it radically different from computers and other machines of artificial intelligence: they are constructed by others, receive their programs from others, and function for others, not for themselves.

Morin also has no problem extending the concept of autos beyond the individual, including progeniture, family, and society.

Chapter 8
Self and Autos

Edgar Morin

8.1 Introduction

Life presents a double countenance: on the one hand, in the form of living beings who appear and disappear discontinuously, on the other hand, in the form of the continuous process of reproduction, which is the temporal propagation of a pattern. "Macroscopically," life presents itself in a manner as paradoxical and "complementary" as physical reality presents itself microscopically, appearing sometimes to be undulatory, sometimes corpuscular in nature. Classical biology has tried to cover up this paradox. During an initial stage, reality was accorded to the notion of species, even though individuals alone are real and the notion of species merely ideal. Individuals comprising the species appeared only as samples or specimens and the *organism* was taken to be the concrete object that permits the study of the species through its individuals. Nevertheless, the duality has continued to reappear with the birth and development of genetics: on the one hand, the germen, on the other, the soma; on one side the genotype, on the other, the phenotype. From a genetic standpoint, the phenotype is merely the genotype after modification by environmental conditions. The term "phenomenal" (the living individual, his behavior) is subordinate to the term "generative," which appears as an anonymous program—produced, so it would seem, by the most anonymous of cosmic actors: chance. Such a simplified and reduced view tends to dodge the disturbing problem of the autonomy of living beings. Those who view the question from this angle never use the prefix "auto."

Nonetheless, the prefix "auto" could have been used in connection with the study of living beings. Instead, the latter were either reduced to the state of organisms, that is to say organization without intelligence, functioning as if by automatic regulation (homeostasis), or else they were considered experimentally, isolated from the concrete conditions of their communicational and/or social life. For decades, they were perceived according to a behaviorist outlook, in which the source of an organism's responses were found not in an autonomous computation but in an exterior stimulus. Only since the development of ethology in the second half of the 20th century have we learned to conceive of these "organisms" as living beings that communicate among themselves and that possess cognitive aptitudes and intelligence. Still, the autonomy of these beings has not been considered from the viewpoint of its organizational foundations.

The double-countenance conception of life, generative (genetic, genotypical) and phenomenal (individual, phenotypical) as autoorganization, is a conspicuous necessity that theoretical work has often attempted to hide in order to construct a simplified conception of life. This attitude is true to the classical conception, according to which determinism is always exterior to objects and therefore to beings as well. The advent of cybernetics was necessary to conceive of the idea of an *endocausality* (Morin 1977), which interacts with exterior causalities (exocausalities) to give rise to and maintain the autonomy of a system. The conception of this endocausality is linked not only with the idea of retroaction, and thus with a retroactive effect allowing it to become causal, but also with the idea of regulation, and thus with an internal cause of constancy in a system. The advent of the informational idea of "program" was necessary if one were to conceive of an endocausality determining finalities suited to a system. This is, however, not at all sufficient, since the pattern applied to the living organization remains an artificial machine (*artifact*), which always receives its program, its materials, its conceptions, and its fabrication from an exterior source, that is, from man. It is nevertheless the impetus of cybernetics and the theory of *automata* that bring to the fore the prefix "auto." The work of von Neumann (1966) on the theory of *self-reproducing automata* was instrumental in emphasizing the idea and theoretical problem of *self-reproduction*. Further, Neumann, in reflecting on the difference between artificial automatons (artifacts) and natural automatons (living beings), paved the way for the idea of autoorganization. If artificial automatons begin to wear out as soon as they start functioning (though made up of very reliable components), while living beings (though made up of unreliable components) hold out against wear for a certain time, it is because the former have no means of regenerating their components and are incapable of reorganizing themselves, while living beings do possess the ability to regenerate their components because they are in a permanent state of autoreorganization. The idea of *permanent autoreorganization* as presented by Atlan (1972) is central in clearing the way for the idea of autoorganization and the idea of autopoiesis. Toward the end of the 1950s researchers began trying to conceive of the organization of life in terms of autoorganizing systems (von Foerster 1960) and autopoiesis (Maturana and Varela 1972), but the question already comes up, What does "auto" mean? It becomes clear that no concept exists that epitomizes the mysterious ability of a being, a system, a living machine, to draw from within itself the source of its particular autonomy of organization and behavior while (in order to fulfill this task) relying on energizing, organizational, and informational nourishment extracted or received from the environment. *What, then, is a living autonomy, autonomous only because, at a different level, it is ecodependent?* To fill this conceptual vacuum I propose the provisional concept of *autos* in order to enable us to consider the problems posed by the prefix "auto."

Varela (1975, 1978) gives the term "self-reference" to the characteristic property of autopoiesis and proposes formally to define *self-reference* as reentry and therefore recursivity. I believe that self-reference, reentry, and recursivity are indeed key notions in the understanding of the living phenomenon. However, as necessary as they are, they alone are not enough because they are too broad in scope: They can account for a great number of physical, self-organizing phenomena that are in no way biological, such as the organization of the atom, of stars, or even of whirlwinds.

I therefore propose to distinguish the notion of *self* from that of *autos*. A whirlwind is *self-organizing* in the very movement by which it establishes its form as a constant circuit, form that is recursive in the sense that final state and initial state are indistinguishable. The birth of the stars, as well as of our sun, is the result of the encounter of implosive (gravitational) and explosive (heat) retroactions, which together close the loop of regulated self-organization. The phenomenon of the self—that is, of being and existence—is a fundamental physical phenomenon since it is around the self that our organized world is formed, made up of atoms and stars. (We can even consider, as does Bogdanski (1978), that waves are self-regulating phenomena.) On another occasion I have developed the theory of self-production (Morin 1977). That is why I consider the autos to be a richer concept than the self, which it contains and incorporates at the same time [in fact, biological autoorganization contains and controls self-organization, which is brought about thermodynamically in and by "dissipative structures" (Nicolis and Prigogine 1976)].

This kind of distinction between autos and self is conventional with respect to the current meaning of these terms: one could call autos what I call self and call self what I call autos. However, if it is admitted that the autos corresponds to the phenomenon of the self with regard to biological complexity, then the autos possesses what is also shared by autoorganization, autopoiesis, autoregulation, and autoreference, and lays the foundations of the autonomy particular to the living being.

8.2   Auto- (Geno-Pheno-) Organization

First of all, let us be careful to avoid any definition of the autos that would mask one of life's two countenances, either the generative (which is crystallized in the notion of species) or the phenomenal (which is crystallized in the notion of the individual). Generally speaking, genetic theories tend to subordinate the phenomenal to the generative, while theories of autoorganization tend to subordinate the idea of autoreproduction to the idea of autoproduction (Maturana and Varela 1974). What is needed is a complex conception capable of revealing the unity in this duality and the duality in this unity.

It is necessary to speak of uniduality within autoorganization. Because of its recursive nature, this double organization is in fact one. As has often been observed, "the cell is at the same time producer and product which incorporates the producer" (Varela 1975); in other words, autoorganization is an organization that organizes the organization necessary to its own organization. Generative organization (which biology reduces, reifies, and unidimensionalizes in the idea of genes carrying an organizing "program") and phenomenal organization (which biology considers as metabolism and homeostasis) can neither be considered as two distinctive organizations nor reduced to an indistinct recursive entity. Distinction and indistinction are both present. Distinction appears as the necessary translation of the language, made up of four signs, of the genetic code into the 20-"lettered" language of amino acids. A heterogeneity appears between even the concept of species and the concept of individual, which seem to belong to different worlds. Indistinction exists in the fact that all of these terms are joined in recursive loops where the conjugation of generative and phenomenal constitutes autoorganization. Therefore, generative and phenomenal should be viewed as two poles: on one side, the generative pole, representing permanent regeneration and reorganization and periodical reproduction, on the other side, the phenomenal pole, the pole of praxis of living beings, of the organization of their interactions and behavior in a *hic et nunc* environment. At one pole, reproduction—that is, the temporal survival of the "species" that the other pole, metabolism, what's happening now, eating, action—that is, the act of living. Geneticists think that we live to survive, in other words, in order to reproduce. Jacob tells us that "the dream of a bacterium is to make another bacterium." Common sense seems to tell us that we eat to live, we don't live to eat. But in fact, we survive also in order to live, that is, to metabolize, that is, to "enjoy" life, and we also live to eat. Means and end are not disconnected, but belong to the living circuit where all is end and means at the same time:

$$\text{live} \rightleftarrows \text{survive}, \qquad \text{eat} \rightleftarrows \text{live}.$$

Any theory of the autos must then contain a theory of auto- (geno-pheno) organization. Any development of the autos must allow for the development and the complexification of the uniduality of the *genos* and of the *phenon*. Thus in the tremendous development of phenomenal individuality characteristic of vertebrates, we see two "epigenetic" apparatuses taking form, at once dissociated and communicating, one devoted to reproduction (sexual apparatus), the other devoted to the organization of phenomenal existence (neurocerebral apparatus).

8.3 Communicational–Computational Autoorganization

Any theory of the autos must also necessarily allow for the idea of communicational–informational organization, and by the same token, the idea of computation. That much would seem clear since the principal aim of the "biological revolution" initiated by Watson and Crick is to apply a cybernetic–informational plan to the organization of the cellular being, and to conceive of this organization as a communicational pattern (DNA–RNA–proteins). Nevertheless, this cybernetic theory and biological conception lack the idea of apparatus. DNA is seen at the same time as memory and as the pure program of a "machine," which is the cell. However, if we look closely we see that the prokaryote cell is almost indistinctly both a machine and a computing apparatus. Indeed, the bacterium computes internal and external data and makes its "decisions" according to the outcome of its treatment of the data. Here again we see the difference between self-organization, which is only physical (stars, whirlwinds, atoms), and autoorganization, which while remaining physical becomes biological. Genophenomenal duality is known to self-organizations, and they possess no communicational–informational organization endowed with a computing apparatus. They form and maintain themselves "spontaneously," whereas in the genophenomenal autoorganization, "prigoginian" spontaneity is set off, controlled, and supervised by the computational–informational–communicational organization.

Here it should be strongly emphasized that no living process, neither the process of metabolic organization nor the process of the organization of reproduction, is conceivable without the action of at least one computing apparatus (and in the case of ontogenesis of a polycellular being, without the interactions of the computing apparatuses of the cells that multiply by mitosis). It should be said that the idea of computation is essential in allowing us to gain an understanding of the logically original nature of the autos.

To conceive of this nature, a double deficiency must be overcome: first, the deficiency of classical biological theory and, second, the deficiency of the theory of self-reference. Classical biological theory, whose influence continues to dwell in the unconscious of biologists, tends to minimize individuality to the benefit not only of genericity, but also of generality. The axiom "any real science is a science of the general" tends to mask the remarkable characteristics of the living individuality: the existence of singular beings, each with its empirical difference, each unique for itself, and each computing its own existence in terms of itself and the being for self (*pour-soi*).

8.4 Being for Self and Autocentrism

Here the idea of autoreference demonstrates its utility. The definitions of autoreference hitherto put forward (Varela 1975) have the great merit of being formalizing definitions, but are nonetheless insufficient. Autoreference should be conceived of as an aspect of the multidimensional reality—at one logical, organizational, and existential—of the autos. In order to understand autoreference, the computational organization of the living being must be considered. All living beings, even the least complex, are individuals endowed with a computational apparatus. This apparatus is radically different from the artificial "computers," which are constructed by others, receive their program from others, and function for others. On the other hand there is in the unicellular being *computation of the self, by the self, and for the self.* Such a computation is not only autoreferring, even though it is fundamentally "egocentric." In the same way that an autoorganizing system is at the same time necessarily an autoecoorganizing system since it needs the environment for its own autopoiesis, an autoreferring computation is necessarily eco-referring; that is, it must be capable of examining, processing, and transforming into information the data or events that come to it from the environment. What is most important, though, is that the data be processed as "objects" by the computation, precisely because the computing being establishes itself as *subject,* insofar as it computes, decides, and acts *of itself and for itself. What is essential then, is the distinct ontological affirmation, unique and privileged, of the self and for the self, that characterizes all living beings.*

This sort of ontological affirmation necessitates the defense of the *identity* (autos = the same), which necessarily supposes the distinction between self and nonself, and as a result the rejection of the nonself into the exterior (immunology). As Varela rightly states, immunology is a property of the total system and not merely a quality particular to certain defense agents. The ontological affirmation of the self and for the self is manifested by the "egotistical" computation that determines actions finalized by and for the self. It is therefore not simply a question of an objective behavior, but also of *ethos*—that is, the behavior performed by a subject for itself. (That is why there is progress when behavior science becomes *ethology*.) The egotistical being for self (*pour-soi*) is not necessarily limited to the individual. Autoreference includes, in a manner sometimes concurrent and antagonistic, as part of its principle of identity, not only the individual, but also the process of reproduction it carries with it, and the circle of the autos can expand to include progeniture, family, and society.

However, even in those cases where it acts for "its own" (family, for example), the living being, from the bacterium to homo sapiens, obeys a particular logic according to which the individual, though ephemeral, singular, and marginal, considers itself the center of the world. All others are excluded from the individual's ontological site, including homozygous twins, congeners, fellowmen. According to a principle of exclusion that brings to mind Pauli's principle, this egocentricity, which excludes from its own site all other beings, this computation and ethos *for the self* furnish the logical, organizational, and existential definition of the concept of *subject*. Being for self, autoreference, and autoegocentrism are so many traits that allow us to formulate and recognize the notion of subject. The opposition self–nonself is not only cognitive, it is ontological: it creates a duality between the valorized realm of the self–subject, centered and finalized, and the exterior universe of objects, which may be dangerous or useful. The duality subject–object is born of this same dissociation. Thus theoretical efforts, which began with the idea of autoreference, should, if they are to be consistent, continue through the idea of autoe-coreference to arrive at the concept of subject, within which are linked the notions of being for self, autocentrism, autologic, ethos, and egotistical computation.

We have been too much accustomed to reducing the notion of subject and subjectivity to contingency, affectivity, and sentimentality. In fact, we should think of the notion of subject in terms of a fundamental logical and organizational category that characterizes the living individuality and is inseparable from the autogenophenoorganization.

Individual subjectivity, although it considers itself to be the center of the universe, is ephemeral, peripheral, punctual. Yet it is precisely this center "point" that brings together the organizing processes and permits the characteristics of life to emerge. In this perspective, it is possible for the point to be richer than the processes that intervene in it, since it is the focal point of the emergences. Individuals (subjects) are the beings that emerge in phenomenal reality. It is in the individual (subject) that all processes of reproduction take place. Thus the concept of subject is not to be considered as an epiphenomenon; it should rather be ontologically registered in our notion of "life."

I shall even attempt to show that the concept of reproduction and the concept of subject share a fundamental common ground. Let us consider the individual (subject) from the standpoint of its egotistical computation: it recognizes the self of the nonself and organizes its own molecular trans-formation and regeneration not only in detail, but also globally, as a whole.

Considering this, we could say that the power of autocomputation, both in detail and in its globality, is at the same time a power of autoreflection. All confusion should be avoided with what we call reflective conscience, conscience of the conscience, which, precisely, supposes a conscience. The computing subject recognizes, knows, computes, decides, but is not "conscious" of itself. Even the human subject is in the unconscious (Lacan 1977). In what terms, then, should we speak of autoreflection, that is, the ability to split the being in two, to consider oneself subject and object at the same time as in the trite sentence of Piccardo, which nonetheless illustrates the ego structure very well, from a human language standpoint: "I am me," that is,

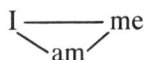

$$I \overset{\textstyle \longrightarrow}{\underset{\textstyle am}{\diagdown \nearrow}} me$$

This idea of autoreflection would remain gratuitous if it weren't for the existence of autoreproduction. What is cellular autoreproduction? It is a process by which, starting with a chromosomal scission, the cell divides itself in two, each half reconstructing of itself its absent half, a process that leads to the formation of two cellular beings. This means that in the very structure of the being (subject) there is a potential duality that leads it to split in two and to multiply by two. This capacity for duplication, which we do not find in the cerebral apparatus (except in the ability to recall representations and images), does exist in the generative memory as the capacity to split the self, practically, physically, organizationally, and biologically. If the ego is capable of creating an *ego-alter*, that is to say, a different self, it is also capable of conceiving of itself as an *alter-ego*, that is, a self that is different. ("I am the other," wrote the poet Rimbaud.)

Let us consider the two egos-alter resulting from mitosis. They are genetically identical and almost identical phenomenally. Nevertheless each one excludes the other from its subjective site and henceforth each computes and acts for itself. The possibility of communication does exist, however, through identification between the two congeners, whence the possibility of inclusion in associations that may take the form of organisms or, for polycellular individuals, of a society. So each living being carries within it simultaneously *a principle of exclusion of others from its subjective site as well as a principle of inclusion of the congener in the broadened circuit of its subjective autos.* The possibility of communication between congeners does not consist solely of an exchange of signs according to a common code, but is also the possibility of intersubjective communication, which, with developments of the living organization, can take the form of a communion or a coorganization. Whence the possibility, through transsubjective interactions (between individual subjects) of the formation of macroindividual subjects of the second order (polycellular beings) or of the third order (societies). Thus it can be seen that the concept of subject, far from being epiphenomenal, may be considered as a hinge between genetic processes of reproduction and phenomenal processes of communicational organization between cells (organisms) and polycellular individuals (societies).

This leads us to an unexpected mental revolution. Classical scientific method forced us to turn away from the notion of subject, including from ourselves, observers–conceivers. Here we not only consider the notion of subject, but also expand it and acknowledge it as something that belongs to all living creatures. Both the genetic code and subjectivity are common to all creatures, from the bacterium to the elephant.

With regard to the preceding, we see that autopoiesis and autoorganization are key notions, but only on condition that they be enveloped and developed in a theory of the *autos*. The autos sums up the conditions of existence and the reproduction of life, and takes the form of the principle of autogenophenoorganization (which is itself included in an incompressible paradigm of autogenophenoecoreorganization). As for the living being, it takes on the characteristics of the individual subject. The notions of *autos* and subject, which recursively refer to one other, lead to a logical and ontological mutation. This is a decisive break with conceptions that look for an explanation in one key term or in one guiding principle, such as the DNA program or behavior.

References

Adler, J., and Wung-Wai Tso (1974), "Decision" making in bacteria: chemotactic response of Escherichia coli to conflicting stimuli, *Sciences* 184, 1292–1294.

Atlan, H. (1972), Du bruit comme principe d'auto-organisation, *Communications* 18, 21–35.

Bogdanski, C. (1978), L'être vivant dans ses aspects dimensionnels et temporels, *Bull. Biol. Fr. Belg.*

Foerster, H. von (1960), On self-organizing systems, and their environments, *Self-Organizing Systems* (M. C. Yovitz and S. Cameron, eds.), Pergamon, New York.

Gunther, G. (1962), Cybernetic ontology and transjunctional operations, in *Self-Organizing Systems* (M. C. Yovits, Jacobi and Goldstein, eds.), Spartan Books, Washington, D.C.

Lacan, J. (1978), *Le Séminaire II*, Le Seuil, Paris.

Morin, E. (1977), *La Méthode, I. La Nature de la Natures*, Le Seuil, Paris.

Morin, E. (1980), *La Methode, II. La Vie de la Vie*, Le Seuil, Paris.

Neumann, J. von (1966), *Theory of Self-Reproducing Automata* (A. W. Burkes, ed.), Univ. of Illinois Press, Urbana.

Nicolis, G., and Prigogine, I. (1976), *Self-Organization in Non-equilibrium Systems*, John Wiley, New York.

Piccardo, O. G. V. (unpublished), *Egostructures*.

Varela, F. (1975), A calculus for self-reference, *Int. J. Gen. Systems* 2, 5–24.

Varela, F., Maturana, H., and Uribe, R. (1974), Autopoiesis: The organization of living systems, its characterization and a model, *Biosystems* 5, 187–196.

Vaz, N. M., and Varela, F. (1978), Self and non-sense, an organism centered approach to immunology (to appear in *Medical Hypothesis*).

PART II
CONVERSATION

Give me a fruitful error any time, full of seeds, bursting with its own corrections. You can keep your sterile truth for yourself.

Vilfredo Pareto

In the second part of this volume we have collected the papers that explore the interfaces of the concept of autopoiesis with the existing and well-established theories of general systems and cybernetics.

Most of these chapters are at least skeptical with respect to autopoiesis. Yet, each represents an attempt to deal with the new concept—either through its incorporation or through its exclusion. The reader might ask why this section should appear *at all* in a book on autopoiesis. Should a theory and its critique be presented within a single volume? Yes, it should.

Autopoiesis is not a dogma. It actually thrives on different interpretations (or even misinterpretations), applications, and subjective assessments. It is still flexible and it can still be molded. It is more like good building material than a final product or creation. To emphasize this quality we expose readers at least to a sample of how the concept of autopoiesis is being "worked over" by more traditionally minded scientists. This provides an excellent opportunity for comparing the two parts and confronting them when they address identical issues.

Without any intention or design on the part of the editor, the two parts also present a good comparative base for studying the cultural differences among the two groups of authors.

We start the section with the most direct critique of autopoiesis written so far—that of Brian Gaines. As president (in 1979) of the Society for General Systems Research, Gaines is aware that his opinions carry a lot of weight and essentially determine the relationship between autopoiesis and general systems theory, at least for the time being.

Gaines formulated eleven questions that, according to him, *should* be answered before the concept of autopoiesis can claim legitimacy. None of the proponents of autopoiesis considered it necessary to answer Gaines's questions explicitly. Some of the questions are clearly irrelevant because they are being asked within the framework of a different and nonintersecting paradigm, some of them were perceived as unjustified "ultimata," and a few of them were based on insufficient information. But several of them are well formulated and obviously addressed to the fields of general systems and cybernetics.

Since such soul-searching has never been carried out within these fields, it would have been nice if the proponents of autopoiesis had set an example and accepted the questions as asked.

Many authors and reviewers have prodded me as editor to provide some answers to these Popperian questions (see Section 9.2). This would require a paper of its own, a careful reasoning, and patient explanations, often with "lessons" on the paradigmal nature of science, subjective aspects of models, hypotheses, and theories, and so on—even more so since Gaines finds most of the literature on autopoiesis "informal and obscure."

I have agreed, instead, to provide a short and personal *response* to these questions, whatever it might be worth to the reader. I am sure that intelligent readers can develop a similar response of their own, perhaps quite different and yet equally valid.

The first question asks about the content and the objectives of the proposed theory of autopoiesis. This seemingly important question is marred by the surprising qualification, ". . . that *we* would regard as a reasonable objective . . ." (the italics are mine). The uncertainty about who might be allegorized by "we" and what "they" would regard as "reasonable" makes this question essentially unanswerable. Nevertheless, among the many "objectives" of the theory of autopoiesis the following may be mentioned: to provide a characterization of the organization of living systems and to describe how the phenomenology of life can be derived from it; to devise models of complex autopoietic systems, mathematical, computer, and physical, in order to test the predictions implied by the theory; to provide an alternative paradigm that would expose the one-sidedness of the components-based reductionistic approach; to describe the phenomenology of the living without evoking the anthropomorphic metaphors ascribed to its functions. No claims of reasonability, success, acceptability, or completeness with respect to these objectives have ever been made. Of course, science has no "objectives"—except those of its own transcendence.

The answer to the second question could very well be "both." Any self-respecting theory of life must attend to its ontological *and* epistemological dimension at the same time.

The third question is hardly relevant. The ability to formalize something is a function of the observer, not of the observed object. But the answer is "yes." Autopoiesis can be formalized, computerized, quantified, and empirically analyzed, usually to a higher degree than customarily experienced with other theories. Yet, the verbal expositions are considered at least as powerful as the mathematical ones, and the complementarity of both modes is surely recognized.

Questions 4 and 5, asking what unity and organization are, have been answered by Gaines himself through quoting their definitions. These are fundamental concepts of autopoiesis, endlessly repeated and refined in the relevant literature, including this volume.

Question 6 refers to the ongoing discussion about autopoiesis and its relationship to dissipative structures, spontaneous social orders, catalytic hypercycles, and the like. We need a comparative synthesis of these approaches; they seem to have much more in common than was originally suspected.

In question 7, Gaines essentially asks about the history of the *idea* of autopoiesis. The recent AAAS symposium volume, *Autopoiesis, Dissipative Structures and Spontaneous Social Orders,* contains a rather detailed discussion of the precursors to the idea. Actually it was autopoiesis that brought also into focus a keen awareness of the historical underpinnings of general systems and cybernetics. It is these latter fields that are generally poor in their appreciation of history. Gaines complains that Weiss's work and the Alpbach Symposium are not referenced in the autopoiesis literature. There are lengthy essays based on it in the aforementioned volume and elsewhere. In fact, it is the general systems literature that mostly ignores Weiss's seminal contributions; and those of Trentowski, Bogdanov, Smuts, Leduc, Menger, Gerard, Waddington, and von Hayek, of the past, as well as those of Prigogine, Eigen, Jantsch, and others of more recent times.

On question 8 the students of autopoiesis differ. The notion of self-production is compatible with the notion of a sharp boundary as well as with that of a continual transition between the nonliving and the living. This is a difficult situation because whatever demarcation we observe today did not necessarily exist at the times of the origins of life (and of autopoietic systems).

Concerning question 9, there is again a difference of opinion, as is reflected also in this volume. Maturana and Varela do view autopoiesis as a biological theory at the cellular level. All other researchers are ready to extend the notion to other levels as well. One implication is that autopoiesis is a necessary but not a sufficient characterization of life. Or, as Gaines answers this question, "an obviously crude fit [of autopoiesis] to a social organization would not. . .in any way undermine the theory as applied to cellular organization."

With respect to question 10, it is proposed that most if not all phenomena of life are derivable from a systems autopoiesis—a system must be autopoietic before it can exhibit the phenomenology of life. That is, autopoiesis is both a primary and necessary phenomenon of the living, but it is not sufficient; the circumstances, the environment, the space, and the interactions under which the autopoiesis takes place are of utmost importance.

The answer to question 11 is "maybe."

Gaines proposed 11 questions to be asked of students of autopoiesis (he refers to them as "propounders"), or of any field of inquiry for that matter. This editor has responded to some of the questions in a short personal comment while introducing this section of the book.

What is interesting about Gaines's approach is that he not only poses questions, but actually attempts to supply some answers. The reader will find his answers at least as good as, if not better than, his questions.

Gaines's critique of autopoiesis can be characterized as constructive, involved, and reasonably informed. A great service could be rendered to the field of autopoiesis if its "propounders" became more receptive to Gaines's comments. Compare the silence that permeated general systems and cybernetics after the famous "sound from Puget Sound" (David Berlinski) launched his attacks.

What is Gaines's view of autopoiesis? He seems to have an intuitive appreciation of its potential explanatory power, its appealing computer modeling possibilities, and the elegance of its underlying holism. On a more rational level, Gaines is obviously turned off by a certain flamboyance and the nontraditional style of the "propounders," the "bandwagon" effect of autopoiesis, unsubstantiated claims, not enough respect for what he considers the best tradition of scientific papers. His unhappiness about the latter simply overwhelms his natural propensity toward the former.

Gaines sensitively affirms that we cannot expect Maturana's fundamental work on the characterization of life to fit tidily into the current frameworks of biology. Nor should one expect it to fit the current versions of general systems and cybernetics. But ignoring it or claiming ignorance of it would be a case of self-inflicted blindness.

Gaines elevates autopoiesis to a theory of systems, yet the propounders

Brian R. Gaines was until 1979 Chairman of the Department of Electrical Engineering Science at the University of Essex, Colchester, England, where he continues to hold an honorary appointment as a Professor. He is currently Deputy Chief Executive and Technical Director of Monotype Holdings Limited, a company concerned with a wide variety of information systems. Dr. Gaines is President of the Society for General Systems Research, a Fellow of the Institution of Electrical Engineers, and Editor of the *International Journal of Man-Machine Studies. Address:* The Monotype Corporation Limited, Salfords, Redhill RH1 5JP, England.

of autopoiesis seem to be unmoved by the honor. They are singularly absent from most general systems and cybernetics meetings, their publications are scattered "elsewhere," and there is no indication that the situation will change in the near future. It is felt by many that this theory can stand fully on its own, communicate directly with individual scientific disciplines, possibly linking up during the 1980s with the theories of order through fluctuation, catalytic hypercycles, and spontaneous social orders.

Chapter 9
Autopoiesis:
Some Questions

Brian R. Gaines

9.1 Objectives

The concept of autopoiesis as a characterization of that which makes a system living, as proposed by Maturana and Varela (1973), has aroused a great deal of interest and derived work, but is has also aroused controversy.[1] This paper is concerned with some aspects of the grounds for criticism of presentations of autopoiesis to date. It is not an attempt to answer criticism, or to give definitive elucidations of unclear aspects of the theory. However, it does attempt to make the problems clear and delimit the forms of acceptable, and of unacceptable, answers. I hope it will serve to stimulate others into providing detailed expositions of the matters raised.

9.2 Some Questions

The controversy stems from several sources. To begin with, the theory so far advanced (in Maturana and Varela 1973; Varela, Maturana and Uribe 1974; Maturana 1975; Zeleny 1977) is informal and obscure, giving

---

[1] Many of the points discussed in this paper arose in the frequent interchanges within continuously changing groups that were such an important feature of the conference on Applied General Systems Research that George Klir organized at Binghamton in August 1977. The session on autopoiesis commenced with an exposition by Nicolas Varela and a demonstration of the computer simulations developed by Milan Zeleny, followed by presentations of several other papers in this volume. It roused considerable interest, but mixed feelings, in the large and varied audience present. In particular, several biologists expressed to me views that prompted some of the questioning in this paper.

rise to the following questions:

1. Has the theory any content? That is, does it explain anything, predict anything, or in some other way provide something that we would regard as a reasonable objective for proposing a theory?
2. Does the theory give a characterization of life in itself, or is it an explanation of how we come to characterize life? That is, is it an ontological theory or an epistemological one?
3. Can it be formalized? That is, can the verbal expositions be given a logicomathematical framework together with a system of interpretation?

I should state immediately that these are not intended to be positivist strictures and that there are a variety of answers possible to each question that might not be acceptable—for example, in terms of the received view (Suppe 1974)—but would satisfy many current scientists. The motivation of the presenters of autopoiesis is both foundational and biological; clearly there is basis for the argument that the science of biology is not amenable to the framework of the science of physics (or perhaps its caricature as conventionally accepted in much of the "philosophy of science" literature). However, at the very least the role of the above three questions in the science of autopoiesis ought to be established, and preferably the answers to them also.

A second source of the controversy is that the theory, in so far as it is well defined,[2] makes at least a two-part characterization of living systems in terms of their unitary organizations and its maintenance through recursive self-production. Both these concepts seem comprehensible, if not well defined; but they also seem at rather different systemic levels, and similar concepts of organizational attributes alone, or self-production attributes alone, have been used in the past to characterize certain aspects of living organisms. The new insight is supposedly concerned with the essential interplay of these two features, and again raises questions:

4. What is unity? That is, is there a decision procedure whereby one can determine that a particular system constitutes a unity whereas another does not?
5. What is organization? That is, is there a system of description by which we may say "this specifies the organization of this system?"

---

[2] Such qualifications in this paper are not intended as the usual rhetorical trick of sowing doubt by insinuation. For many people, including myself (on Mondays, Wednesdays, and Fridays!), the theory is reasonably well defined. What I sometimes wonder is whether all our definitions agree, and whether we have a rich enough common framework of terminology, belief, and experience to seek out and determine the form and nature of disagreement.

A related question concerns the nature of structure, since again the presenters of autopoiesis make a careful distinction between the role of organization and that of structure but do not present an equally careful distinction between these two terms themselves. Perhaps this, as a criticism, is too pedantic since we "all know" what is meant by "organization" and "structure." However, as soon as such terms take on technical denotations in terms of a new theory, the previous connotations that we "all know" become potentially misleading.

6. What systems have the unity of organization characteristic of autopoiesis but are not so because of their lack of recursive self-production? And, vice versa, what systems exhibit self-production but lack unity of organization? In other words, let us have not only intensional definitions but let us also enrich the context with extensional, discriminatory ones.

This, informally and in part, has been done in some of the literature but it needs further examples. I also suspect, from discussions between different workers in the field, that there would be differences of opinion on some possible exemplars that would themselves be illuminating.

A third source of the controversy is that apparently similar considerations as to the organization and nature of life have been advanced by others in the past (e.g., Paul Weiss 1940, 1962, 1969), and one question that has been raised many times in discussion is the degree and type of novelty involved:

7. How does the concept of autopoiesis differ from other frameworks for the characterization of life proposed in the past (e.g., in essential terms or in greater precision, or in terms of new system-theoretic tools now available better to express previous concepts, etc.)?

Fourth, it is reasonable to ask whether any attempt to characterize living systems is not creating a distinction without a difference. Would a biologist or biochemist nowadays be prepared to advance a clear decision procedure that divides the living from the nonliving?:

8. What are the grounds for supposing that the living can be discriminated from the nonliving? Note that the implied metaquestion is "does the discrimination essentially involve autopoiesis?" (i.e., are we dealing with an explanatory theory of an already well-defined phenomenon, or a new definition of a previously ill-defined phenomenon?). Either could be useful, but the basis for evaluation is different in each case.

Fifth, the derivative literature on autopoiesis (e.g., Zeleny 1977, and papers in this volume) has a common pattern of commencing with biological problems at a cellular level and then switching to commentaries on societal organizations. It is again reasonable to ask whether we are to evaluate the theory over such a broad spectrum (in which case, why the emphasis on "living"—the concept that a social group has "life" is not unreasonable but it is radical scientifically to equate this in a foundational theory with the life of a cell), or whether the social examples are to be taken as analogies only to clarify certain concepts but are not essential to the status of the theory:

   9. Is autopoiesis essentially a biological theory at cellular level or does it apply to a wide spectrum of "autopoietic unities"? In other words, are the social examples applications of analogical reasoning from well-established biological domains to less-known social situations, a possibility for much biological theory, or are there "real" autopoietic systems at a noncellular level?

Sixth—and the answer to the previous question is clearly very relevant to the legitimacy of this one—there have been many previous characterizations of life at varying levels, in terms of reproduction, self-organization, adaptation, learning, intentionality, self-awareness, and so on, and it is reasonable to ask whether these definitions are implied by, imply, or are correlated with those based on autopoiesis. Again, this question is touched on in the literature but mainly in terms of the autopoietic characterization somehow preceding any other.

   10. What is the relation of characterizations of living organizations based on autopoiesis with those based on other features of living systems? That is, are any derivable from the others? If not, what are examples of systems satisfying one definition but not another?

Finally, there is one question that, apparently, should have been asked first:

   11. Is it worth it? Do we really want a characterization of living organizations? This is not just a flippant remark, since the full corollary to a "yes" answer must be "why is it worth it?" That is, what are we going to do with the characterization once we have it?

I suspect this is where much of the elucidation lies; definitions are the tools of science and it is the use to which they are put that ultimately determine their legitimacy.

9.3 Some Possible Forms of Answer

If the questions in the previous section had been rhetorical I could now go through them again giving the "correct" answers. However, this would destroy the dialectical intent of this paper. The (diverse) answers of others more closely associated with the field will be more interesting and relevant than mine, and examining other chapters of this book for answers to the questions posed will itself be informative.

Before we can evaluate new system-theoretic or scientific concepts we must decide on the basis for such evaluation (and be prepared to change it since that may be part of the proposal). When life decides to reflect upon itself we have a situation analogous to that of a television camera looking at the picture at its own output: forms may flow and boundaries may shift. We cannot expect fundamental work such as Monod's (1972) on the role of chance and necessity in developmental biology, Dawkins' (1976) on the dynamics of evolution in social biology, or Maturana's (1975) on the characterization of life in foundational biology to fit tidily into current frameworks. Biology is itself going through an introspective phase of examining its own fundamentals and presuppositions (Ayala and Dobzhansky 1974; Hull 1974; Lewis 1974),[3] and to pose the right questions about the nature of life is a key problem. The formulation of a framework, a language, in which to pose these questions is a formidable task: If the terms used correspond too closely to existing vocabularies we may dismiss the discussion as clearly ill conceived, whereas if the terms themselves reflect their new content then we may dismiss the discussion as vague and metaphysical.

Maturana's 1975 paper represents an extended attempt to define clearly and communicate the terms used in formalizing the notion of autopoiesis. He states the aim:

To explain the organization of living systems by describing the organization that constitutes a system as an autonomous unity that can, in principle, generate *all the phenomenology* proper to living systems if all the historical contingencies are given [Maturana 1975, p. 317].

The italics are Maturana's and suggest that the answer to my question 10 is that we should, by viewing the effect through autopoiesis of outside influences, be able to generate the other, correlated, manifestations of life.

Maturana also gives definitions of some of the terms I have queried, for example, unity:

A unity is an entity (concrete or conceptual) separated from a background by a concrete or conceptual operation of distinction. A unity may be treated as an unanalysable whole endowed with constitutive properties, or as a composite entity with properties as a unity that are specified by its organization and not by the properties of its components [Maturana 1975, p. 315].

He emphasizes the observer-generated nature of actual "unities" as re-

---

[3] Similar moods of massive introspection have occurred in physics (Korner 1957) and psychology (Royce 1970). They are probably an essential component of the maturation of any science.

sults of ostension that makes a distinction, and notes (Maturana 1975, pp. 325–326):

Different operations of total distinction separate different kinds of unities because they define different kinds of boundaries and, therefore, imply different organizations.

This suggests in answer to my question 2 that the theory is basically epistemological and that "unities" are entities hypothesized by observers in the way that they view the world. Such a relativistic view of the basic phenomena of life corresponds to similar views of the nature of possibly related phenomena such as "adaptation" (Gaines 1972).

However, Maturana (1975, p. 326) later states, "The organization and structure of a unity specify all the operations of distinction through which it can be separated from the background." Now, either this is trivial in that any arbitrary distinction made by an observer must, by definition, be possible for the unity, or it is intended as a strong ontological statement restricting the freedom of the observer to ostend only such unities as are "actually there." This exemplifies a fundamental problem for those attempting to propound a formal theory of autopoiesis: they have to decide which terms are undefined and open to unrestricted application subject only to the defined constraints of the theory. It is thereafter confusing to argue for the "reality" of such terms; some uses of the definitions will clearly be more fruitful than others, but that relates to a stage of the theory that we have not yet reached with autopoiesis, its evaluation (e.g., in terms of internal coherence or external correspondence).

This dilemma is no new one, but a basic conflict between epistemology and ontology, between knowing and being. As a biological example from a previous generation of theories of life, take the dialogue between Koestler and Weiss on "hierarchical organization":

As Paul Weiss said yesterday: "The phenomenon of hierarchic structure is a real one, presented to us by the biological object, and not the fiction of a speculative mind." It is at the same time a conceptual tool, a way of thinking, an alternative to the linear chaining of events torn from their multidimensionally stratified contexts [Koestler 1969].

Replace "hierarchic structure" with "autopoiesis" and you might have a direct quote from recent literature.

I am tempted to quote here from propounders of *the* ontological argument, such as Descartes: his power of rhetoric in stating St. Anselm's proof of the necessary existence of God (Plantinga 1968) is a salutory reminder of the difference between the arts of persuasion and the pursuit of science. On the other hand, we have also to accept that the ultimate, undefined terms of any theory are accepted as "acts of faith," not that they be true (for the word is meaningless in this context), but at least that they be potentially useful. If we go looking at the world in terms of unities, recursive self-production, and autopoietic organization, then we shall find a certain kind of world: It is not yet clear why we should want to find it or what we will do with it when we have it.

To summarize the preceding discussion, I should conclude currently that the answer to my question 1 is still lacking; we do not have a clear statement of the objectives of a theory of autopoiesis. We have an aim certainly, to characterize the organization of living systems, but no clear definition of objectives whose satisfaction may be used to determine whether our aims have been successfully carried out. This basic lack is probably the prime cause of further difficulty in answering later questions. The theory is not yet formalized, although the papers to date, particularly Maturana (1975), sketch some informal definitions that might be formalized given appropriate tools.[4] However, question 3 can only be answered in the light of clear answers to questions 1 and 2—then question 11 might also be evaluable.

Turning now to question 7, I am not concerned with priorities of discovery but instead with the apparent novelty of the concepts of autopoiesis. Maturana (1975) has only three references, none to other authors. Zeleny (1977) has a considerable bibliography of past, related studies, but no detailed statement of what the relationships are (e.g., which are previous attempts to characterize the organization of living systems, which failures, or successes, when concepts are the same or different). The comparison is all the more difficult because Zeleny has widened the scope of "autopoiesis" to social system organization.

The intended novelty of their work is stressed by Varela, Maturana, and Uribe (1974, p. 187):

All living systems must share a common organization . . . . At present there is no formulation of this organization, mainly because the great developments of molecular, genetic and evolutionary notions in contemporary biology have led to the overemphasis of isolated components.

This insight and the reaction against the stress on the components brought about by the tremendous advances in molecular biology in the past 20 years is no new one (see Goodfield 1974 for an excellent survey). In particular, the Alpbach Symposium, *Beyond Reductionism* (Koestler and Smythies 1969), was primarily concerned with just this problem, and the names of the participants—Weiss, Waddington, Bertalanffy, Hayek, and so on—read like a roll call of honor for those who for many years have emphasized the role of "organization" in biology and its systemic for-

---

[4] It would be a book in itself to discuss what are "appropriate tools." Varela (1977) has been developing an extended form of Brown's (1969) logic, which is particularly attractive as its basic semantics is one of "making a distinction"; the extension consists of allowing self-referential systems within the calculus, essentially of using Russell's paradox to generate new descriptions within the existing logical framework.

mulation. In this symposium will be found not only the overall insights that led to the formulation of the concept of autopoiesis, but many of the detailed criticisms of the relevance of "self-reproduction," and so forth. It is singularly unfortunate that this volume is omitted as a reference from all the key papers on autopoiesis, not because of (scientifically irrelevant) claims for priority of publication, but because it illuminates so much of the thinking that lies behind the concept of autopoiesis.

The foregoing does not imply that there is nothing new in the work on autopoiesis, only that much of the groundwork has already been done and many of the insights existed before. The novelty is best seen, however, in the divergence between the *emphasis* in, for example, Weiss's work on *hierarchical* organization and that in Maturana's work on *self-producing* organization. As bald paraphrases, Weiss sees the whole hierarchy as characterizing life, while Maturana sees a particular phenomenon at a given level of it characterizing our right to apply the term "living" at that level. I have personally tended to view the formulation of autopoiesis as a refinement of the analytical tools necessary to formalize the earlier insights, and Maturana's emphasis on the characteristic organization at a single level seems precisely the form of introduction of isolated, "automic" concepts, analyzable in their own right, that characterizes the "precisiation" stage of a science (Carnap 1962).

Questions 8–10 are related and all difficult to answer without some clear statement of objectives in reply to question 1. However, it is in relation to these three questions in particular that very different, yet totally legitimate, answers may be given. For example, in reply to question 8 we may adopt the extreme relativistic view that there is no absolute distinction—many systems can be characterized as either "living" or "nonliving," according to the viewpoint adopted (e.g., the distinctions one ostends). Also, one does not expect the middle to be excluded: under any definition there will always be systems whose inclusion in either category seems artificial. This is a problem of any program of precisiation and it is possible to deal with it in a variety of ways (Gaines 1976). It does not necessarily indicate that false distinctions have been drawn, although that too is possible (Popper 1976, p. 24).

The answer to question 9 can be indeterminate in general, although one must make a specific answer in any particular context. In formal terms, the theory of autopoiesis cannot be confined to biology: it is a theory of systems and it is unlikely, although not impossible, that the only extant systems satisfying the specified constraints are those at a particular biological level; on the other hand, an obviously crude fit to a social organization would not, for example, in any way undermine the theory as applied to cellular organization.

Finally, it is to question 10, in the light of the answers to all the preceding questions, to which we may turn in future for the most fruitful developments arising out of a theory of autopoiesis. In the first quotation of this section Maturana's aim is to generate *all the phenomena* proper to living systems. The prima facie plausibility that an adequate theory may allow this is the major attraction to the study of autopoiesis. Clearly, even if this is done, it does not rule out alternative starting points; it seems likely that we discover connections rather than equivalences between the many phenomena associated with life. However, the challenge now is to formulate clearly and consistently a characterization of living organizations that allows the derivation of as many of these connections as possible.

9.4 Conclusions

The 11 questions in Section 9.2 are intended to stand apart from the remaining material. Even if the rationale given for them is demonstrably false or based on misunderstanding, the questions themselves should still be answerable, and answered. I should not rule out meta-answers of the form, "at this stage in the development of the theory it is not possible to give a clear answer to that question"; at least the questioner would know what level of discourse to adopt. To some of the questions also no short answer may be given, and the "answer" has to take the form of a dialogue since the preconceptions of the questioner are as important as those of the answerer.

I hope that the questions here will serve as tools to the reader of this book, helping him to prize much that is of value from the main papers on autopoiesis itself.

References

Ayala, F. J., and Dobzhansky, T. (eds.) (1974), *Studies in the Philosophy of Biology,* MacMillan, London.

Brown, J. S. (1969), *Laws of Form,* Allen and Unwin, London.

Carnap, R. (1962), *Logical Foundations of Probability,* Univ. of Chicago Press, Chicago Ill.

Dawkins, R. (1976), *The Selfish Gene,* Oxford Univ. Press, Oxford.

Gaines, B. R. (1972), Axioms for adaptive behaviour, *Int. J. Man–Machine Studies* 4(No. 2), May, 169–199.

Gaines, B. R. (1976), Foundations of fuzzy reasoning, *Int. J. Man–Machine Studies* 8(No. 6), November, 623–668.

Goodfield, J. (1974), Changing strategies: A comparison of reductionist attitudes in biological and medical research in the nineteenth and twentieth centuries, in *Studies in the Philosophy of Biology* (F. J. Ayala, and T. Dobzhansky, eds.), MacMillan, London.

Hull, D. (1974), *Philosophy of Biological Science,* Prentice-Hall, Englewood Cliffs, N.J.

Koestler, A. (1969), Beyond atomism and holism—the concept of the holon, in

Beyond Reductionism (A. Koestler, and J. R. Smythies, eds.), Hutchinson, London.

Koestler, A., and Smythies, J. R., eds. (1969), *Beyond Reductionism,* Hutchinson, London.

Korner, S., ed. (1957), *Observation and Interpretation in the Philosophy of Physics,* Dover, New York. USA.

Lewis, J., ed. (1974), *Beyond Chance and Necessity,* Garnstone Press, London.

Maturana, H. R. (1975), The organization of the living: A theory of the living organization, *Int. J. Man–Machine Studies* 7(No. 3), May, 313–332.

Maturana, H. R., and Varela, F. G. (1973), De Máquinas y Seres Vivos, Ed. Universitaria, Santiago de Chile. English version: *Autopoietic Systems,* Biol. Computer Lab. Res. Report 9.4, Univ. of Illinois, Urbana.

Monod, J. (1972), *Chance and Necessity,* Collins, London.

Plantinga, A., ed. (1968), *The Ontological Argument,* MacMillan, London.

Popper, K. (1976), *Unended Quest,* Fontana, London.

Royce, J. R., ed. (1970), *Toward Unification in Psychology,* Univ. of Toronto Press.

Suppe, F., ed. (1974), *The Structure of Scientific Theories,* Univ. of Illinois Press, Urbana.

Varela, F. (1975), A calculus for self-reference, *Int. J. Gen. Systems* 2(No. 1), 5–24.

Varela, F. G., Maturana, H. R., and Uribe, R. (1974), Autopoiesis: The organization of living systems, its characterization and a model, *Biosystems* 5, 187–196.

Weiss, P. (1940), The problem of cell individuality in development, *American Naturalist* 74, 34–46.

Weiss, P. (1962), From cell to molecule, in *The Molecular Control of Cellular Activity* (J. M. Allen, ed.), McGraw-Hill, New York.

Weiss, P. (1969), The living system: Determinism stratified, in *Beyond Reductionism* (A. Koestler, and J. R. Smythies, eds.), Hutchinson, London.

Zeleny, M. (1977), Self-organization of living systems: A formal model of autopoiesis, *Int. J. Gen. Systems* 4(No. 1), 13–28.

Andrew is attempting to reconcile the autopoietic viewpoint with the teleological implications of using a negative-feedback control system as a model of the living phenomena. This effort has seemingly been seconded by Varela (in this volume), who argues that autopoiesis and allopoiesis are complementary characterizations of a living system.

This view is also reminiscent of Gaines's implication that the behavior of an autopoietic system is most naturally described in terms of goals. Roughly, if a system is autopoietic and it exhibits behavior, and that behavior has been phylogenetically molded to preserve autopoiesis, then the behavior of the system is most compactly described in teleological terms.

Yet, Andrew is aiming at something different: to show that the classical language of cybernetics can be used to describe the observed phenomenology of autopoietic systems. This seems to be self-evident: an observer can treat the autopoietic system as an allopoietic system (i.e., consider its perceived inputs and outputs, and thus describe it *as* an allopoietic system). The autopoiesis itself, however, would not be affected by the chosen mode of description.

It should be recalled here that organizational *closure* of autopoietic systems is meant in the sense of circularity and has nothing to do with the "openness" of such systems with respect to environmental "inputs" or perturbations. Structure refers to the realization of the organization in a particular space of components. Thus, structure is *not* only a result of a system's interaction with its environment, but, more fundamentally, of the underlying organization.

There are, however, many interesting points in Andrew's paper. For example, by considering how the autopoietic viewpoint can be enriched by everyday observation, Andrew evokes the idea of a possible enrichment of our everyday observation by the autopoietic viewpoint.

He emphasizes one idea especially: the fact that many animals and humans engage in activities that are not conducive to their survival (altruistic behavior) is taken as a phenomenon that is at variance with the autopoietic viewpoint. Andrew actually refers to such kinds of altruism as nonautopoietic behavior. It is interesting to contemplate why autopoiesis should be equated with survival. No known autopoietic individual has ever survived indefinitely; disintegration of an autopoietic unity can be viewed as a natural outcome of its own autopoiesis, as was often demonstrated through the published computer models of autopoietic behavior. Death or a disintegration of a component (an animal) is necessary for the maintenance of an autopoietic whole (society). Yet the fact that some systems are capable, consciously or subconsciously, of interrupting their own autopoiesis is interesting and intriguing.

There are some unexpected comments in the Conclusions about Ashby's treatment of state-determined systems, as it was later adopted by Maturana. In 1962 Ashby expressed his conviction that a state-determined model is an adequate description of living systems and that the term "self-organizing system" could be very well dropped from usage. Yet this is where modern cyberneticians part even with Ashby.

Alex M. Andrew, born in England in 1925, received B.Sc. in Mathematics and Natural Philosophy (1949) and Ph.D. in Physiology (1957), both from Glasgow University. During 1954–1955 he worked in McCulloch's group at M.I.T. His other appointments include National Physical Laboratory at Teddington (1957–1962); SIGMA (Science in General Management), headed by Stafford Beer (1962–1964); and Biological Computer Laboratory (with Heinz von Foerster) at the University of Illinois (1964–65). Since 1965 he has been a lecturer in Cybernetics at the University of Reading. Dr. Andrew serves on editorial boards of *Information Sciences* and *International Journal of General Systems*. He is currently writing a book on *Self-Organizing Systems* soon to be published. *Address:* Department of Cybernetics, University of Reading, Whiteknights, Reading, RG6 2AL, England.

Chapter 10
Autopoiesis–Allopoiesis Interplay

Alex M. Andrew

10.1 Introduction

Maturana (1975) and his co-workers (Varela et al. 1974) have introduced the term *autopoiesis* to designate what is arguably the most fundamental characteristic of living organisms. It is also characteristic of larger living systems, including human institutions. *Autopoiesis* is the capability of living systems to develop and maintain their own organization: The organization that is developed and maintained is identical to that performing the development and maintenance. Autopoiesis denotes a bootstrapped or self-referential quality of living systems, reminiscent of Warren McCulloch's observation that a theory of the working of the brain, if correct, must be written by itself.

Stafford Beer (1975a) has described the self-maintaining property of human institutions by likening them to physical or chemical systems obeying Le Chatelier's principle. Such systems react to an outside disturbance of moderate severity by allowing changes in systemic variables sufficient to counteract the disturbance, the system remaining qualitatively unchanged.

The autopoietic viewpoint recognizes the organization that appears autonomously within the system; characteristics resulting from interaction with an environment are termed *structure* rather than *organization*. With this distinction, autopoietic systems are organizationally closed and can be described without reference to inputs or outputs. Systems that are not organizationally closed are termed *allopoietic*.

Maturana further claims that

Descriptions (of living systems) in terms of information transfer, coding and computations of adequate states are fallacious because they only reflect the observer's domain of purposeful design and not the dynamics of the system as a state-determined system (Maturana 1975).

Presumably goals and feedback loops can be added to the list of proscribed terms.

This paper considers how the autopoietic viewpoint is to be reconciled with the everyday observations that the behavior of people and animals is very readily and satisfactorily described in terms of goals and attempts to achieve them and that many physiological systems conform extremely closely to the engineer's idea of a feedback-control system. The early literature on cybernetics abounds with examples of the apparent use of negative-feedback control by living systems. One of the most interesting is the pupillary response of the human eye to light, since it affords a convenient nonsurgical way of opening the feedback loop in order to measure its characteristics (Stark 1959).

It is clear that living systems embody subsystems that are readily describable in allopoietic terms. These subsystems are to be regarded as structure rather than organization but they are nonetheless vitally important to the maintenance of autopoiesis. They have come into being as byproducts of a parent autopoietic system. They can in fact be described in either autopoietic or allopoietic terms.

The partitioning of a system into structure and organization requires justification. The distinction is needed for precise formulation of the autopoiesis hypothesis. Clearly, those parts of a system that react to environmental changes cannot be said to be maintained constant in a circular process. One must either interpret "maintained" in a weaker sense than "maintained constant," or else divide off some parts of the system from the essential core that is maintained by autopoiesis. This core is the organization, and the parts that are conceptually divided off are the structure. Even when this partitioning is allowed, the problems are not fully resolved; the organization cannot remain absolutely constant. If it were it could not implement the circular process that maintains it. This is the crux of the problem of self-reference, which has stimulated Varela's elegant extension of Brown's Laws of Form (Brown 1969; Varela 1975) and his later work related to Scott's treatment of fixed points of algebraic expressions (Varela and Goguen 1977).

The relationship between the two modes of description of the subsystems constituting *structure* is not a simple matter, and depends on the view taken of the nature of biological learning and evolution. This point will be discussed further in Section 10.2, where the two modes will be presented as alternatives, with no real discrepancy or conflict between them.

Some phenomena, however, appear to be at variance with the autopoietic viewpoint. It is not difficult to demonstrate that many animals, particularly man, engage in activities that are not conducive to survival. A way of resolving this apparent contradiction will be given in Section 10.4.

10.1.1 Top–Down and Bottom–Up Analysis

The development of the autopoietic viewpoint has a strong top–down character, in that a large amount of theory is made to flow from a few central truths. It is hardly necessary to observe that any top–down approach must be developed with extreme caution. The difficulty is not so much in discerning the central truths as in deciding what their consequences are. In this paper the central hypotheses of autopoiesis are not disputed, but a new view of some of their consequences is developed.

The terms *top–down* and *bottom–up* are widely used in connection with compilers (Gries 1971) and it may be useful to pursue a loose analogy between compilation and biological research. The terms are used to refer to the method by which the compiler analyzes the user program (in ALGOL, FORTRAN, etc.), which it is required to translate into machine language. In a bottom–up method the smallest constituents of the program are recognized and combined into larger fragments and so on until a complete program is recognized. A top–down method utilizes, from the start, the knowledge that the input should be a legal program in the source language. The recognition of the constituent fragments is then facilitated because the range of possibilities is narrowed.

Top–down methods of compilation are more efficient (for the recursively defined ALGOL-like languages, at least). To find a description of a pure bottom–up approach it is necessary to go to rather early publications, such as Ledley's admirable exposition (1962).

There is, however, a significant difference between the study of biological systems and the analysis of a program in a high-level language. If the program fails to conform to a precise set of rules the compiler is expected to terminate with the message "compilation failure." Biologists are not permitted any such escape and must be wary of a pure top–down approach to their subject. On the other hand, they will not get very far unless they are prepared to think in top–down fashion for some of the time; scientific thinking normally proceeds by a succession of alternate episodes during which a top–down or a bottom–up approach is used.

10.1.2 The State-Determined System

It is important to be clear about the sense in which a description of living systems may be said to be fallacious when expressed in allopoietic terms. Any such description constitutes an empirical theory, based on observations, and as such it may in time be supplanted by a new theory. The autopoietic viewpoint, on the other hand, has a nonempirical logical basis. The allopoietic "explanation" must therefore be demoted to the status of a description. Instead of saying, "This is a feedback loop," we should say, "This behaves like a feedback loop." Used as a descriptive analogy, the allopoietic explanation is unobjectionable. Also, if the same allopoietic explanation proves useful in describing a wide variety of biological systems, it is reasonable to suppose that the correspondence is not fortuitous and that these biological systems share some property that is reflected in the allopoietic explanation.

Ashby's concept of the state-determined system (1960) provides a useful way of talking about systems in general and allows the derivation of some important results. In the context of one particular system it provides nothing more than a descriptive framework, which may or may not be the most convenient to use for some purpose. To describe a given system within this framework it is necessary to plot behavior lines in the phase space, on the basis either of observations or of some analysis of the system.

10.2   Interplay of Ideas

This section contrasts the alternative descriptions of living subsystems in autopoietic and allopoietic terms. Since there is no essential conflict, what is being considered is an interplay of ideas.

The organizational closure of autopoietic systems implies that their organization is unaffected by the unpredictable disturbances the system receives from its environment.

An observer of an autopoietic system might, however, attribute its stability in the face of disturbances to one of the following mechanisms. (The list is not exhaustive, nor are the mechanisms mutually exclusive.)

1. The system may utilize redundancy in the manner generally understood by "error-correcting coding." Von Neumann (1956) postulated duplicated channels and majority elements as constituents of reliable nervous systems. Since that time, more efficient ways of utilizing redundancy have been devised (Peterson 1961) and related to neural computation (Winograd and Cowan 1963). Other studies have treated the particular case where the unpredictable disturbances influence neural thresholds (McCulloch 1959; Blum 1962; Andrew 1970).
2. The system may allow self-repair by keying in randomly diffusing elements. The model described by Varela et al. (1974) could be described in these terms.
3. The system may embody subsystems that are in fact negative-feedback control systems operating to maintain steady values of parameters that would otherwise be disturbed by environmental effects.

4. As an extension of (3), subsystems may use feedforward or antici-
patory control, the *directive correlation* of Sommerhoff (1950). A
person who steps out of the path of an approaching motor car uses
a visual feedforward of information to avoid disintegration of the
living system perceived as self. If only feedback control were used,
evasive action would not be taken until the car was actually deflect-
ing some part of that person's body.

The perception of any of these mechanisms in a living system (or in an
autopoietic model) is a construction formed by the observer. Nevertheless
it may constitute a very satisfactory description of the system's behavior,
especially where the mechanism attributed is (3) or (4). It is relevant and
legitimate to consider how behavior that is describable in these terms can
arise in the structure of an autopoietic system.

10.2.1 Iterative Versus Recursive Evolution

The autopoietic viewpoint implies deemphasis of mechanisms such as
those listed above (1)–(4). Reference to such mechanisms is a descriptive
expedient only, and the mechanisms are features an observer attributes
to the *structure* associated with an autopoietic system.

To study living systems in detail their structure must be described in
some way. Some presentations of the autopoietic viewpoint subsume an
implicit assumption that detailed study of the systems (and hence the
description of structure) is in some sense unnecessary. The implication
is that the systems are adequately understood once their *organization* is
known; the evolution of *structure* by the coupling of the organization to
the environment is held to be understood in principle, even though a good
deal of tedium would be involved in working out the details.

Just how much of the detailed working of a system has to be unraveled
before it is held to be adequately understood is of course a matter of
opinion. It seems reasonable to regard the autopoietic explanation as adequate if evolution is held to be a relatively simple single-stage process,
as its discussion in terms of state-determined systems tends to suggest.
If, on the other hand, evolution is functionally recursive, in that evolutionary changes set up mechanisms facilitating further evolution, it seems
necessary to examine much more than the basic autopoietic organization
in order to have any real understanding of a living system.

Maturana and others have discussed recursion in living systems, but
they are referring to a form of recursion that is structural rather than
functional, though this distinction cannot be a firm one. The type of recursion that is more closely connected with structure (in the everyday
sense of the word) is referred to by Beer (1975b) with reference to socioeconomic systems, where enterprises are nested in an industry and
industries within a country's economy, with essentially the same form of

organization at each level. In living systems, individual cells are nested in larger organisms, which are in turn nested in social or ecological systems, and so on.

There is, however, another sense in which evolution is recursive; that is, it may be interpreted as evolving the means of permitting further evolution. If learning and other forms of ontogenesis are considered to result from evolution over generations, this would be an outstanding example of the evolution of the means of further development. As Ashby (1960) points out, the brain may be thought of as an organ evolved to facilitate further adaptation. In the terminology of Locker and Coulter (1977) the same idea may be expressed by saying that teleogenic systems have evolved.

Apart from this, there is a good deal of evidence that both learning and evolution can usefully be described as recursive processes. The term *learning-to-learn* is accepted by psychologists (Hilgard and Atkinson 1967). Biologists are quick to postulate special mechanisms evolved to facilitate further evolution. Even if we reject the tentative suggestion of Hardy (1965) that something in the nature of telepathy plays a part in evolution, there is evidence that animal behavior (Anonymous 1974) and the acceptance of fertilization by plants (Burnet 1971) follow laws that have been evolved to maintain a level of genetic diversity conducive to subsequent evolution. These are fairly simple examples of "adaptation-to-adapt" but they demonstrate that evolution *can* operate recursively, or at least in a manner that is readily describable as recursive.

The argument that evolution operates recursively does not conflict with the view that living systems have an autopoietic organization associated with an evolved structure. It does, however, support the view that any theory of the living system will be very incomplete unless this structure is examined. For its description there seems to be no real alternative to the use of such allopoietic terms as *goal* and *feedback*. One must use them with care, remembering that they have been introduced as aids to description. These terms therefore have no a priori fundamental significance, though in view of their wide applicability it is natural to try to explain their correspondence to real phenomena on some basis other than fortuitous coincidence.

10.3 Language

It is argued elsewhere (Andrew 1978) that both evolution and ontogenesis are facilitated by a succinct representation of information within the system. What is meant by *language,* in a cybernetic (Arbib 1970) sense of the term, is a system of coding of information allowing succinct representation. The coding process is not reversible since it entails selection. Language, in this very general sense, includes means of communication whose syntax is trivially simple, in contrast to spoken languages and programming languages.

Succinct representation of information is useful to a living system, in the sense that it tends to ensure its survival. Living systems have therefore evolved so that they behave as though they had the goal of representing information succinctly. In the general sense of the term, it is not necessary to have the interaction of two distinct systems to have the appearance of language.

The state of a system at any instant is a function of its history but is not usually such that the history can be reconstructed from it. If the (many-to-one) mapping of histories onto states can be interpreted as preserving features of the history that are important to the system's survival, the system is using a language. Such use of language is an *elementary exemplification* (Andrew 1977) of the hypothetico-deductive scientific method.

10.4 Nonautopoietic Behavior

Biologists have for a long time been worried by the fact that animals, human beings in particular, sometimes exhibit heroic behavior that achieves some goal at the cost of their own life. Obviously a genetic trait encouraging self-sacrifice is likely to be eliminated by natural selection. Haldane (1955) suggested an explanation that is essentially the same as that embedded in a formal theory by Hamilton (1964). The argument is that if the self-sacrificial behavior (or other less drastic form of altruism) improves the chances of survival of individuals closely related to the individual exhibiting the altruism, there may be natural selection of genes that predispose their carriers towards altruistic behavior. The care that animals take of their own young is a form of altruistic behavior readily explained in this way.

It is clear, however, that a great deal of animal and human behavior that is not conducive to survival is inexplicable in terms of the Hamilton–Haldane extension of standard genetic theory. Religious martyrs have died rather than renounce highly abstract principles, and modern men are ready to risk their lives to climb mountains or explore space. Artistic pursuits of all kinds have no obvious survival value and yet they key in to human mental processes in some very powerful and desirable way.

Up to a point, nonautopoietic behavior can be explained by supposing that behavior patterns evolved in one ecological niche are appearing in another. Pask (1971) and others have characterized aesthetic experience as the pleasurable exploration of new environments and the discovery of means of controlling them. Insofar as aesthetic activity conforms to this view it can be regarded as a carryover of curiosity, which is conducive to survival in many environments, into contexts in which it no longer has survival value.

The carryover of behavior patterns from one context to another occurs readily if the information determining the behavior is stored in succinct form. Succinct representation is necessary for generalization and inductive inference, but it may lead to an "analytic continuation" of the behavior into contexts different from that in which it was evolved and in which the behavior may be disadvantageous to the living system and contrary to autopoiesis.

10.5   Conclusions

Along with a number of other papers in the present volume, this paper was presented, in a preliminary form, at a conference held in Binghamton, New York, in August 1977. The polemic tone of Section 10.2 was in response to the very polemic tone of Maturana's paper (1975) and particularly to his use of the term "fallacious." At the conference Francisco Varela made it clear that he believes that much has to be added to the theory of autopoiesis to form any satisfactory description of living systems; in particular, autopoiesis tells nothing about evolution.

To some extent, therefore, the argument of Section 10.2 has been conceded. It remains relevant, however, because the boundary between those aspects of living systems that can be discussed in autopoiesis terms and those that cannot has to be defined. Also, caution is advisable in drawing conclusions from Ashby's treatment (1960) of the state-determined system. Ashby himself has expressed rather different views of the implications of this concept in his writings at different times. The view adopted by Maturana is perhaps most clearly presented in a paper entitled "Principles of the Self-Organizing System" (Ashby 1962). In this paper Ashby claims that the description of living systems in terms of phase-space and state-determined behavior is adequate and that the term "self-organizing system" could very well be allowed to drop from use. Another discussion (1960) under the heading of "Amplifying Adaptation" presents a somewhat different view. Section 10.2 is essentially a defense of this latter viewpoint and a claim that the apparent implications of Maturana's discussion should be modified in accordance with it.

Varela has commented that in the discussion of nonautopoietic behavior in Section 10.4 autopoietic systems at two different levels are being confused. To explain altruistic or other self-destructive behavior it is necessary to consider autopoiesis of the social group to which the individual belongs. The fact that the altruism is nonautopoietic for the individual viewed in isolation is irrelevant since the individual does not live in isolation.

However, this point does not resolve the paradox discussed by Haldane (1955) and Hamilton (1964). Since this paradox is only seen in the context of evolution, an explanation in autopoietic terms should not be expected, a limitation of the applicability of autopoiesis theory that should be noted. The operation of a neuron in a nervous system is analogous to that of an individual in a social group. One must therefore consider not only autopoiesis, but also a complex interplay of autopoietic and allopoictic processes, if one hopes to throw any light on what is surely the most fundamental problem of neurophysiology: This is the problem of how the largely autonomous neurons come to act in concert. How, for instance, does it come about that neurons in my motor cortex, cerebellum, and elsewhere in my body choose to collaborate to control the raising of a glass of beer to my lips, when they themselves exist in a regulated environment in which they can have no experience of thirst or of the flavor of beer? We are still a long way from the answer.

ACKNOWLEDGMENTS
I am indebted to Francisco Varela and Gloria Guiloff for helpful discussions.

References

Andrew, A. M. (1970), A note on logically stable neural nets, *Automatica* 6, 615–620.

Andrew, A. M. (1977), Cybernetics and artificial intelligence, in *Modern Trends in Cybernetics and Systems* (J. Rose and C. Bilciu, eds.), Springer, Berlin.

Andrew, A. M. (1978), Succinct representation in neural nets and general systems, in *Applied General Systems Research: Recent Developments and Trends* (G. J. Klir, ed.), Plenum, New York.

Anonymous (1974), Inbreeding to preserve genetic diversity, *Nature* 252, 345.

Arbib, M. A. (1970), Cognition—a cybernetic approach, in *Cognition, a Multiple View* (P. L. Garvin, ed.), Spartan, New York.

Ashby, W. R. (1960), *Design for a Brain,* 2nd Ed., Chapman and Hall, London.

Ashby, W. R. (1962), Principles of the self-organizing system, in *Principles of Self-Organization* (H. von Foerster and G. W. Zopf, eds.), Pergamon, Oxford.

Beer, S. (1975a), The cybernetic cytoblast: Management itself, in *Platform for Change* (S. Beer, ed.), John Wiley, London.

Beer, S. (1975b), Fanfare for effective freedom, in *Platform for Change* (S. Beer, ed.), John Wiley, London.

Blum, M. (1962), Properties of a neuron with many inputs, in *Principles of Self-Organization* (H. von Foerster and G. W. Zopf, eds.), Pergamon, Oxford.

Brown, G. S. (1969), *Laws of Form,* Allen and Unwin, London.

Burnet, F. M. (1971), Self-recognition in colonial marine forms and flowering plants in relation to the evolution of immunity, *Nature* 232, 230.

Gries, D. (1971), *Compiler Construction for Digital Computers*, John Wiley, New York.

Haldane, J. B. S. (1955), Population genetics, *New Biol.* 18, 34–51.

Hamilton, W. D. (1964), The genetical evolution of social behavior, *J. Theoret. Biol.* 7, 1–52.

Hardy, A. (1965), *The Living Stream*, Collins, London.

Hilgard, E. L. and Atkinson, R. C. (1967), *Introduction to Psychology*, 4th Ed., Harcourt, Brace and World, New York.

Ledley, R. S. (1962), *Programming and Utilizing Digital Computers*, McGraw-Hill, New York.

Locker, A. and Coulter, N. A. (1977), A new look at the description and prescription of systems, *Beh. Sci.* 22, 197–206.

McCulloch, W. S. (1959), Agathe Tyche: Of nervous nets—the lucky reckoners, in *Mechanisation of Thought Processes*, Her Majesty's Stationary Office, London.

Maturana, H. R. (1975), The organization of the living: A theory of the living organization, *Int. J. Man–Machine Studies* 7, 313–332.

Neumann, J. von (1956), Probabilistic logics and the synthesis of reliable organisms from unreliable components, in *Automata Studies* (C. E. Shannon and J. McCarthy, eds.), Princeton Univ. Press, Princeton, N.J.; also in Taub, A. H., ed. (1963), *John von Neumann—Collected Works*, Vol. V, Pergamon, Oxford.

Pask, G. (1971), A comment, a case history and a plan, in *Cybernetics, Art and Ideas* (J. Reichardt, ed.), Studio Vista, London.

Peterson, W. W. (1961), *Error-Correcting Codes*, MIT Press, Cambridge, Mass., and John Wiley, New York.

Sommerhoff, G. (1950), *Analytical Biology*, Oxford Univ. Press.

Stark, L. (1959), Stability, oscillations and noise in the human pupil servomechanism, *Proc. I.R.E.* 47, 1925–1939.

Varela, F. J. and Goguen, J. (1977), The arithmetic of closure, in *Progress in Cybernetics Research* (R. Trappl, G. Klir, and L. Ricciardi, eds.), Vol. 3, Hemisphere, Washington.

Varela, F. J. (1975), A calculus for self-reference, *Int. J. Gen. Systems* 2, 5–24.

Varela, F. J., Maturana, H. R. and Uribe, R. (1974), Autopoiesis: The organization of living systems, its characterization and a model, *Biosystems* 5, 187–196.

Winograd, S. and Cowan, J. D. (1963), *Reliable Computation in the Presence of Noise*, MIT Press, Cambridge, Mass.

Ben-Eli's paper is written in the best tradition of the early cybernetic notions of Ashby and it complements well the preceding paper by Andrew. The reader will also notice a rather peculiar, machinelike format of ordered assertions, a style that probably takes its origin in the writings of Gordon Pask (also in this volume).

Ben-Eli's notion of evolution is based on such interpretative tools as equilibrium, stability, regulation, control, and sequential causality. Again, this editor believes that such exposure of the older and more traditional viewpoint does belong in a book on autopoiesis. It provides an immediate and purified contrast to the autopoietic perspective and allows readers to draw their own comparisons of paradigms within the confines of a single volume.

Stability and equilibrium as a purpose of evolution can also be contrasted with the recent notions of dissipative structures, order through fluctuation, and the self-organizing properties of systems far from equilibrium.

Ben-Eli perceives the world as a complex hierarchy of differentiated structures and considers evolution a process proceeding toward the higher orders of complexity. Evolution is regarded as a regulatory "strategy" for achieving stability in a dynamic environment. This is succinctly exemplified in his assertion that "evolution is a type of stability" (see Section 11.3).

Using the notion of regulation leads Ben-Eli to invoke the concept of a "goal" or "criterion" as well as that of a desirable equilibrium state and a continuing maintenance of its stability. His notion of self-organization entails such teleological perceptions as "improvement," "better achievement of a goal," and so forth, all strictly implied by the theory of regulation. The concept of "improvement" becomes crucial to Ben-Eli's perception of evolution. It is interesting to follow Ben-Eli's efforts to make this relativistic concept more objective, as if asking, Can there be any observer-independent consensus about what might constitute an improvement?

Ben-Eli insists that evolution involves more than maintaining autopoiesis: it involves a consistent, selective improvement, in a nontrivial sense. It seems, at least to me, that it is the complexity of the system's organization that is finally offered as being the elusive criterion of progress.

The useful notion is that organizational closure is not only maintained but also consistently expanded or "improved"; this is in contrast to the protoautopoietic concept of strict organizational invariance and a view of evolution as a process of historical structural change through the continuous structural coupling with the medium. Autopoiesis can account for "improvement" by pointing out that the perceived complexity relates to the structure, and its effectiveness to the structural coupling with the medium.

But at the end, Ben-Eli states that the least unit of evolution is an autopoietic system and that evolution only *appears* to be directional, and the reader finally realizes that what is meant by organization in his context is really structure (Section 11.6).

Michael U. Ben-Eli, born in Israel, received his first degree from the Architectural Association in London. During the years at the Association he closely collaborated with R. Buckminster Fuller, working on projects related to Fuller's World Resources Inventory, the World Game, and experiments with low-cost geodesic structures. Subsequently he received a Ph.D. from Brunel University, Department of Cybernetics, under Professor Gordon Pask. Dissertation: "The Cybernetics of Stability and Regulation in Social System." In 1976 he founded a systems-management consulting firm, Cybertec, which he currently heads. Dr. Ben-Eli is Associate Professor at the Department of Administrative Medicine of Mount Sinai Medical School and an Adjunct Associate Professor at the Department of Health Care Administration of Baruch College in New York. His major interests include the dynamics of evolutionary processes, the characteristics and underlying mechanisms of "stability" in complex systems, and management-organization problems in complex human organizations. Dr. Ben-Eli lives in New York with his wife Marcia and daughter Gabrielle. *Address:* 345 East 86th Street, New York, N.Y. 10028, U.S.A.

Chapter 11
Self-Organization, Autopoiesis, and Evolution*

Michael U. Ben-Eli

11.1 Introduction

The purpose of this chapter is to examine some thoughts pertaining to evolution with reference to the concepts of self-organization and autopoiesis and with emphasis on some peculiarly cybernetic notions concerning regulation for viability. The latter in particular, it is suggested, lends the subject matter of evolution a novel focus, providing for an intriguing interpretation of problems involving the question of the direction of evolutionary processes and of what characterizes evolution as a general phenomenon. Thus, unlike some of the original expositions of autopoiesis (Varela, Maturana and Uribe 1974; Maturana and Varela 1975), this chapter stresses an evolutionary viewpoint and employs the theory of regulation (Ashby 1958a, 1966; Conant and Ashby 1970) as a useful interpretative tool for examining the concept of evolution.

One important consequence of this cybernetic perspective is that evolution is viewed as characteristic of a particular kind of dynamic behavior in systems, reflecting a particular aspect of a logic of mechanisms (in the sense suggested by Ashby 1958b). From this viewpoint, evolution corresponds to a specific type of regulation and, as a process, it is embodied in a particular type of organization (of which Pask's iconic representation of learning processes, as in Pask et al. 1973, is an example). This broad statement is significant insofar as it stresses that evolution is a general type of stability (steady state being a special case) and that as such it is

* A paper presented at the NATO International Conference on Applied General Systems Research: Recent Development and Trends, Binghamton, New York, August 1977.

a general condition typical to environments that are subject to the operation of a particular set of constraints.

In a more specific vein, this chapter presents an attempt to relate the concept of evolution to the law of requisite variety and to the possibility in principle of amplifying regulation. In this context, it is argued, a complex dynamic environment, consisting of a multitude of interacting organizations, would put a definite selective premium on the possibility of local increases in the potency of regulation capabilities. This condition in itself would suffice, it seems, to explain the persistent evolutionary tendency, observable in the terrestrial environment, to form stratified organizations of increasing complexity. Only through such a stratification can an increased advantage in regulating capabilities be obtained.

Accordingly, a perception of the world as a complex hierarchy of structures, differentiated by discontinuities and characterized by an increasing order of complexity and organization (Bronowsky 1970), can receive a specific interpretation. Such a hierarchy can be regarded as a stratified organization of controllers interacting such that across its levels regulation is amplified (Ben-Eli 1976, 1978).

Evolution, by conjecture, is the process through which such an amplification—qua complexification or increase in regulation potency—is achieved. In this sense, it can be regarded as an essential "regulation strategy" for achieving stability in a dynamic environment in which the context for stability is changing.

11.2 Basic Premises

Although specific embodiments will not be dealt with explicitly, the context of the comments that follow is that of stability in viable organizations. The term "viable organizations" is meant to designate the class of complex dynamic systems typically dealt with in biology, psychology, and sociology. The focus is on processes or active systems, involving properties and mechanisms associated with various manifestations of life, consciousness, and sociocultural phenomena. Insofar as the discussion of such processes can be reduced to problems concerning equilibrial characteristics of polystable organizations (Ashby 1966), it is assumed that a fundamental organizational isomorphism exists between these processes on the level of mechanism of regulation.

Early contributions to cybernetics (notably Wiener 1948) have pointed out the existence of such an isomorphism between various man-made control apparatus and processes encountered in physiology. The very same correspondence exists between cognitive processes and the processes that mediate stability in social systems (Pask 1973), and it extends to processes involving conscious experience and the operations that characterize evolutionary phenomena (Pask 1969).

On the appropriate level of abstraction concerning the working of regulation mechanisms, this isomorphism transcends the detailed characteristics of entities involved—their materiality, for example—and it can therefore serve to highlight a general logic of mechanism in Ashby's sense (Ashby 1958b). In other words, there are common organizational features underlying the dynamic behavior of systems, whichever the specific nature of their components (Beer 1969).

The particular characteristics of mechanisms through which stability is mediated vary with specific identifications. It is when we focus on an actual "real-world" organization that such mechanisms assume specific identities, coinciding with specific embodiments and related to specific processes in the "flesh," the "metal," or otherwise.

11.3 Organizational Closure and Autopoiesis

The concept of self-organization offers a convenient starting point for discussing current thinking regarding the nature of stable processes and the condition for stability in dynamic systems. The crucial point is that the designation "self-organizing system" embodies two key ideas: the idea of a consistent identity and the idea of dynamic variations essential for its continuous viability. These two aspects of systemic behavior are entirely compatible, expressing the facts that notions of stability and adaptive behavior revolve around the problem of maintaining a balance between constancy and steady-state on the one hand and between change variability and reorganization on the other.

The first aspect bears upon contemporary thinking of what a system "is" in the first place; in other words, of what constitutes an autonomous entity distinguishable from its background as a conceptualizable unity called a system, an organization, a stable process, an individual, a pattern integrity, and so forth. The second aspect emphasizes the dynamic dimension underlying the entity "system" having to do with variations in the operations that constitute "it" and, perhaps even more significantly, with optimization of its self-assertive qualities. This aspect bears upon the concept of evolution.

The dynamic aspect underlying the activities of a self-organizing system can receive various expressions. One relates to McCulloch's formulation of the concept of redundancy of potential command (McCulloch 1952). Another relates to von Foerster's notion of a self-organizing system as a system in which the rate of change of redundancy is positive (von Foerster 1960). Yet another is Pask's notion of an indeterminacy characterizing the interaction with a self-organizing system, so that both the observers' frame of reference and their description of the system have to be changed as observation goes on (Pask 1965).

Thus, for example, evolution can be viewed as a time-dependent characteristic of a self-organizing system (Pask 1961). In other words, an observer interacting with a self-organizing system is faced with a sequence of steps, or systems, that become a single self-organizing system by virtue of a consistent topic the observer has in mind. Under certain conditions such a sequence may generate a trend interpretable as evolutionary.

Note the relativistic connotations immanent in the concept of self-organization. The concept calls for a distinction between three interacting entities: a system, its environment, and an observer.

In contrast to classical science, the participatory and reflective role of the observer has been emphasized in cybernetic literature (for example, Pask 1975, 1978; Maturana 1970; von Foerster 1976). The observer-dependent qualities of fundamental systemic notions have thereby been brought to the fore. The observer provides criteria for distinction relative to which events occur and systems are identified. The observer's perspective determines a frame of reference with respect to which observations are made and conclusions concerning observations are reached.

Units of observations are embodied in sets of relations, as is the idea of stability. For the purpose of description, both may require a second-order resolution into what may be called relations of relations. In the process of a system's activities (interaction with its environment), some of these relations may be modified. The concept of evolution entails a specific type of modification in such relations and relations of relations. Evolution is, therefore, a type of stability. It is characteristic of a particular kind of dynamic behavior in systems reflecting a general condition typical to environments that are subject to the operation of a particular set of constraints.

Concepts pertaining to the general definition of "regulation" provide the specific criterion for the type of modifications in relations (as we have just described), which are interpretable as evolutionary. The argument runs as follows.

"Organization" entails conditionally (between parts) and regularity of behavior (Ashby 1962). In this context, "self-organizing" suggests the following possibilities (Ashby 1962): (1) the initially unorganized may become organized or (2) an organization may become a "better" organization. From the viewpoint of a theory of regulation, the notions of organization and viability bear upon the concept of survival in its generalized sense. The key issue is expressed by the idea of a number of parts so related as to maintain stability around a focal condition (Sommerhoff 1950), also called a goal, an assigned equilibrium, or an acceptable outcome—all observer-related concepts. Accordingly, judging by the criterion of its viability, an organization is deemed effective or "good" when it acts so as to maintain a system's essential variables within their "admissible" range (Ashby 1958a). This notion forms the core of a concept of regulation.

Regulation is mediated through organization. It entails specific structures of relations among interacting entities, such that a particular outcome is brought about.

In the course of self-organizing, an organization may become better in achieving a goal, or generally, in maintaining a stability under more variable conditions. It improves!

A perception of improvement depends to a great extent on the observer; nevertheless, in the strict sense implied by the theory of regulation, the concept of improvement is crucial to a nontrivial concept of evolution. This notion will be developed further in Sections 11.4 and 11.5. In order to improve, there must be an organization in the first place, and we are back to the conceptually convenient, although, perhaps, not absolutely prerequisite starting point, of the idea of a system as an autonomous stable identity.

Reviewing the notion of whole, or stable systems, Varela has drawn attention to the reciprocal interactions and fundamental circularity that characterize viable organizations, and he coined the term "organizational closure" as a characterization of system-wholes (Varela 1976). In this context, the concept of a system as a stable organization has become synonymous with that of organizational closure (see also Pask 1976a, 1977). Stable processes (systems) constitute unambiguous units of reality. They are organizationally closed, and "organizational closure" stands for stability in a dynamic and generalized sense. "Unambiguous units of reality" refers to identifications and/or to conceptualizable identities. Fuzzy systems, processes, concepts, boundaries, and the like (as in Zadeh 1973) are not thereby excluded.

The concept of organizational closure (and related formulations), even when not so named has been inherent, in current ways of thinking about stability. It is intrinsic not only to all homeostatic mechanisms, in their original as well as in the generalized sense, but also to the various contexts of processes involving recursive computations, self-reference, self-stabilization, and so forth. Thus, for example, it is part and parcel of the underlying circularity of von Foerster's finite function machines (von Foerster 1970), as well as of von Neumann's theory of self-reproduction (von Neumann 1969). It is an essential aspect of Pask's theory of conversation and his concept of P individuation (Pask 1976b). It is the basic concept underlying Fuller's notion of a stable structure (Fuller 1975), and more. In a biological context, it has received a definite expression in the concept of autopoiesis (Maturana 1970; Varela, Maturana, and Uribe 1974; Maturana and Varela 1975).

Autopoiesis has been advanced as the nonreducible organizational characterization of living systems that emphasizes the recursive nature of processes involved in their self-realization as stable identities. An autopoietic system is defined by a network of recursive processes for the production of components that constitute and specify the system itself as a unity. Thus,

an autopoietic machine continuously generates and specifies its own organization through its operation as a system of production of its own components and does this in an endless turnover of components under conditions of continuous perturbations and compensation for perturbations [Maturana and Varela 1975].

An autopoietic system is, accordingly, a homeostatic system with its own organization as the essential variable (Maturana and Varela 1975). As such, autopoiesis articulates a "second-order" metaconcept of an overall organizational homeostasis (closure), to the maintenance of which, in any given living organization, all structure-specific homeostatic and other related processes are subjected.

The concept of autopoiesis is couched in mechanistic terms applicable to a general logic of machine behavior. On this level, the concept is independent of the specific characteristics of particular identifications. It therefore embodies the abstract attributes of organizational closure as a general criterion for stability: Autopoietic systems are organizationally closed and organizational closure is a characteristic of autopoiesis.

This statement is significant insofar as it may help resolve an argument whether the concept of autopoiesis is applicable to domains other than that of living organizations in the strict biological sense. If by autopoiesis the concept of organizational closure and the general logic of organizationally closed processes are emphasized, it certainly is applicable. If the term is restricted to mean organizational closure as manifest specifically in the processes of biology, then it is not. Whichever the case, such an argument should not detract from the importance of the underlying generality of the idea of organizational closure, serving, as it does, to unify a concept of stability pertinent to biological entities, cognitive processes, and sociocultural phenomena.

11.4   Autopoiesis, Evolution, and Regulation

Formulations involving (specifically or by resemblance) the notion of organizational closure employ basically similar conceptual constructs. In spite of the underlying similarity, however, different considerations are emphasized in different contexts. As a result, pertinent notions are embodied in a variety of paradigms. Some put emphasis on an abstract characterization and involve simulations of closure (Loefgren 1972, 1975); others emphasize a concept of structure and the dynamics of underlying mechanisms (Pask et al. 1973; Pask 1975; Pask 1977); still others provide a descriptive interpretation of aspects of closure in a specific context— a society for example (Rappaport 1968). Although different in emphasis, such formulations are not mutually exclusive. They provide a range of conceptual tools useful for a variety of needs. The particular paradigm employed depends on the context and purpose. The concept of autopoiesis is one among the currently available paradigms.

In the case of autopoiesis, the emphasis is on a characterization of that organization common to all living systems. This characterization stresses the organizational features that define a living system as a unity. The fundamental circularity inherent in the process of realizing such a unity is captured by the definition of autopoiesis. It also finds expression in the way the concept is employed. Thus, a living organization is defined by its autopoiesis, and autopoiesis, in turn, is an essential and a sufficient characterization of a living system. All other dynamic features involving the continuous maintenance of autopoiesis in the face of perturbations become subordinate to the integrity of the autopoietic condition itself. The power of the concept is in its metasystemic nature, which allows for collapsing a variety of features, all secondary to a concept of a living system as a unity, into one global statement.

Two immediate consequences follow: One is that evolution is treated as a secondary aspect in the phenomenology of living systems. The other is that the concept of regulation, and hence the structure and dynamics of pertinent mechanisms, are deemphasized.

For example, in the context of the characterization of a living organization as an autopoietic organization, evolution can be regarded as a sequence of historical changes through which autopoiesis is sustained. Maintenance of autopoiesis is the overriding consideration insofar as, whatever an evolutionary change entails, it must ultimately satisfy the requirement for autopoiesis.

This view is indubitably correct and entirely consonant with the definition of autopoiesis and the conceptual metaframework within which the concept is conceived. Nevertheless, it is inadequate for a comprehensive characterization of life, since the emphasis on the invariant and unitary characteristics of a living organization as an individual living system obscures the otherwise essential features of evolution as a general underlying process. Such features emerge as a consistent pattern generated by the historical sequence of individual autopoietic realizations and modifications in such realizations.

Beyond considerations of local closure and specific adjustments in the way local closure is attained, a concept of evolution must accomodate a notion of systemic, cumulative characteristics unifying all successive single realizations. As a consistent cumulative trend, such characteristics belong to a domain different from that of specific realizations and their embodiments. It requires a higher level logic than the one employed in the description of any single living organization as an individual unity.

This consideration is further compounded by the fact that evolution is itself a recursive process that continuously conditions and reinforces its own further realization. (There is a definite sense, in this respect, in which we can talk about the "evolution of evolution.") Although autopoiesis is of necessity maintained through the evolutionary process, evolution involves more than maintaining autopoiesis. In a nontrivial sense, evolution involves a consistent selective improvement in autopoietic organizations and in the means by which autopoiesis is attained. Not only is the constancy of closure maintained across a sequence of evolutionary realizations, but closure is expanded (Ben-Eli 1976; Ben-Eli and Tountas 1977). There is a second-order constancy involved that is embodied in a consistent dynamic trend. In the terrestrial environment such a trend is manifest in the persistent emergence of ever more complex organizations.

Such arguments concerning evolution do not conflict with the characterization of living organizations as autopoietic systems. They serve to point out, however, that the concept of autopoiesis as it stands may acknowledge but not provide a complete explanation of those consistent aspects of living systems involving the notions of complexification, increase in organization, and self-improvement. The question of which aspect should be emphasized in principal—the unitary characteristics of individual autonomy or the evolutionary considerations—is misconstrued. Ultimately, both are "behavioral" manifestations of stability in complex dynamic systems. Which of the two is emphasized depends on the observer's viewpoint, purpose, and temporal frame of reference.

Similarly, as a metasystemic statement concerning the characterization of living systems as autonomous unities, the concept of autopoiesis sheds little light on the specifics of structures and underlying mechanisms. Detailed considerations of such mechanisms may not be essential for a broad characterization of organizational viability; they are indispensable, however, for a thorough understanding of how autopoietic systems work.

The concept of autopoiesis identifies a class of organizations and defines the global condition for their viability. It does not comment on the structural details of processes that mediate such viability. By accepting the concept that autopoietic or organizationally closed systems constitute fundamental units of reality, we may dispense with the question of *why* there are such systems (Pask 1977). Nevertheless, we may still need to know more about the *how* of their underlying processes. In the context of the *how,* paradigms that emphasize the structural aspects of organizational closure (Pask's iconic representations of conversation theory, for example) become especially useful.

By stressing the question of mechanisms that maintain dynamic organizations invariant, such paradigms bring the concept of regulations to the fore. The crucial notions are universally those of a purposive or goal-directed system and the resolvability of complex organizations into purposive components. Such notions served tremendously to increase scientific understanding of behavior in dynamic systems. They have been successfully employed in diverse fields, ranging from engineering, physiology, and biology to ethology, anthropology, psychology, and sociology. The suggestion that the idea of purpose and notions concerning regulation are descriptive ploys employed by observers and not "objective" features of observed domains does not detract from their usefulness in helping to understand, control, and synthesize natural processes. One particularly useful application involves the use of the cybernetic concept of regulation as an interpretative tool for a characterization of evolution. The basic argument is briefly developed in the following section.

11.5   Amplifying Regulation Through Evolution

By assuming that realizable configurations imply the fulfilment of thermodynamic and energetic requirements, and by emphasizing organizational aspects that are independent of material considerations, the cybernetic theory of regulation offers a view of evolution that is generally applicable to different systemic domains. The general applicability is obtained by a formulation of principles that concern behavior in dynamic systems and bind the concepts of regulation, control, information, and communication to the general notions of survival and viability.

These formulations are conceived on a level of abstraction that makes their transfer across systemic boundaries particularly convenient. The approach can, therefore, contribute to a unified view of evolution and it provides for a consistent interpretation of both the dynamics of special-case evolutionary processes and the overall evolutionary tendency toward forming organizations of increasing complexity.

The emphasis is on the close relation between the concepts of adaptation and regulation. From the cybernetic viewpoint, the adaptive processes underlying evolution are subject to the laws of control, specifically to the law of requisite variety, and to the possibility, in principle, of amplifying regulations through linkages in hierarchical organizations of interacting controllers (Ashby 1964). Evolution, from this viewpoint, is characteristic of a particular type of dynamic behavior in systems reflecting the operation of a particular set of constraints. It corresponds to a specific type of regulation, embodied in a particular kind of organization.

In this context two ideas are crucial:

1. In any complex dynamic system subject to consistent constraints, some properties will be more resistant to change than others. These will tend to survive and gradually to dominate their environment, appearing to be particularly well adapted to its demands (Ashby 1962).
2. The active interactions of coexisting organizations continuously alter the properties of the medium in which they occur. As initial

constraints are modified, and with them the norms of "survival success," new needs and conditions for further evolution are being continuously established. The whole cumulative process proceeds with 'new possibilities and challenges created at each evolutionary step.

In the process of seeking a local viability, various organizations and modes of behavior arise, subject to satisfying the condition for stability under existing constraints. Favorable organizations and modes of behavior will be allowed to persist (survive, maintain closure) and those that entail an improvement will be encouraged to develop, thus generating a trend that an observer would deem evolutionary.

Specifically, a viable system that is adapted to its environment can be regarded as a successful regulator (having its own stability as an open-ended goal), in that the repertoire of its actions matches effectively the variety of the disturbances "threatening" its stability. Selection, accordingly, entails a process that operates to encourage an appropriate match between a regulator's variety and the variety in its environment. In a complex dynamic world, it will favor the formation of high-variety regulators, those effective in securing a viable closure under a wide range of dynamic events. This is to say that variety "mismatches" between interacting controllers will produce a pressure for local variety supplementations. Accordingly, a dynamic environment of high variety can be expected to put a definite premium on the possibility of local increases in the potency of regulation capabilities. This condition in itself is sufficient to explain the persistent evolutionary tendency to form stratified organizations of increasing complexity. Only through such an organization can an increasing advantage in regulating capabilities be achieved.

The selective process that mediate the match of varieties operates simultaneously on two levels:

1. On one level it operates to increase regulation potential of specific organizations, producing a better match between such organizations and the variety of their environment.
2. On a higher metalevel, however, it operates not only by selecting particular organizations but by encouraging variability in general.

The result, as manifest in organic evolution on earth, is a local increase in the range and variety of adaptive possibilities and a consistent general trend characterized by a succession of progressively more capable regulators. A view of the world as a hierarchy of structures characterized by an increasing order of complexity and organization obtains, accordingly, a specific meaning. Such a hierarchy can be regarded as a stratified organization of controllers interacting such that across its levels regulation is amplified. Each level in this hierarchy corresponds to a class of regulators and these become more potent as they ascend the scale of complexity.

Evolution is the process through which such a complexification is achieved. It can be regarded as an essential regulation "strategy" for maintaining closure in a dynamic environment in which the context of stability is shifting.

As a process, evolution involves relative changes in variety, especially relative increases and local amplifications of variety among interacting organizations. While variety is locally amplified, closure is maintained. At the same time the means of ensuring closure gain in potency, thus expanding not only the closure itself but the boundaries of the domain in which it can be sustained.

Incorporating additional variety from the environment brings about the formation of a new and more complex unity. The latter corresponds to a new level of systemic integration that is marked by a relative expansion of regulation capabilities and is subject to selection for some specific survival advantage.

In this regard, it is important to note the following:

1. Other than in the special case of the isolated steady state, the context of autopoiesis is an evolving domain.
2. The least unit of evolution is an autopoietic system.
3. Evolution is regenerative; it is unlimited for as long as sufficient diversity is generated locally, and sufficient distinction is maintained, among interacting autopoietic organizations.

11.6 Syntropoiesis: A Characterization of Evolution

Evolutionary processes can be depicted by a dynamic activity in a redundant network of interacting regulators (each of which has to secure its own closure) involving the selective formation of linkages between initially independent loci of control. The overall context of stability will shift and actual realizations of stable local configurations in such a network continuously change as linkages are formed and reformed.

While such changes depend on various "chance" events, they are not entirely random. They are "directionally" biased by selection processes that reinforce particularly "survival-worthy" patterns, thus altering the condition probabilities of their own further realizations. As a result evolution appears to an observer as a consistent trend involving a directed increase in organization and suggesting a built-in drive for consistent self-improvement. Both are fundamental to a perception of life as a process that not only maintains but also improves itself (Szent-Gyoergyi 1974). As essential features of evolution, they merit a special name.

The term *syntropy,* coined by Fuller to describe the cohesive antientropic tendency of forming structures of increasing order and organization (Fuller 1975) seems particularly appropriate. Accordingly, I would like to suggest the designation *syntropoiesis* as an identification of this general aspect of viable existence.

Syntropoiesis involves autopoietic systems such that (1) the closure of their circular processes of self-realization is maintained in spite of changing underlying circumstances and (2) their sequential realizations not only maintain but also produce selective improvements on previous autopoietic norms.

The evolutionary sequence of realization and continuous improvements of realizations that results appears to an observer to be a coherent trend characterized by a consistent dominant feature. We are now in a good position to characterize this dominant feature specifically: it involves the consistent production, through variety amplifications, of ever more general regulators fitted for an increasingly more comprehensive niche.

At any instant, the question of how the next amplification, how the next expansion of closure will be obtained, is left open.

References

Ashby, W. R. (1958a), Requisite variety and its applications for the control of complex systems, *Cybernetica* 1 (No. 2), 83–99.

Ashby, W. R. (1958b), General system theory as a new discipline, *General System Year Book* III, 1–6.

Ashby, W. R. (1962), Principles of the self-organizing system, in *Principles of Self-Organization* (H. von Foerster and G. W. Zopf, eds.), Pergamon, Oxford.

Ashby, W. R. (1964), *An Introduction to Cybernetics*, Methuen–University Paperbacks, London.

Ashby, W. R. (1966), *Design for A Brain*, Chapman and Hall, London.

Beer, S. (1969), The prerogatives of systems, *Management Decision*, Summes, pp. 4–12.

Ben-Eli, M. (1976), *Comments on the Cybernetics of Stability and Regulation in Social Systems*, Ph.D. thesis, Brunel Univ., Uxbridge.

Ben-Eli, M. (1978), Amplifying regulation and variety increase in evolving systems, *Proc. 4th Europ. Mtg. on Cybernetics and Systems Res., Linz, Austria*.

Ben-Eli, M., and Tountas, C. (1977), The evolution of complexity in a simulated ecology, *Proc. SGSR Symposium*, Denver.

Bronowski, J. (1970), New concepts in the evolution of complexity: Stratified stability and unbounded plans, *Zygon* 5 (No. 1), 18–35.

Conant, R. C., and Ashby, W. R. (1970), Every good regulator of a system must be a model of that system, *Int. J. Systems Sci.* 1 (No. 2), 89–97.

von Foerster, H. (1960), On self-organizing systems and their environments, in *Self-Organizing Systems* (M. C. Yovits, and S. Cameron, eds.), Pergamon, New York.

von Foerster, H. (1970), Molecular ethology, in *Molecular Mechanisms in Memory and Learning* (C. Ungar, ed.), Plenum, New York, pp. 213–248.

von Foerster, H. (1976), *Collected Publications of Biol. Computer Lab*, Univ. of Illinois, Urbana.

Fuller, R. B. (1975), *Synergetics*, Macmillan, New York.

Loefgren, L. (1972), Relative explanations of systems, *Trends in General Systems Theory* (G. Klir, ed.), John Wiley, New York.

Loefgren, L. (1975), On existence and existential perception, Univ. of Lund, Sweden.

Maturana, H. R. (1970), Neurophysiology of cognition, in *Cognition: A Multiple View* (P. Garvin, ed.), Spartan Books.

Maturana, H. R., and Varela, F. (1975), Autopoietic systems—a characterization of the living organization, Biol. Computer Lab. Res. Report 9.4, Univ. of Illinois, Urbana.

McCulloch, W. S. (1952), Finality and form in nervous activity, in *American Lecture Series* No. 11 (C. C. Thomas, ed.), Springfield, Ill.

McCulloch, W. S. (1958), Agatha Tyche: Of nervous nets—the lucky reckoners, in *Mechanization of Thought Processes*, Symposium at National Physical Laboratory, November.

von Neumann, J. (1969), in *Theory of Self-Reproducing Automata* (A. Burks, ed.), Univ. of Illinois Press, Urbana.

Pask, G. (1961), The cybernetics of evolutionary processes and self-organizing systems, *Proc. 3rd Cong. Int. Assoc. of Cybernetics, Namur Belgium* (Gauthier Villars, ed.).

Pask, G. (1965), Comments on an indeterminancy that characterizes a self-organizing system, in *Cybernetics of Neural Processes*, National Res. Council, Rome.

Pask, G. (1969), The Cybernetics of Behaviour and Cognition Extending the Meaning of 'Goal' ", System Research Ltd., paper delivered at the Int. Congr. of Cybernetics, London.

Pask, G. (1973), Models for social systems and for their languages, *Instructional Science* 1 (No. 4), 395–445.

Pask, G. (1975), *The Cybernetics of Human Learning and Performance*, Hutchinson, London.

Pask, G. (1976a), Revisions in the foundations of cybernetics and general systems theory as a result of research in education, epistemology and innovation, *8th Int. Congr. on Cybernetics, Namur, Belgium*.

Pask, G. (1976b), *Conversation Theory*, Elsevier, Amsterdam.

Pask, G. (1977), Organizational closure of potentially conscious systems, in *Proc. NATO Int. Conf. on Appl. Gen. Systems Res., Binghamton, N. Y.*, August.

Pask, G. (1978), A conversation theoretic approach to social systems, *4th World Congr. of Cybernetics and Systems, Amsterdam*.

Pask, G., Scott, B. C. E., and Kallikourdis, D. (1973), A theory of conversation and individuals (exemplified by the learning process on CASTE), *Int. J. Man–Machine Studies* 5, 443–566.

Rappaport, R. (1968), *Pigs for the Ancestors*, Yale Univ. Press, New Haven, Conn.

Sommerhoff, G. (1950), *Analytical Biology*, Oxford Univ. Press.

Szent-Gyoergyi, A. (1974), Drive in matter to perfect itself, *Synthesis* 1 (No. 1.), 12–24.

Varela, F. G., Maturana, H. R., and Uribe, R. (1974), Autopoiesis: The organization of living systems, its characterization and a model, *Biosystems* 5, 187–196.

Varela, F. G. (1976), The arithmetic of closure, in *Proc. 3rd Europ. Mtg. on Cybernetics and Gen. Systems Res., Vienna* (R. Trappl, ed.).

Wiener, N. (1948), *Cybernetics*, MIT Press, Boston.

Zadeh, L. A. (1973), Outline of a new approach to the analysis of complex systems and decision processes, *IEEE Transactions on Systems, Man, and Cybernetics*, SML-3 (No. 1), 28–44.

Atlan works with another classical concept of cybernetics—the information theory of Shannon. His concern about complexity and its measurement provides a useful complement to the preceding paper.

The earlier proposition of Von Foerster of order from fluctuations (disorder + noise = order) gains a new significance in Atlan's careful treatment. Two classical limitations of the information theory are being overcome: (1) the creation of information can be explained without contradicting the theorem of the noisy channel, and (2) an approach to the meaning of information is provided by applying the order through noise principle to hierarchical systems.

But the major concern of Atlan is self-organization. He emphasizes the *randomness* of stimuli, fluctuations, or perturbations, as opposed to the programmed ones, as providing the main characteristic of self-organization. Thus he states that the system's organization must be open to such random perturbations (compare with Varela's concept of organizational closure). The reader would probably surmise that Varela could well agree on the complementarity of the two different approaches to self-organization, especially after reading his article in this volume.

The complexity of a system to Atlan is a mechanistic and quantitative attribute, measurable by logarithmic functions. He does distinguish between the complexity and complication and relates them to only partially known natural systems and to fully described artificial systems, respectively. But the complexity is still assumed to be expressible through a unidimensional index, a single number.

As this paper is not related to autopoiesis in any direct way, the reader might also question its appropriateness for this volume. There are several connections. The theory of autopoiesis implies that the notions of coding, programming, and transfer of information become misleading if used as explanatory notions at the cellular level. A good, up-to-date exposure to the theory of information appears to be useful for comparative purposes and as a reference. Atlan provides an interesting discussion of the order from noise principle and its implications for self-organization and evolution: if we know all the internal production mechanisms, self-organization cannot exist! That is, self-organization is an observer-dependent phenomenon, based on perceived randomness or incomplete knowledge and ignorance of the observer. The reader should compare this with the similar conclusions derived by Uribe elsewhere in this volume, although through a different way of reasoning.

Atlan's definition of self-organizing systems takes the role of the observer explicitly into account. Randomness, information, and noise are viewed as being relative to the observer, or more precisely, to the operations of measurement and control. With Uribe, Atlan could agree that a system can be random (i.e., unknown) with respect to the observer, while being deterministic or ordered (i.e., known) with respect to itself (as if the functioning parts were "known" to the whole).

Atlan's main results seem to be of great value to the current problems with DNA redundancy: self-organization is a unity of continuous disorganization and subsequent reorganization with more complexity and less redundancy. A system should be periodically "recharged" with redundancy in order to be able to evolve.

Henri Atlan was born in Algeria in 1931. He received his M.D. and Ph.D. from the University of Paris Medical School and Faculty of Sciences. His appointments include Professor of Biophysics at the Rouen University Medical School (1966), University of Paris VI Medical School–University Hospital Broussais (1973), and most recently, Professor of Medical Biophysics and Head of the Department of Medical Biophysics at the University Hospital Hadassah in Jerusalem (since 1975). Dr. Atlan was also Research Associate at the NASA Ames Research Center (Moffett Field) in California (1966–1968) and Visiting Professor at the Weizmann Institute of Sciences in Rehovot, Israel (1970–1973). His experimental work in cell biology contributed to the fields of the biology of aging, radiation biology, and cellular cybernetics (membrane-induced regulation of protein synthesis and cell metabolism). His theoretical works have dealt with various approaches to the logic of the integrated biological organization, leading to new developments in information theory and network thermodynamics. Dr. Atlan has published two books (in French) on *Biological Organization* and *Information Theory*. *Address:* The Hebrew University–Hadassah Medical School, Department of Medical Biophysics and Nuclear Medicine, P.O.B. 12 000, 91120 Jerusalem, Israel.

Chapter 12
Hierarchical Self-Organization in Living Systems
Noise and Meaning

Henri Atlan

12.1 Introduction

Shannon's formulas for the functions H and R (respectively, information content and redundancy) have been used extensively to define what order, complexity, and organization are—a need apparently foreseen by von Neumann. He compared the definition and clarification of these concepts to that of energy and entropy in the development of the 19th-century natural sciences.

The order-from-noise principle (von Foerster) and its later development as a principle of information or complexity from noise (Atlan) have helped us to understand in the language of and without contradicting the Shannon theory how information can be created. It has further been used as a basis for a theory of self-organization.

Self-organization has been described as a property of a system in which novelty and changes in organization can be observed. In order for these changes to be real the organization of the system must be open. Otherwise the changes would be contained in the already existing law of organization, which is self-contradictory (cf. Ashby 1962). What makes one call the system self-organizing (although the sources of newness come from elsewhere) is the randomness of the stimuli coming either from the environment or from internal fluctuations in the system (e.g., temperature). In other words, the organization—as a process—of a system implies reactions to stimuli. These stimuli can be either programmed or random. In the latter case the system is called self-organized.

The purpose of this article is to show how this formalism may add something to the understanding of hierarchical self-organizing systems. In addition, the same formalism can give a hint regarding the problem of the meaning of information in such systems.

However, the concepts of complexity, complication, order, and organization are used with different, at times contradictory, implications in the literature. Therefore, before proceeding further, some clarification is needed.

12.2   On Complexity and Related Concepts

12.2.1 Complexity and Complication

It is well known that Shannon's H function represents the missing information that would be needed in order to describe a natural system completely. As such it rightfully measures the system's *complexity* if we realize that complexity is a negative quality: We shall say a system appears *complex* when we do not know how to specify it completely although we know enough about it to recognize it and to call it a system. In this respect, complexity must be distinguished from what we may call *complication*: The latter only expresses a high number of steps necessary to describe or specify or build a system. In this sense, complication should be the attribute of artificial, man-made systems, and its measure, computable from actual blueprints, could be, for example, the minimum number of steps that a Turing machine would need to describe it. In fact, very often such measures are given by means of the computer time needed to achieve some task: The more time, the more complicated the task for the same computing facilities.

12.2.2 Complexity and Coding

As a measure of the information content of a system, H implies the existence of a channel from the system to the observer. Function H is in fact transmitted information in this channel, or the information content of the message output in this channel (Gatlin 1966; Atlan 1968). As information transmitted from the system to the observer, it is indeed transmitted information in the Shannon sense: The coding or decoding of the meaning of the messages at the input and output is taken for granted but is not taken into account.[1]

The observer of the natural system receives the output of a channel without knowing the code that would allow understanding of the messages received and their translation into a specific and explicit description of the system. This imperfect observation at least allows measurements by the observer of the need to know the system completely; it is this meas-

---

[1] As is well known, Shannon theory is limited to Weaver's level A (transmission of signals within the channel) at the exclusion of levels B and C, which imply the emitter and receiver operations of coding and decoding and have to do with the meaning and efficiency of the message.

urement that is transmitted in the channel to the observer by means of the only possible observations of the system; that is, the probability distribution of the constitutive parts. At the same time, it is this same measurement that allows estimation by the observer of the complexity of this system, whose code (the internal order) is unknown because it is not understood.

12.2.3 Complexity and Levels of Organization

An important consequence, often considered an additional flaw of the classical concept of information content, is that the observation level, although critical, is seemingly left to the arbitrary choice of the observer. The estimates for H can be very different and depend upon the choice of the constitutive elements: elementary particles, atoms, molecules, macromolecules, organelles, cells, organs, organisms, production and consumption units, societies, and so on.

As will be considered further in Section 12.4, the meaning (for the system itself) of the information (that the observer is lacking) is to be found at the articulations between these different observation levels. For example, if the observer chooses to describe a system in terms of its constitutive atoms, information is available concerning only the different kinds of atoms to be found in a statistically homogeneous ensemble of identical systems and their probability (or frequency) distribution, whereas H will measure the missing information necessary to specify the system. Obviously this deficient information is very large compared to the case of the same system described in terms of molecules: In that case we would already make use of additional information that we have, or assume we have, on how the atoms are associated to build molecules, and so much complexity then disappears.

Another, and very spectacular, example is that of the information content or complexity of a living organism, which is considerably reduced when we assume that the organism is completely determined by the informational structure of its genome. The organism's complexity is then reduced to that of its genome, that is, of its constitutive DNA (Dancoff and Quastler 1953).[2] The building of the cell proteins from their constitutive atoms and monomers is no longer considered as a priori undetermined (uncertain) once we know the genome and the mechanisms by which it determines the protein structure. The knowledge of these mechanisms reduces our missing information, that is, the apparent complexity

---

[2] This widespread idea should be accepted as a working hypothesis or a metaphor and not taken literally, since DNA as a genetic "program" needs the products of its reading and execution (RNA and regulatory proteins) to be read and executed. In living cells, the "program," if any, seems to be identified with the whole cell. The DNA looks rather like the memory where parts of this program are stored.

of the cell system, showing the constraints between constitutive elements. In other words, this reduction is an application of the well-known opposition between H and R: The information and redundancy functions are related by $H = H_{max}(1 - R)$. H measures the complexity because it is the information that we lack; R measures the simplification because it is information that we have, at least in part, about the internal constraints, in the form of the conditional probabilities.

From this point of view a very complicated artificial system—one whose complication would be measured by a large number of instructions to a Turing machine—may have its complexity reduced to zero and redundancy maximum ($= 1$). In effect we may choose to describe such a system by a function H with only one constitutive element, namely, the whole system already described in a deterministic way by its known blueprint. Our a priori deterministic knowledge of this blueprint allows us to choose the whole system as an elementary constitutive part, just as a priori knowledge of the association of atoms in molecules allows us to consider the molecules as elementary parts instead of the atoms. Again we meet here the basic difference, often forgotten, between well-known artificial complicated systems and natural complex systems.

12.2.4 Measures of Complexity

The feeling of complexity can come first from the large number of constitutive elements. It is measured by their "variety" in the sense of Ashby or $H = \log_2 N$ and does not take into account any repartition of the different elements. An approach to (the imperfect knowledge of) such a repartition is given by the element's probability distribution and leads to a different, lower value of H; this is coherent since the probability distribution already represents some additional knowledge and therefore reduces the complexity. Thus, we can recognize three kinds of complexity, according to the three different expressions of function H. The first, which is trivial (and maximum), is the variety of elements given by $H = \log_2 N$; the second has to do with disorder or statistical homogeneity and is given by

$$H = - \sum_i p(i) \log_2 p(i).$$

Finally, as seen in the previous section, the third is a measure of the lack of knowledge concerning internal constraints (or redundancy) of the system and is given by $H = H_{max}(1 - R)$.

12.2.5 Complexity and Disorder

A system appears to be *ordered* to a given observer if the latter can see some internal articulation and can understand or guess the code that governs the arrangement of the elements. An ordered complexity is no longer complex, but it may be complicated. The relationship between complexity and disorder appears clearly when we realize that it is a statistically homogeneous structure that is most complex for somebody who wants to reproduce it exactly, distinguishing between molecules having different locations. In fact, maximum disorder (or entropy) looks like maximum homogeneity to the observer, who is unable to distinguish between elements. For the observer, the actual microscopic state cannot be determined since it cannot be distinguished from other states.

Thus, while a state of maximum thermodynamic disorder is seen as maximum macroscopic homogeneity, it is in fact a maximum heterogeneity at a level where we cannot observe it, that of its individual constitutive particles. Thus maximum disorder or homogeneity is a state of maximum microscopic complexity, a state in which we measure maximum missing information.

12.2.6 Order and Redundancy

Known internal constraints within the system can be measured by the redundancy function, which reduces H. Since the latter is a measure of the system complexity and disorder, the former is a measure of simplicity and order. Thus, within the framework of this theory, what we call order appears to be in fact redundant or repetitive order. It need not be physically repetitive in the simple sense of one element repeated many times. But it is redundant in the sense of deductively repetitive: knowing one element gives some information on the others, and this is the source of the appearance of order.

12.3   Order-from-Noise and Complexity-from-Noise Principles

12.3.1 Beneficial Effects of Noise

Weaver had already noticed that the effect of noise on signals in a channel increase the message information content at the output, since its uncertainty has increased. This looked to him a paradoxical "beneficial effect" of noise that could not be accepted within the framework of a communication theory, where the goal is to transmit information with as few errors as possible.

However, the situation is different when one is interested not in the output of a given communication channel, but in the information content of a system containing this communication channel as a constitutive part. It is then easy to show that Weaver's first intuition was right and that it can be the basis for a solution of the creation-of-information problem within the framework of Shannon theory (Atlan 1968). In addition, we shall show that it helps somehow to clarify the profound unity of Weaver's three levels of information starting from the narrow definition limited to level A. This unity was postulated by Weaver immediately after he distinguished among them.

12.3.2 The von Foerster Magnet Model

More generally, von Foerster (1960) related the properties of adaptation and evolution to the capability not only to resist but also to utilize the effects of noise and he coined the name "order-from-noise principle." He gave a qualitative example in the form of his famous box of magnetized cubes randomly shaken (Figure 1). However, the quantitative treatment was limited to a simple case in which the ordered structures that appear as a result of the shaking were assumed to be pairs of cubes. According to this treatment his order from noise appears clearly as a repetitive order, or redundancy from noise. On the other hand his qualitative example, where the produced structures are not known—and cannot be known in advance—appears to be a nonrepetitive order, that is, variety or complexity from noise.[3]

More precisely, the whole reasoning in the von Foerster qualitative model is based on the assumption that we do not know that the cubes are magnetized and cannot forsee the structures produced. This is why the system looks to us "self-organizing," it is for this reason that the organized structures that appear as a consequence of the random shaking look to us more complex than the initial random mass. On the other hand, the increase in redundancy calculated for the case that these structures are pairs of magnets is based on our specific knowledge about the structures that appear (namely, that these are pairs). Therefore the assumption is equivalent to knowledge of the internal mechanism (the magnetization) by which the structures are produced, the opposite of the assumption in the qualitative case. It is therefore obvious that it is the complexity that diminishes, as in the example of an organism determined by its genome (Section 12.2.3); indeed, what increases is the repetitive order, as in a crystal formation. The noise allows the constraints potentially contained in the attraction forces to be realized, so that the constructed system will correspond exactly to our a priori knowledge regarding the mechanisms by which it comes about.

---

[3] Now the situation is different from that discussed in Section 12.2.5. Here the absence of structure (or "disorder") of the initial random state of the magnets is seen not as a state of maximum complexity, but as a random mass in which we do not assume any constraint. The apparent disorder or homogeneity appears to represent maximum complexity only if we have some reason to consider it not random but rather the result of some (yet unknown) constraints. Only this assumption can justify the hypothesis mentioned in Section 12.2.5 that the system must be described in terms of its microscopic state. This shows that our intuitive idea of order and disorder is not clear, but based on implicit assumptions regarding both our actual and potential knowledge of the system.

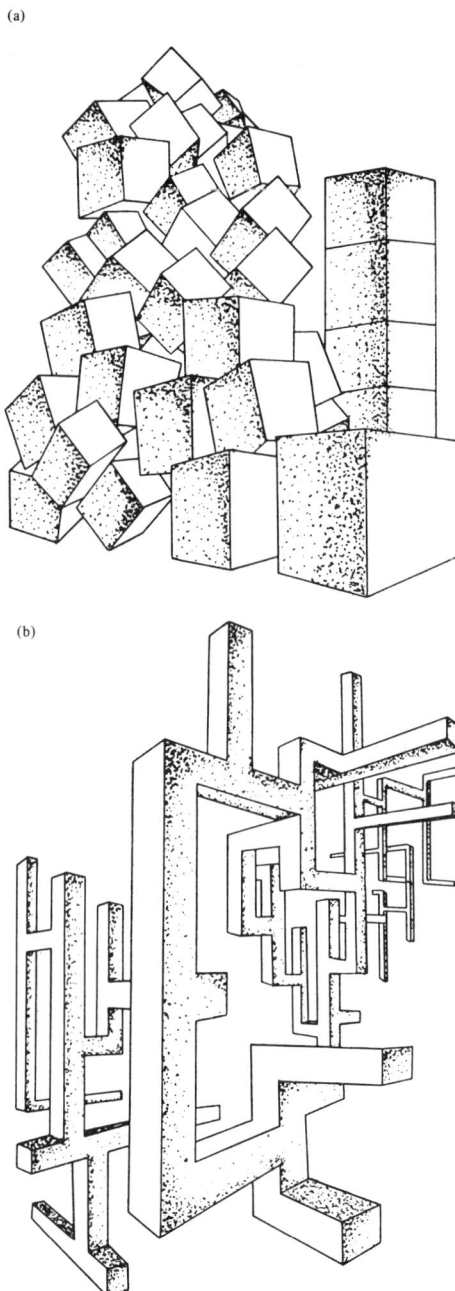

Figure 1. (a) Magnetized cubes (three faces in one direction, three in the other) in a box before shaking. (b) The same cubes during random shaking of the box form geometrical figures subject to change, thus giving the impression of self-organization to an observer who opens the box from time to time.

The great interest of the qualitative von Foerster magnet model, however, lies in the assumption that the observer does not know these mechanisms. This feature makes them a model for systems that look (to us) self-organizing, while it implies at the same time that in the absolute sense (i.e., if we were to know everything about such systems), self-organizing systems cannot exist. Only under this assumption do the structures appear to us more complex than the initial structureless mass, so that our missing information—that is to say, the function H—is larger. More precisely, again under this assumption of ignorance of the internal initial constraints (initial magnetization), geometric structures appear more complex than the initial structureless state in the following way. Let us decompose the space occupied by the system into pieces, that is, small elementary volumes to which a probability of being occupied by a magnetized cube can be assigned. The perception of the random mass as unorganized implies that each volume can be replaced by another with corresponding probability of being occupied without producing a sensible change in the global form, which is still perceived as random and structureless. This means that the number of different elements that we would need to specify in order to reconstruct a statistically similar mass (i.e., the number of different volumes with their probability of being occupied) is very small, since important changes in these probabilities for most elements would not affect our perception of the mass as random and structureless. On the other hand, a geometric form (which has appeared seemingly by itself) implies that every cube has a well-defined location in space, which means that volumes and their probability of being occupied by a cube cannot replace one another. In that case the number as elements to be specified is thus much larger, which amounts to a larger value of H.

While studying the beneficial effects of low doses of noise-producing factors (ionizing radiation, heat, time of aging) on living systems (1968), we tried several time to give this principle mathematical expression (1972, 1974, 1975). At that time, the necessary distinction between these opposite kinds of order was not so clear. We used the same von Foerster expression of order from noise, while we expressed it as an increase of the H function, that is, an information- or complexity-from-noise principle.[4]

---

[4] In light of the above developments clarifying the identity of complexity and disorder, this amounts to disorder from noise, which seems quite trivial. In fact, the interest of this operation lies in that this complexity or disorder is measured by the H function for a natural, not a well-known system. In other words, as already mentioned, the intuitive notion of order or disorder is relative to the knowledge and understanding of the observer.

12.3.3 Complexity from Noise and the Bourgeois Graphs

The transmitted information $T(x;y)$ in a channel is given by

$$T(x;y) = H(y) - H(y/x),$$

where   (1)

$$H(y/x) = - \sum_{i,j} p(i)p(j/i) \log_2 p(j/i).^5$$

The function $H(y/x)$, called the *ambiguity,* is an increasing function of the noise in the channel. It measures the independence of y with respect to x and is therefore a loss in the transmitted information. We call it a *destructive ambiguity.*

However, if one is interested in the total information content of x and y, this same function bears a plus sign, as in

$$H(x,y) = H(x) + H(y/x).$$ (2)

Here, the independence of y with respect to x measures the decreased redundancy or the increased information content of a system containing x and y. We call it *autonomy producing ambiguity.* Therefore, the sign of this function depends upon the position of the observer: negative when he sits at the output of the x-to-y channel, positive when he sits at the output of a channel from a system that contains x and y as constitutive parts. Thus, we can see how apparently paradoxical positive effects of noise on the information content of a system can be formulated.

These two possible measures of the ambiguity (Atlan 1968) appear most clearly in the graphic representations conceived by M. Bourgeois and shown in Figure 2a. This diagram assumes conservation of that which is flowing, namely, uncertainty. Thus, by specifying that the algebraic sum is zero at each node, one immediately obtains all of the well-known Shannon equations.

We can see that at node C [for channel $(x;y)$] we obtain the transmitted information

$$T(x;y) = H(x) - H(x/y)$$ (1a)

in the channel. Similarly, at node y we get

$$T(x;y) = H(y) - H(y/x).$$ (1b)

On the other hand, at node O (for observer of the whole channel, i.e.,

---

[5] The $p(i)$s are the probabilities of the x_i in the input message x. The $p(j/i)$s are the conditional probabilities of having the symbol y_j in the output message y at the place corresponding to x_i in the input.

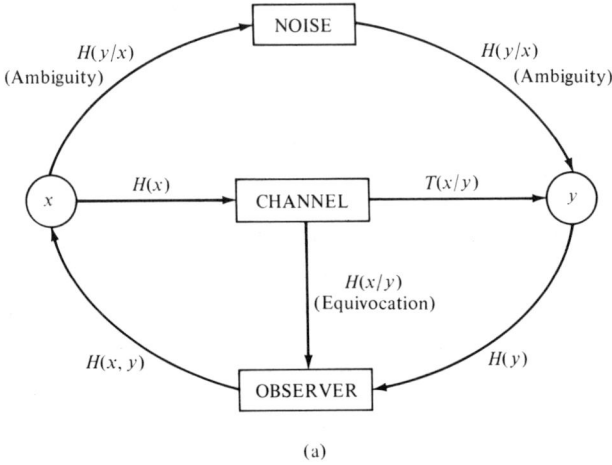

(a)

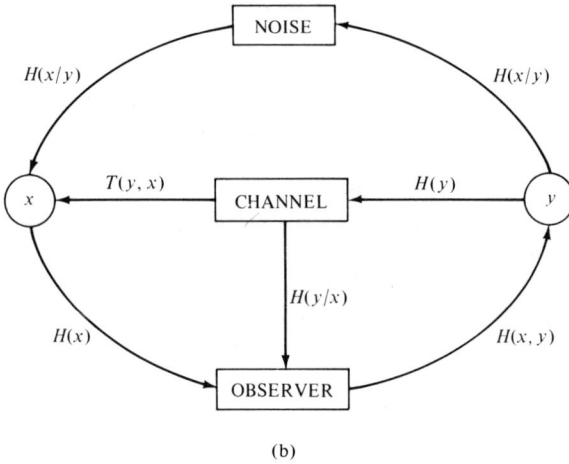

(b)

Figure 2. The Bourgeois graphic representation of the two measures of ambiguity (a) assuming conservation of uncertainty and (b) after interchange of input and output.

of x and y) we obtain

$$H(x,y) = H(y) + H(x/y), \qquad (2a)$$

and similarly at the node x

$$H(x,y) = H(x) + H(y/x). \qquad (2b)$$

If we interchange the variables by setting y as the input and x as the output, the graph is seen to have the same symmetry as the function T (see Figure 2b).

Apart from its value in clarifying these different kinds of ambiguity, the great advantage of the Bourgeois graphic representation is to underline the implicit role and position of the observer in the definition of the Shannon functions. The observer is defined by its knowledge of the plain and conditional probabilities that make up these functions. Thus the existence of a communication channel, with a flow of information in it, implies another flow of information toward the observer. Moreover, the feedback from the observer to the source x, necessary to close the graph and to provide the correct equations, means that the quantity conserved is the total uncertainty $H(x,y)$, which divides into various parts after reaching the source. Thus the real source of uncertainty, which feeds the source of the channel, is none other than the observer. This is another way of noting that the definitions of the information functions depend upon the operation of observing and the level of observation.

This kind of graph can be extended to communication networks with more than two vertices as in the work of W. J. McGill (1954).

12.4   Hierarchical Self-Organization

12.4.1 Complexity from Noise and Levels of Organization

The complexity-from-noise principle has been used as the basis for a theory of (self-) organization (see the Appendix). As such, it may prove to be useful in biology (Nagl) and the social (Morin) and psychological sciences (Ganguilhem, Serres). Our previous observation regarding the two different kinds of ambiguities is based on an analysis of differences in the position of the observer (see Section 12.3).

However, this distinction has a more profound meaning related to a hierarchical organization (Atlan 1975) with different levels of integration of the kind observed in living organisms (molecular, supramolecular, cellular, organ, and physiological systems).

The question of hierarchical organization is a very complex one and not yet solved from a logical and mathematical point of view. When we describe a dynamic system by means of a set of differential equations, the boundary conditions are in general imposed from a different, higher level of integration: For example, the cell, as a set of molecular species that undergo chemical transformations and transports of various kinds, may be described by such a dynamic system. Its boundary conditions will be defined by a state of the cell imposed by the supracellular organization constraints—either the in vitro cell culture conditions or the in vivo constraints of the organ.

Thus it is easy to understand, from a mathematical point of view, how a higher level of organization can determine the structure and behavior of the lower. However, the opposite is much more difficult to formulate, since it would imply a dynamic system, the boundary conditions of which should be determined by the differential equations themselves. A relatively close analogy to this situation may be found in the analysis of a vibrating sphere, in which the boundary conditions of a piece of the sphere are determined by the state of the adjacent piece and vice versa. Although this situation seems unusual and difficult to formalize from a classical mathematical point of view making use of system dynamics, it is the basis of the day-to-day physicochemical approach to biology, in which the functioning and structure of the organized system is assumed to be determined by the properties of its parts.

Now, we can see that the complexity-from-noise principle can contribute to an understanding of the logic of such hierarchical systems when we realize that the differences in the position of the observer mentioned above can also be viewed as differences in the level of organization of the system itself. In effect, when we consider a communication channel within a system, the output can be viewed as the functioning of the system itself (or at least part of it), inasmuch as the transmission of information in this channel will affect the system's structure and behavior. On the other hand, the global point of view of an observer standing at the output of the implicit channel from the system to the observer can be viewed as that pertinent to the higher level of organization for which the whole system acts as a subsystem, that is, as a constitutive part. Thus the observer for which the ambiguity bears a plus sign is not merely a logical postulate, but also the description of the effects of a lower level of organization on a higher one. As an example, noise in the communication channel between DNA and proteins in a cell has a negative effect when it is felt by the cell itself, in the form of false proteins with nonproper enzymatic properties different from those required by the present state of the cell metabolism. However, these false proteins may have new properties that would make them suitable for new adaptive reactions to a new environment. From the point of view of the organ or physiological apparatus, this same noise has the effects of creating variety and heterogeneity among cells, which allows them more adaptability. Therefore, up to a certain point, and providing the redundancy of the cell is large enough so that these false proteins are not going to impair the cell function, the same effects of the noise on the channel within the cell that are viewed as detrimental by the cell itself can be viewed as beneficial by the organ. Thus the change in the sign of the ambiguity may be understood as a consequence of the change in the level of the observation, which is itself related to a change in the level of a hierarchical organization.

The graphic representation of Bourgeois can help visualize this point. The simplified system of a nonspecific source x and a channel $(y_1;y_2)$ (Figure 3), dealt with in our previous work (1968), is represented in the graph of Figure 4.

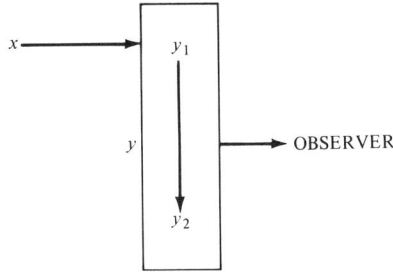

Figure 3. A simplified system consisting of a nonspecific source x and a channel $(y_1; y_2)$.

Our previous equation for the information content of this system

$$H(S) = H(y_1) - H(y_1/x) + H(y_2/y_1) \qquad (3)$$

is now obtained by summing up the flows at the node y_1, where we have

$$H(y_1,y_2) = T(x;y_1) + H(y_2/y_1)$$
$$= H(S). \qquad (4)$$

In other words, what was called $H(S)$ is nothing else than $H(y_1\ y_2)$ under the condition that the input $H(y_1)$ in $(y_1;y_2)$ is replaced by $T(x;y)$. This amounts to considering the input in the channel $(y_1;y_2)$ as the transmitted information in the previous channel $(x;y_1)$, that is to say, to take into account the ambiguity $H(y_1/x)$ due to the noise in this channel. Thus, because of the asymmetric role of x as a nonspecified source, the communication among x, y_1, and y_2 must not be seen as a communication network with three vertices. Rather, it is a link between two separate channels. These channels remain closed on themselves because they represent two separate hierarchical levels, which means also two different

Figure 4. The graphic representation of the system of Figure 3.

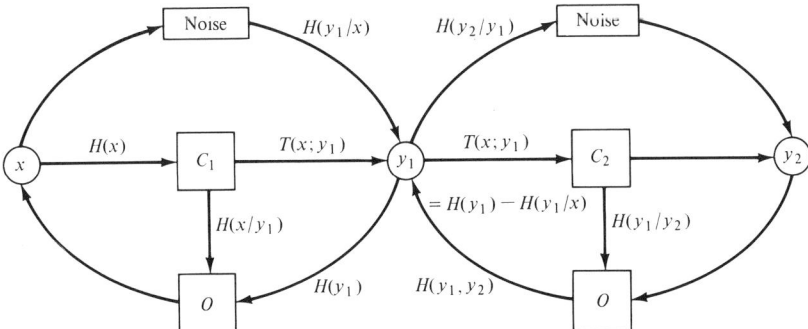

levels of observation. This is why the link between them does not cancel their individual closure by means of two separate "observers." This link takes place by means of a noiseless transmission of what was transmitted at the output of the first to the input of the second. (A noisy transmission between one level and the other would imply an additional decomposition of y_1 from C_1 and C_2 through an additional noisy channel between them.)

It is worth noting that Eq. (3) relies on the assumption that x is an unspecified source, unknown to the observer, whereas only y_1 and y_2 are accessible to observation. (A possible physical meaning of the unspecified source x can be the state of the system at a time $t - dt$ immediately preceding the time t of observation.) If it were not so, we would have to consider a channel $(x;Y)$, where Y stands for y_1 and y_2 (Figure 3) in a different, more classical way, making use of double conditional probabilities. The analysis would proceed as follows:

$$T(x;Y) = H(Y) - H(Y/x). \tag{5}$$

$H(Y) = H(y_1) + H(y_2/y_1)$ is substituted into $H(Y/x)$, which gives

$$H(Y/x) = H(y_1/x) + H(y_2/y_1,x).$$

By substituting $H(Y)$ and $H(Y/x)$ in (5), we get

$$T(x;Y) = H(y_1) + H(y_2/y_1) - H(y_1/x) - H(y_2/y_1,x). \tag{6}$$

Writing Eq. (3) instead of (6) means that we assume $H(y_2/y_1,x) = 0$. This assumption is in fact a consequence of the unspecified character of x. In effect, it means that the uncertainty in y_2 is zero if y_1 and x are specified. In other words, from the point of view of the system itself, there is no uncertainty at all as to y_2 when y_1 and its source x are given. The uncertainty in y_2 is only relative to the observer, as a consequence of the fact that the observer does not have access to x. Since the creation of additional uncertainty in y_2 by decreasing the observed constraints of the system (i.e., those between y_1 and y_2) still allows the system to be in a functioning state (complexity from noise), this new state must be admissible from the point of view of the real internal constraints governing the unknown organization of the system, that is, those between x (the unspecified, unknown source) and Y (which is observed).

12.4.2   Noise and Meaning

Transmission from One Level to the Other. One further step can be taken by considering the transmission of information from one level to another within a hierarchically organized system in terms of the information's meaning, which we shall define, in a restrictive way, as the efficiency of transmission.

We have already seen how the use of the complexity-from-noise principle implies a change in the sign of the effects of noise according to whether one is looking at the transmission within a single level or from one level to another.

Again, so far as the meaning is concerned, we must now distinguish between the transmitted information as it is seen by the system itself and what can be grasped from this by the observer. For the system itself, the transmission of information from one level to another must imply the existence of all that is needed for a channel of communication to function, namely, the three levels of Weaver, including that of the meaning of the information.

The Operational Definition of Meaning. Before going further, let us realize how the meaning of information can be understood for a natural (non-man-made and nonhuman) system, on an objective basis different from our introspective feeling or linguistic experience. We propose to define the observed meaning of information as its observed consequence on the receiver. In other words, we suggest unifying Weaver's levels B and C (semantics and efficiency), although we know that in our linguistic experience no such unity exists.

A good example of this is given by the functioning of the genetic code and the meaning of genetic information. On the one hand pure Shannon theory for the transmission of information restricted to the A level can be applied to analyze the protein–synthesis machinery as a noisy channel between the DNA nucleotides sequences (written in a four-letter alphabet coded in words of three letters) and the protein amino acid sequences (written in a 20-letter alphabet) (Atlan 1968, 1972). In such an analysis, the question of the meaning of the genetic information is disregarded when the only problem is that of the accuracy of the DNA expression into the protein structure. However, at the output of the channel this meaning manifests itself through the enzymatic activity of the proteins, and more generally through their function in the cell metabolism. Thus it appears that the meaning of the genetic information is to be found in the cell functions that the genetic message stimulates or inhibits when it is transmitted in the (DNA or protein) channel. Meaning as a cell-function efficiency may be viewed at different levels, that of the cell itself, for example, or that of the organism.

At the level of the species, the meaning of the genetic information may have to do with the selective value of the given species within a given environment. A precise definition of the selective value of information carriers has been proposed by Eigen (1971), who has studied a process of self-organization of matter by means of replications and interactions of such carriers (biomacromolecules). Here again, when these carriers are duplicated an optimum nonzero percentage of errors in the transmission of information is necessary for the process of increase in information content to take place. This percentage appears as one of the parameters by which the selective value is defined. The selective value of information carriers is a particular case of a meaning of information where the receiver for which the information is meaningful is the natural selection system. Thus the value of the information for this system (i.e., its meaning) is related to the existence of noise in the replication channels.

As imperfect as it is from a psychological, introspective point of view, this proposed operational definition may be applied to, and include, the human linguistic meaning as a particular case. Here, the observed systems are the human cognitive and acting systems. From this point of view, the meaning of the messages received by these systems will be defined as the correlated observed changes that they produce either on the internal state of the cognitive system or on the observed output (i.e., behavior) of the acting system, or both.

Meaning of Information Within a Hierarchical System. Now, we as observers cannot in fact know the meaning of information transmitted from one (lower) level to another (higher) more integrated one. However, this meaning is what makes the structure and function of the system itself at its higher level of integration. If we knew it we would be in a position to understand the hierarchical organization of the system in a specific and complete way and would not need this probabilistic nondeterministic approach. At the same time, the system itself at its higher level obviously "knows" the meaning of this information, since it receives and reacts to the information in an efficient way that is manifested in its mere existence as an organized functioning system. On the other hand, although we do not understand the organization on a deterministic basis, what appears to us outside as noise within the system appears to have beneficial effects on this organization insofar as the complexity-from-noise principle is applicable. As Piaget rightly noticed, the very fact that the system is able to utilize noise to improve its organization means that this noise is not noise for the system. For us, however, this noise cannot be anything else, since it appears to us to be made up of random perturbations with no logical articulations relative to the organization of the system as we can know it on a deterministic basis.

We are therefore led to suggest that what appears to the observer as an organizational noise acting in the channels connecting different hierarchical levels is, in fact, for the system itself the meaning of the information transmitted in these channels. It is because we are using a formalism (the Shannon theory) from which the concept of meaning is absent that its expression appears as paradoxical, as the information-from-noise principle. In other words, the positive effect of noise appears as the consequence of the negation of a negation: What is destroyed by the noise is information, but since it is information whose meaning we do not know, its negation can be used indirectly to appreciate the effects of the positive, meaningful (to the system) information that is still unknown (to the observer).

Thus it is our opinion that the complexity-from-noise principle may have far-reaching consequence: the possibility of widening the scope of the Shannon information theory without losing the benefit of its well-established rigor. This is achieved by taking seriously the role of the observer and applying a Shannon-like analysis to a noisy channel toward the observer. For this purpose, attempts like that of Bourgeois's graphic representation (see Section 12.3.3) may be fruitful.

12.4.3 Artificial Systems, Natural Systems, and Observer Systems

We have already mentioned the distinctions to be made between man-made, well-known artificial systems and imperfectly known natural systems. With regard to the latter, the existence of the observer and his position outside the system have appeared to be critical—a fact taken into account in definiting the concepts and quantities related to such a system.

For artificial systems, a choice is sometimes possible between deterministic and statistical methods: the position of an unacquainted external observer of the system may be adopted for operational reasons, because total knowledge, although accessible, is less convenient than a statistical approach on the basis of a partial knowledge, as if the system were natural. In the case of natural systems, though, there is no choice but to resort to statistical methods like those of thermodynamics and information theory.

However, distinction must also be made between observed natural systems and human natural systems, where the observer is inside the system as a part (of a social system) or as a whole [an individual as an "assimilating" system, with senses both of biological and of cognitive assimilation (Piaget 1974)]. When we apply the above analysis to human systems we take the position of someone who would observe them from outside. This is a very particular and somehow artificial point of view. From it we observe things as if we do not intend to know the meaning for ourselves of what we ourselves are living—either as organized individuals or as an element of the social system. However, this point of view is merely a consequence of the extension of the scientific method to the phenomena of our life, in the form of the postulate of objectivity. We can see how this postulate takes the shape of a prejudice that neglects an important, perhaps essential part of the information available to us. Here we reach the limits of this method transposed to human systems analysis: Although we have a direct access to the information that the system receives from itself, including its meaning, we neglect it because it is in the realm of what we call subjective knowledge.

As an example, it is worth analyzing briefly attempts to apply these ideas on organization to understanding psychic organization (Atlan 1973a,b; Ganguilhem 1975; Serres 1976). The theory of self-organization has been used to analyze the roles of memory and self-organization processes in the constitution of the psyche by learning and, more generally, by what Piaget calls assimilation—in both its biological and psychological sense. Some interesting conclusions on the nature of biological and psychological time can be drawn (Atlan 1973 a,b).

In its most general form, the complexity-from-noise principle has helped to provide a cybernetic understanding to the seemingly paradoxical death pulsion of Freud (Ganguilhem 1975). In its application to the problem of meaning in hierarchical organization, this principle has helped to suggest how what appears as noise and nonsense to the observer of the conscious level is in fact meaningful messages from the unconscious level (Serres 1976). However, it is at this point in the psychoanalytic approach that our remark must find its place: in effect the observer of the conscious and unconscious levels together is at once psychoanalyst and patient. The patient is rightly called the subject (Lacan 1966), being in the grammatical position of the one who is talking, that is, sending messages from the whole psychic system, while seemingly under objective observation by the psychoanalyst. This is why a very special status with respect to its scientific character has been attributed to modern psychoanalysis. While it has always aimed to be differentiated as a science from magic and religion, one is forced to differentiate it from science as well because of the special status it accords its object, which is that of subject! The epistemologist M. Foucault (1966) called it and ethnology ''antisciences'' for this reason. The psychoanalyst J. Lacan tried to express the problem by resorting to a primordial status of linguistic rules in the creation of the psychic organization as a whole. His first example of how these rules can be the basis of a self-generated symbolic reality appears to us with hindsight as another application of the complexity-from-noise principle: A random series of plus and minus signs can give rise to a set of symbols with very specific rules when it is seen at a different level of integration where the units are made of groups of signs.

Moreover, in social systems, the observer is not only an element of the system but also a metasystem: The system is contained by the observer inasmuch as the observer views it from outside. We have seen above that the transmission of information from one level to another implies a transmission of meaning (i.e., of the codes) from one level to the other. It is the lack of knowledge of these codes for the external observer that gives rise to the apparent paradox of the information-from-noise principle.

In the case of social systems, individuals are contained in the system and bear their own coded information to be transmitted to the higher, social level of organization and vice versa from the social system code to the individuals. However, all this holds only from the point of view of an objective information, that is, if we forget that they themselves are the observers. In fact their positions as observers transform their individual codes of meaning into something more general than the social code. The latter is contained in the former inasmuch as the observed is contained in the observation. This is a source of difficulty, together with a new wealth of organization, specific to the social systems. The individuals who make up the system use and process an information with a meaning that acts at the same time at two different extreme levels: the elementary level of the constitutive parts and the most general level of a metasystem containing the society (and even the universe!), namely, the observer-cognitive system.

The order-from-noise principle, in a vague qualitative form, has been applied to social systems as a principle of permanent reorganization after disorganization due to the tensions between the individuals and the society: the individuals are viewed as sources of random perturbations for the social organization.

However, the peculiarities of the social systems just mentioned make more subtle the analysis of the tensions between individuals and societies. These tensions, which can culminate in what is called a "crisis," may be understood as a bad transmission of the meaning of information within the system itself, between the individual and the society. That is to say, the individual and collective codes of significations are not the same any more.

Two mechanisms can be imagined by which such a crisis can be avoided without being solved, thus being transformed into a prolonged state of latent crisis. It is suggested that this state has been reached by the developed societies, which provide us with two (extreme) examples of these mechanisms. The first mechanism is to impose the meaning of the social code on the individual in such a way that the individuals' codes will be repressed, as it is done in all totalitarian societies. The second possible mechanism is the opposite: to project the meaning of the individual codes and impose it on the social reality. This latter is done in the so-called consumption societies, where everyone wants to think that the only goal of the social organization is the satisfaction of individual wishes. In fact, the social organization resists somehow, at least because of the contradictions between the individual needs and desires. However, the individual's situation of being at the same time contained and containing allows him to make such a projection. Similarly, in the totalitarian society, this situation allows the social code to be more or less "introjected" in the form of an ideology, which "convinces" the individuals—willingly or unwillingly.

In both cases this peculiar situation allows one code to be more or less repressed by another and maintains the system in this state of avoided but nonsolved crisis. There is still bad communication between the code of the individuals and that of the social system. According to our development above, this amounts to a functioning opposite to the complexity from noise. That is to say, this state amounts to a decrease in complexity, possibly corresponding to an increase in redundancy. Interestingly enough, in both symmetrical extreme cases, we can notice an increase in redundancy in the form of a tendency to an uniformization of the individual wishes in what is called today the "masses." Usually this uniformization is related to that of the mass media as the overwhelming means for social "communications." However, might it be that the development of these media at the expense of other means of communication—possibly more meaningful from an internal, individual life point of view—was necessary to avoid the explosion of societies in crisis?

Appendix
The Formal Theory of (Self-)Organization in Living Systems

The possibility for noise to reduce the redundancy of a system is the basis for a theory of organization able to account for (1) the opposite intuitive features of organization, namely, redundancy and variety, (2) organization as both a state and a dynamic process, and (3) the possibility of self-organizing properties and the conditions for their appearance (Atlan 1968, 1972, 1974).

The various proposed definitions of organization found in the literature follow two major trends, which contradict one another (Dancoff and Quastler 1953; Linschitz 1953; von Foerster 1960; Rothstein 1962; von Neumann 1966; Ashby 1967; Theodoridis and Stark 1969, 1971; Eigen 1971; Mairlot and Dubois 1974). By organization is meant either constraint between parts (i.e., redundancy) or nonrepetitive order (i.e., variety and inhomogeneity, as in Shannon's information content). In fact, these two intuitions correspond to two extreme views, according to which a model for the best-organized system would be either a perfect crystal or an apparently random but functioning mass of elements. Interestingly enough, neither of these extreme views fits completely our intuitive notion of what organization is.

Compromise Between Complexity and Redundancy:
The Library Metaphor

When we look at real natural organized systems we are dealing in fact with some intermediate situation, an example of which is provided by a cultural system contained in a library. The culture exists as an intermediate situation between complete independence of the books from each other (no constraint) and mere repetition (maximum redundancy). The latter case would reduce the culture to the content of our one single book. The former would not be compatible with the existence of a culture either, since each of the books would be an isolated system. The cultural system is made of what is common to all the books in the form of quotations, references, allusions, deductions, inferences, and so on. All this means some degree of communication between the books, although not perfect (i.e., with some ambiguity), which prevents them from being mere repetition.

Thus optimum organization must be viewed as a compromise between maximum information content (i.e., maximum variety) and maximum redundancy (Atlan 1968, 1974).

In a review on the recent experimental findings about the DNA organization in the nucleus of living cells, W. Nagl observed that this definition "fits the organization of the cell nucleus surprisingly well. Its structure and function are based on both variety (unique units of genetic information) and repetitiveness (reiterated units that are noninformative in a genetic sense)."

Mathematical Expressions of the Organization

A simple way to express the organization of a system as a dynamic process is to look at it as a time course of change in its information content H. In addition, the function H is split into a maximum information content and a redundancy function according to the classical definition of the redundancy:

$$H = H_{max}(1 - R).$$

In this equation H is a measure of the information content of a system with internal constraints between parts measured by the redundancy R. H_{max} is the maximum potential heterogeneity computed by not taking into account the constraints—that is, by assuming complete independence of the parts. Thus H, being an increasing function of H_{max} and a decreasing one of R, appears to be the required compromise between them. Differentiating H versus time (again with the assumption that time means accumulated effects of noise-producing factors), we get

$$\frac{dH}{dt} = (1 - R)\frac{d\,H_{max}}{dt} + H_{max}\left(-\frac{dR}{dt}\right). \tag{7}$$

(This assumption ignores the evolution in time determined by a sequence of instructions in a program imposed by the designer and builder of the system from the outside. In a programmed system, the organization is that of the program and its measure may be given by the computer time necessary to have it run, or by the number of steps in a Turing machine as mentioned above for the estimate of the complication of an artificial man-made known system.)

The two terms on the right-hand side of (7) can be identified with the two effects of noise previously described. The first term has the meaning of a destructive ambiguity that destroys H_{max} (i.e., the total information transmitted from the system to the observer, counted without taking into account the constraints). It is the classical disorganizing effect of noise. Because of decrease in the constraints, dR/dt is negative, and the second term has the meaning of an autonomy-producing ambiguity that, as explained above, produces a decrease in the redundancy of the system. In other words, as the accumulation of errors acts by decreasing both H_{max} and R, the first term is negative and the second positive.

Now, dR/dt and dH_{max} are themselves two different functions of time that express the kinetics of the effects of noise on the system. It has been proposed that these two functions, together with equation (7), express the overall organization of the system, both structural and functional.

A particular case, limited to a channel between two substructures $(y_1;y_2)$ of a system, and a nonspecified source x emitting to the observer, was treated in detail. The following equation was proposed (Atlan 1968), where the two different ambiguities act together on the information content $H(S)$ of the system:

$$H(S) = H(y_1) - H(x/y_1) + H(y_2/y_1). \tag{8}$$

Its time derivative can be expressed by means of the generalized Yockey's equation (Atlan 1968) and integrated. Under certain conditions H in equations in (7) and (8) can be increasing in time.

Self-Organization by Decrease in Redundancy

We proposed to call self-organization a process where the change in organization with increased efficiency is not directed by a program but occurs under the effects of random environmental factors. A condition for self-organization (i.e., increasing H in time, at least up to a certain time) amounts to a minimum initial redundancy to start with.

In effect, according to this view a self-organizing system is a system redundant enough and functioning in such a way that it can sustain a decrease in redundancy under the effects of error-producing factors, without ceasing to function. This decrease in redundancy leads to an increase in information content or variety, which allows for more possibilities in regulatory performances as shown by Ashby (1958). In other words, self-organization appears as a continuous disorganization constantly followed by reorganization with more complexity and less redundancy. Additional mechanisms can be imagined at various levels of organization by which a system can regain redundancy in order to be able to evolve again. There is little doubt that DNA redundancy (in the form of additional chromosomes or else) must have played a central role in the species-evolutionary processes. Our cognitive system is another example of self-organizing systems, and here the paradoxical sleep and dream have been proposed as such hypothetical mechanisms for regaining redundancy (Atlan 1973a,b).

References

Ashby, W. R. (1958), Requisite variety and its implications for the control of complex systems, *Cybernetica* I (No. 2), 83–89.

Ashby, W. R. (1962), Principles of the self-organizing system, in *Principles of Self-Organization* (H. von Foerster and G. W. Zopf, eds.), Pergamon, Oxford.

Ashby, W. R. (1967), The place of the brain in the natural world, *Curr. Mod. Biol.* 1, 95–104.

Atlan, H. (1968), Applications of information theory to the study of the stimulating effects of ionizing radiation, thermal energy and other environmental factors: Preliminary ideas for a theory of organization, *J. Theoret. Biol.* 21, 45–70.

Atlan, H. (1972), *L'Organisation Biologique et la Theorie de l'Information*, Hermann, Paris.

Atlan, H. (1973a), *Conscience et desirs dans des systemes autoorganisateurs, in L'Unité de l'Homme* (Morin and Piattelli-Palmerini, eds.), Le Seuil, Paris.

Atlan, H. (1973b), Le principe d'ordre a partir de bruit, l'apprentissage non dirigé et le reve, in *L'Unite de l'Homme* (Morin and Piattelli-Palmerini, eds.), Le Seuil, Paris.

Atlan, H. (1974), On a formal theory of organization, *J. Theoret. Biol.* 45, 295–304.

Atlan, H. (1975), Organisation en niveaux hierarchiques et information dans les systemes vivants: Reflexions sur de nouvelles approaches, in *l'Etude des Systemes,* CNSTA, Paris, pp. 218–238.

Bourgeois, M. (1977), personal communication

Dancoff, S. M., and Quastler, H. (1953), The information content and error rate of living things, in *Information Theory in Biology* (H. Quastler, ed.), Univ. of Illinois Press, Urbana.

Eigen, M. (1971), Self-organization of matter and the evolution of biological macromolecules, *Naturwissenschaften* 58, 465–523.

Foerster, H. von (1960), On self-organizing systems and their environments, in *Self-Organizing Systems* (M. C. Yovitz and S. Cameron, eds.), Pergamon, New York.

Foucault, M. (1966), *Les Mots et les Choses,* Gallimard, Paris.

Ganguilhem, M. (1975), Vie article in *Encyclopedia Universalis,* Paris.

Gatlin, L. L. (1966), The information content of DNA(I), *J. Theoret. Biol.* 10, 281–300.

Lacan, J. (1966), *Ecrits* (esp., "Seminaire sur la Lettre Volée"), Le Seuil, Paris.

Linschitz, H. (1953), The information content of a bacterial cell, in *Information Theory in Biology* (H. Quastler, ed.), Univ. of Illinois Press, Urbana.

Mairlot, F. E., and Dubois, D. M. (1974), Basic criteria in cybernetics: communication and organization: advances in cybernetics and systems research, *Proc. Europ. Mtg, Vienna,* Transcripta Books.

McGill, W. J. (1954), Multivariate Information Transmission, *Psychometrika* 19, 97–116.

Morin, E. (1977), *La Methode,* Le Seuil, Paris.

Morin, E. (1973), *Le Paradigme Perdu: la Nature Humaine,* Le Seuil, Paris.

Nagl, W. (1976), Nuclear organization, *Ann. Rev. Plant Physiol.* 27, 39–69.

Neumann, J. von (1966), *Theory of Self Reproducing Automata* (A. W. Burks, ed.), Univ. of Illinois Press, Urbana.

Piaget, J. (1974), *Adaptation Vitale et Psychologie de l'Intelligence,* Hermann, Paris.

Rothstein, J. (1962), Information and organization as the language of the operational viewpoint, *Philosophy of Science* 29 (No. 4), 406–411.

Serres, M. (1976), Le point de vue de la biophysique, in Critique, Paris, *La Psychoanalyse vue de dehors* (special issue), March, 265–277.

Theodoridis G. C., and Stark, L. (1969), Information as a quantitative criterion of biospheric evolution, *Nature* 224, 860–863.

Theodoridis G. C., and Stark, L. (1971), Central role of solar information flow in pre-genetic evolution, *J. Theoret. Biol.* 31, 377–388.

Weaver, W., and Shannon, C. E. (1949), *The Mathematical Theory of Communication,* Univ. of Illinois Press, Urbana.

Locker's writing is complex, flowery, and often singularly poetic. The reader is invited to read Locker's last paragraph in order to appreciate the editor's task. Yet, I have interfered only minimally because Professor Locker's underlying ideas and reasoning are highly original and a price must be paid for understanding them—by the reader and the author equally.

Locker turns our attention to autopoiesis again. Some of his conclusions have been characterized as "brave," or at least "boldly asserted." It will become obvious to the reader that during the process of writing and rewriting of papers, Locker and Glanville have become intellectually and mutually self-respecting unities.

Essentially, Locker grapples with the problem of the origin of autopoietic systems—through partially vindicating Virchow's *omnis cellula e cellula* with his *omne systema e systema*. Another variation on this ploy (in English) would be Weiss's earlier "System begets system" and even its precursor, *omnis organisatio ex organisatione*. Life cannot originate "by itself."

Locker reinstates the role of a "program," conceived as a creative–generative system comparable to the subject, or "self." The subject has to precede, as a creative system designer, any occurrence of "origin." Locker's attack on scientism, and his inclusion of autopoiesis within its realm, represents a metaview of considerable import. In more simple terms, autopoiesis is only a result of human reflection, a construct of the human mind, a fruit of a self-referential comprehension of ourselves.

It is interesting to follow how Locker "takes on" Eigen's hypercycles, a theory that is becoming central to Jantsch and respected by both Atlan and Varela. Locker insists that *nothing* arises "by itself" and that the objective origin of anything is not only impossible but unthinkable: the cognizing subject is always indispensable for an origin to be recognized, and the creative subject (God?) is always indispensable for an origin to be accomplished.

As Locker refers to the object of his attacks as "scientism," without specifying any of its various meanings, the reader might find it useful to recall its most common definition: an exaggerated trust in the efficacy of the methods of natural science to explain social or psychological phenomena, to solve pressing human problems, or to provide a comprehensive unified picture of the meaning of the cosmos.

The following should be pondered: Locker has used the established methods of natural science (logical reasoning, mathematics, empirical argumentation, and referencing other scientists) to reach rather unitary and systemic conclusions on the questions of cognition, self-reference, consciousness, cosmogeny, language, and origin. If we accept that there *is* something outside the picture derived by the methods of science and scientism, as for example Locker's logically derived implication that "nothing arises by itself," are we then justified in entertaining the opposite idea, that of spontaneous generation?

Alfred Locker was born in Vienna in 1922. He received his doctorate in biophysics from the University of Vienna in 1949. His research appointments include Research Laboratory of the First Medical Clinic and Antibiotics Research Unit at the University of Vienna (1949–1960), Unit of Physiology and Biophysics and Unit of Medical and Biological Radioprotectivity at the Institute of Biology, Austrian Reactor Center (1960–1969); in 1965 he became associated with the Institute of Theoretical Physics at the Technical University of Vienna where he is currently a Professor and Head of the Department of Theoretical Biophysics. Dr. Locker published over 150 scientific papers and edited *Quantitative Biology of Metabolism, Biogenesis–Evolution–Homeostasis,* and *Radioprotection.* He recently published a textbook on *Theoretical Cybernetics. Address:* Institut für Theoretische Physik I, Technische Universität Wien, Karlsplatz 13, A-1040 Wien, Austria.

Chapter 13
Metatheoretical Presuppositions for Autopoiesis
Self-Reference and "Autopoiesis"

Alfred Locker

13.1 Introduction

Autopoietic organization has been defined (Varela et al. 1974) as

unity by a network of productions which (1) participate recursively in the same network of productions of components which produced these components, and (2) realize the network of productions as a unity in space in which the components exist.

Additional defining remarks by the same authors specify "that the realization of an autopoietic organization is the product of its operation," thus separating this kind of organization from the allopoietic one that "characterizes systems in which the product of their operation is different from themselves."

In order to formulate the problem under consideration in relation to autopoiesis, we have to ask, (1) What is autopoiesis? In conformity with the definitions quoted above, we may find, as a preliminary answer, that it is both a process and a state; since the process apparently is characterized through its equality to its own product, the statement can possibly be translated into "equality (or complementarity) of process and state." When we continue to ask, (2) How does autopoiesis come into existence?, we raise a question about a process that can again be regarded as equal to autopoiesis itself. It could be maintained that in addition to the operation within the system—the metaoperation also leading to the occurrence of autopoiesis as a kind of principle out of which autopoietic systems arise—is again equal to autopoiesis. Thus, the recursion invoked in the definition does not only refer to autopoiesis from "inside."

When we ask about the presuppositions required for an autopoietic system to come into existence, we are first confronted with purely scientific presuppositions that do not properly deserve their name, such as theoretical mechanisms that (according to some scientific hypotheses) should have brought about the system that exhibits an autopoietic character. This character is explained in familiar terms.

The moment we realize that the scientific answers with which we have contented ourselves cannot be considered the ultimate ones, we can begin questioning with interest the presuppositions for science itself. For our particular problem the reasons we have to challenge the assumptions generally made in science are the following:

1. The definition of autopoietic systems deals with recursiveness (i.e., a part of the theory of computation).
2. There is no doubt that autopoiesis, as a kind of continuous self (re)production of the system, must be brought into connection with self-reference and hence with ourselves. Thinking about ourselves is the task of philosophy.

When hidden assumptions are made explicit something very astounding may be revealed: apparently the relationship of autopoiesis to self-reference and to consciousness does not occur because of the "emergence" of consciousness due to the evolution and increase of complexity of autopoietic systems (e.g., the brain)—construed even as an epiphenomenon (Varela 1971)—but rather for an opposite reason, namely, that the self-evident comprehension of ourselves has to precede the contrivance of autopoietic systems. What seems to be objectively given emerges as the result of the projection of knowledge of our own properties, as self-referential conscious subjects, onto an object that happens to be, in the very nature of the organism, a subject as well.

Admittedly, there are three available methodologies for tackling the problem of what presuppositions are required for the conception of autopoietic systems: (1) the scientific, (2) the systems theoretical, and (3) the metatheoretical (Bense 1960). Our main concern here is the question of to what extent any result that considers self-reference and the subject nature of systems contributes to an appropriate understanding of autopoiesis. The investigation we pursue is, in parts, polemical and we shall paraphrase the answer several times.

13.2 The Inadequacies of a Scientific Approach to Autopoiesis

Autopoiesis presents a new conception in that, rather than being a mechanistic approach, it confronts an intuitively understood fundamental property of the organism, that is, its existence as a unity or a "whole." However, appropriate recognition of the significance of this new apprehension of a known item necessitates the avoidance of seductive conceptual schemes; in recent years the most coercive one seems to be the evolutionist empiricist scheme, the acceptance of which, after the advent of the cybernetic paradigm, would indeed lead us into an inexcusable relapse. This without doubt would be the case if we were to consider the evolutionary scheme, especially in the most suggestive form as given by Eigen (1971), as an example of the only valid line of treatment. In Eigen's theory, because of the encounter of the two main classes of biomolecules, nucleic acids and proteins, life originated as a pure chance (or probability) event; because of the mutual profit by the features of the constituent biomolecules, the resulting primitive organism, called "hypercycle," exhibits a set of properties indicative of life. Due to competition among several specimens of hypercycles as against the challenge of selection, the organisms are subject to evolution.

Despite the fascination this theory now arouses in public we have to face some difficulties:

(1) Autopoietic organization is *recursive*. A predicament for evolution theory is that recursive functions should become operative in nature via the mechanisms evolution theory propounds. Since it makes use of the concepts of "chance" (i.e., mutations) and "necessity" [i.e., natural laws restraining the outcome of mutations (Monod 1975)], the "emergence" of recursive relations (in time) is unthinkable unless one considers the togetherness (and mutual conditionality) of "chance" and "necessity" itself as an expression of the (atemporal) existence of recursiveness (and circularity). Indeed, astonishingly simple computer algorithms, such as those for the "Garden of Eden" simulation of the evolution of life (Gardner 1971), are based on recursiveness. Thus, the objective occurrence of recursiveness in nature seems to be one of the presuppositions that exceed scientific contexts and of which science generally is not aware. But how could we take notice of recursiveness (or even invent recursiveness as a mathematical theme) if we were not subjects with self-reference?

(2) We have to examine the *language* in which a scientific theory in general and the theory of evolution in particular is being formulated. We find immediately that evolution theory is developed within the context of a language that allows the expression of an "emergence theory" (*E*-theory). Within the confines of this theory one thinks of observing and describing something objectively existing that allegedly arose de novo, but one does not ask about the presuppositions that make this description possible. These presuppositions can be distinctly expressed only if another language (level) is assumed and it is shown that the language in which the theory has formerly been uttered—and the contents of the theory—is valid only relatively. This elucidation is made possible by the language of a "transition theory" (*T*-theory) (Rosen 1973), which naturally comprises *E*-theory. Whereas in *E*-theory some events appear as randomly occurring and thus not predictable, just the opposite holds true for the same event described in *T*-theory. Here, according to laws that have to be sought out and formulated according to the requisite boundary

conditions, the event appears as determined and predictable. However, the laws to be assumed here are also the laws that characterize the subject's organization for obtaining knowledge. Since we have to distinguish here between immediate description on the one hand and consideration of the possibilities of this description on the other, we are inevitably dealing with the cognizing subject.

(3) Highly indicative of this involvement of the subject in the outcome of cognition is the occurrence of *complementarities* (or dual statements). In evolution theory the complementarity between chance and necessity, or between "additive" and "subtractive" processes—in line with the "arched structure" model (Cairns-Smith and Walker 1974)—is overlooked, although any single event is the result of both of them.

Having taken the difficulties of comprehending evolution within the usual theoretical framework seriously, we have to see in them strong hints at the necessity of asserting the precedence of the cognizing subject, that is, the observer and theory builder. Ordinary science, however, tends deliberately towards hiding this precedence. Thus, in criticizing the inadequacies of the scientific approach we affirm the centrality of the subject.

13.3 The Difficulties for a Systems Theoretical Approach to Autopoiesis

13.3.1 On the Systematization of Instrumental Systems

Taking autopoiesis as akin to self-(re)production we have to deal with two problems:

1. What are the possibilities for a system to produce something else (e.g., another system), and finally itself?
2. Which systems theoretical presuppositions are required for this kind of activity?

We understand as systems theoretical presuppositions those suppositions that are antecedent to the acquisition of knowledge about the system's activity as well as to the activity itself, provided the latter can be separated from the former. We attempt now a systematization on the basis of the system's ability to produce something else (as an allopoietic or *aP*-system), or to produce itself (as an autopoietic or *AP*-system).

The criterion for this systematization is the system's function as an *instrument* for some task (or purpose), which recognizes that the system's instrumental character cannot be disconnected from the notion of purpose. We preliminarily limit our consideration of *instrumental* or *I*-systems to the observable and describable body-machine (*b*-machine, comparable to "hardware"); but it will be shown that the mind-machine (*m*-machine, comparable to "software" or program, the latter being called *p*-system) will increasingly gain importance. Our array is the following (Figure 1a–f):

(a) About the *executing* or *E*-system we know nothing more than that it simply "works," that is, executes the instruction given by a *p*-system, although the latter is not explicitly beheld. Therefore, any discussion of purpose cannot be contemplated, and consideration is limited to "what" the system is doing. The dependence on the constraint of the view an observer might choose allows the statement that any arbitrary system may suit the character of an *E*-system. Here a distinction between a producing system and a product system (or process and state) does not make sense.

(b) The *making* or *M*-system is a system that, according to the aspect of the observer, enables one already to obtain a vague specification for the system. In addition it may be stated that the product of the *M*-system does not exceed certain features (e.g., complexity) of the *M*-system itself.

(c) The *producing* or *P*-system has a clearly specified function, that of producing a product. The latter's complexity may be less than, equal to, or greater than the complexity of the *P*-system itself. It is evident that the *P*-system obeys the instructions given by the pertinent program, that is, the *p(P)*-system.

(d) The *reproducing* or *R*-system represents a subclass of the *P*-system system; it "aims" at producing a series of copies of itself, each of which exactly resembles the original without being, of course, identical to it. The products are not connected, and the production goes on in one direction only. In order to understand the activity of the *R*-system, its pertinent *p(R)*-system has to be taken into account.

(e) The *self-reproduction* or *sR*-system represents an *improper* autopoietic system. It appears as a subclass of the *R*-system in that its outcome should, abstractly speaking, be (complementarily or dually) identical to itself[1]; its product, by being connected with its producer, can reversibly (and recursively) assume the role of the producer. Here we are confronted, not with a potentially infinite series of quasi-identical systems, but with a totally new kind of system; it is not characterized through a one-directional relation between constituents that are linearly threaded, but rather through an intrinsic circularity, forming a unity. If we regard the circular relation underlying an *sR*-system abstractly, then it can be called a self-referential relation, the significance of which will be examined.

---

[1] This argument also applies to the definition that the *sR*-system (well understood as improper "autopoietic" system) "emerges, given a domain of processes of production of components, when and only when such processes concatenate in a recursive fashion" (F. Varela, personal communication).

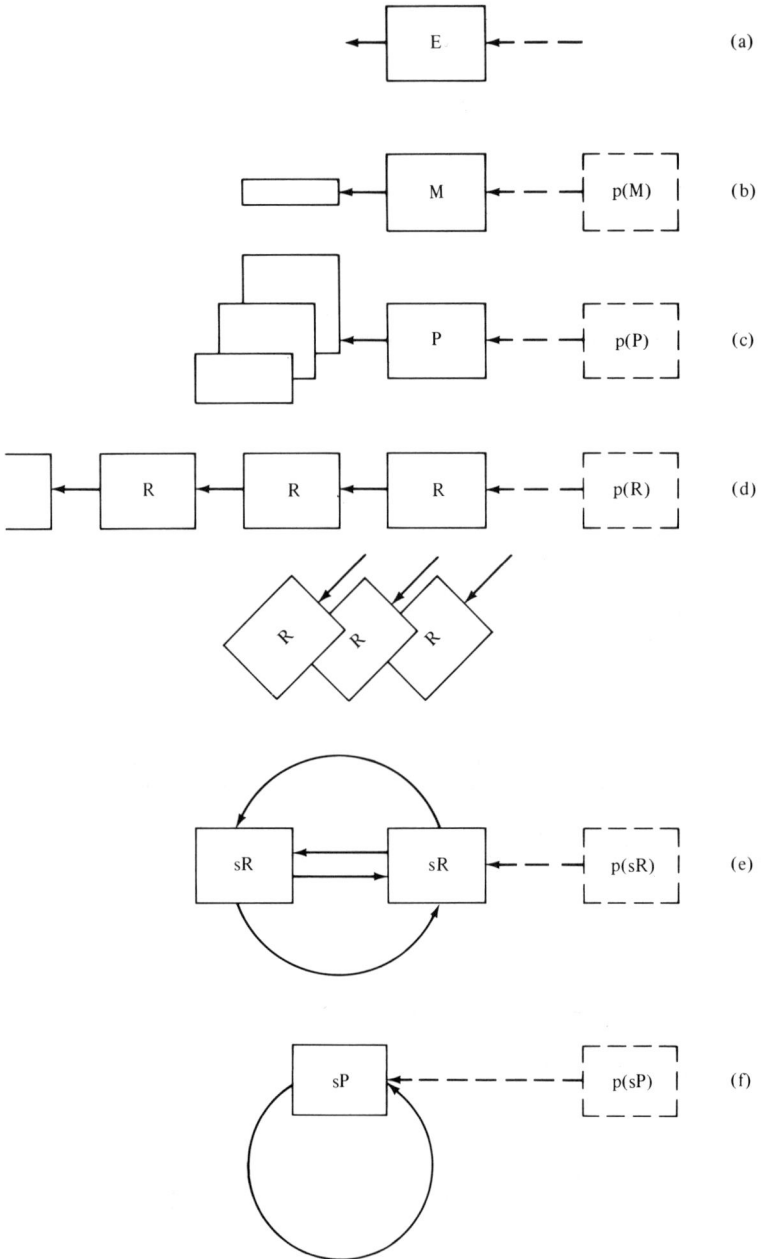

Figure 1. Display of *I*-systems (i.e., *E*-, *M*-, *P*-, *R*-, *sR*-, and *sP*- (≡*AP*)-systems) (center) with their pertinent programs, or *p(I)*-systems (right) and products (left). (d) The distinction between complete reproduction (each product shares the capability of reproduction and continues the series) and incomplete reproduction (the products are sterile). (See text for further discussion of parts a–f.)

(f) The autopoietic or *AP*-system *proper* differs from the foregoing system in that it *self-produces* (therefore also called *sP*-system) in the most astonishing way, (1) by letting the producer's system product become virtually identical with the producer system and (2) by connecting the two in such a peculiar way that they become a new, *"in itself" identical* and autonomous unity. By doing so, the *AP*-system proper (called *AP*-system for short) surpasses even the *sR*-system in a qualitative way; the latter is still dually organized, while the *AP*-system is unitarily built up. No difference between the producer and the produced system exists any more; they become (or better, they are) immediately one with the other at the moment the *AP*-system arises.

The systematization of *I*-systems serves a better understanding of the *AP*-system. In the *AP*-system a concentration (or better, *centration*) of systems characters becomes obvious. The series from the *E*-towards the *AP*-system shows an increasing subjectivization and, more and more, a "becoming itself"; in the *AP*-system's identity, its own connectedness "in itself" is incorporated.

The growing role of the program (i.e., the *p*-system) and the problems of cognition are studied next.

13.3.2 A Systems Theoretical Approach to the Problem of Cognition

The systematization of *I*-systems was carried out without considering the *cognitive domain* (CD); that is, the closed domain in which a system is located and the location of which is taken into cognizance by an observer (*O*-system) equally belonging to the CD (Figure 2). It is conceivable to look at the CD quasi-objectively, as if, in speaking about the CD, we were exempt from an entanglement in it. Then we may perceive how the *O*-system describes, in the CD, the system (here called *S*-system). But we may in addition surmise that the *O*-system is bound to make a hypothesis on how the *S*-system came into existence and to assume that it did so through another system in just the way explained for *I*-systems. For a natural system (i.e., the organism), the whole physical universe is usually accepted as such a system, making entry into existence of the system under observation possible; then the mechanism of evolution is proffering itself as the execution of that system's activity, leading to the system under specification. In order to distinguish in the CD between the *S*-system (proper) and the system that brought the former (instrumentally) into existence, we call the latter the designing system, designer, or *D*-system.

Within the CD the *S*-system can be assessed by the *O*-system only in the form of a *model*. The notion of model is always ambiguous since for one and the same bulk of data (observable about the *S*-system) the *O*-system can formulate quite different descriptions. Depending, however,

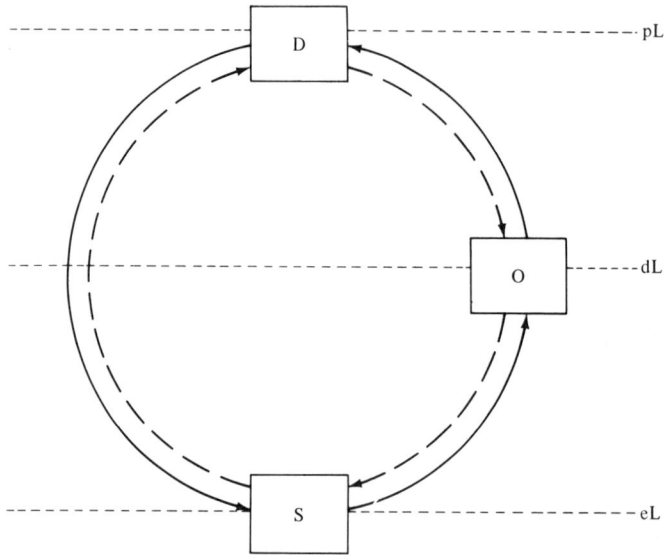

Figure 2. Mutual interrelationships between the *D*-, *O*-, and *S*-systems in the CD_{obj}; the language levels (*pL*, prescription language for program formulation; *dL*, descriptive language for model representation of the *S*-system through the *O*-system; *eL*, executing language of the *S*-system) have not been mentioned in the text.

on whether the *S*-system is viewed quasi-objectively (by us) or subjectively (by the *O*-system in the CD, taking the *S*-system's model character explicitly into account), we have to refine the notion of the CD and to distinguish between (1) the CD_{obj}, consisting of relations between the *O*-system, the *D*-system, and the *S*-system, and of having the other observer (i.e., us) only available as a "hidden parameter"; and (2) the CD_{subj}, consisting of the relations between the *O*-system, the *S*-system, and its model and of having the *D*-system (i.e., "objective" reality as designer for both the *O*-system and the *S*-system) again available as a "hidden parameter." The CD can thus be interpreted as the environment (surroundings) of each of the three systems, which for each of these three systems (because of the complete connectivity between them and closure of the CD) perfectly coincides.

The CD of the *AP*-system poses a different problem. Because of its autonomy the *AP*-system does not share the CD with any other system; the *D*-system, the *O*-system, and the *S*-system fuse within it. As a consequence of this feature the *AP*-system is essentially unrecognizable from outside (e.g., not simply identifiable as an *I*-system) by any *O*-system that wants to pin down the *AP*-system as the *S*-system of the *O*-system's *own* CD. In order to avoid speaking about purely fanciful products or figments, the observer making propositions about an *AP*-system must empathically (or intuitively) share the *AP*-system's properties on the ground of self-understanding, and then project this understanding into (or onto) the system. Other possibilities are not available unless the observer wants to destroy the system in order to explore it and thereby degrade it into systems of lower classes.

13.3.3 On the Significance of Program and Purpose

During the systematization of *I*-systems, the underlying programs (*m*-machines or $p(I)$-systems) became increasingly significant. Although it may be an oversimplification, we equate the notion of program with the notion of purpose and therefore apply some ideas proposed by Pask (1970) (Figure 3). In the CD the *D*-system (i.e., designer), before doing its job, has to formulate a program for its own activity, and this program pre-destinates the *S*-system's production (i.e., the program is the program "for" the *S*-system, in short, *for*-program). When the production of the *S*-system has been performed, the latter possesses the program (as its own $p(S)$-system *in* itself. The third partner in the CD, the *O*-system, must hypothetically infer from the *S*-system's behavior the program underlying

Figure 3. Role of programs in the CD. The $p(S)$-"for"-program, formulated in *pL* by the *D*-system, becomes the $p(S)$-"in"-program of the *S*-system, executed in *eL*. The *O*-system, in order to gain insight into the *S*-system, formulates the latter's $p(S)$-"of"-system in *dL*; the $p(D)$-"of"-program is tentatively formulated by the *O*-system in order to gain access to the *D*-system.

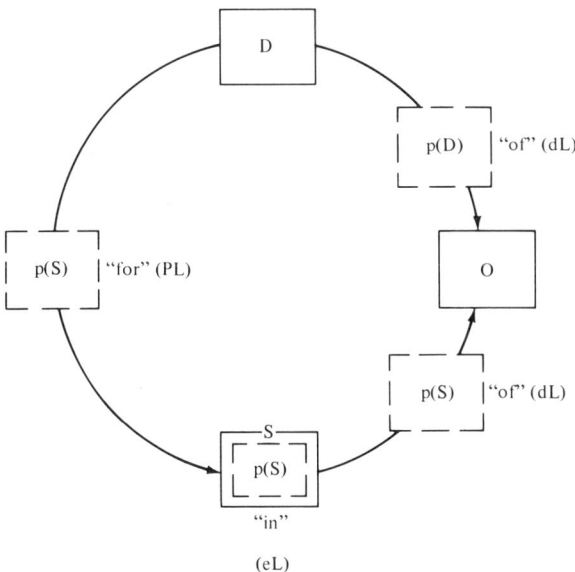

this behavior, which the *O*-system then interprets *post festum* as the program *of* the *S*-system. Thus, the of-program becomes the model of the in-program.

In any *p*-system two major features may be discerned:

(1) A program of an *I*-system (i.e., a *p(I)*-system) can be *realized,* or made an *m*-machine for a *b*-machine, in different ways; several *I*-systems realize one *p(I)*-system (Rosen 1966). Therefore, an individual *I*-system's program exhibits less "content" than the whole *p(I)*-system set. The *p(I)*-system in turn is more complex than the *I*-system, in the sense that it not only represents the in-program, but also (and prerequisitely) the for-program. It contains the instructions for each individual *I*-system, that is, how it may be built (in one of the ways *M*-, *P*-, or *R*-systems indicate), and how it may work (as an *E*-system). In the for-program, the *p(I)*-system comprises the *m*-machine and the *b*-machine, as well.

(2) The *p*-system undergoes a *conceptual movement.* When being formulated ("produced") as a program by a *p[p(I)]*-system, it is first part of that system (as in-program in the *p[p(I)]*-system); then it is conceptually moved (and made operative via the CD to its product system. Furthermore, when the latter now plays the role of the *D*-system, the *p*-system becomes part of that system's program as *p(D)*-system; from there it finally moves to the *S*-system produced (becoming its *p(S)*-system).

Here we note that the connectivity of the CD, predominantly considered with respect to *p*-systems and pertinent *b*-machines, is conterminous with actional connectivity. Anticipating metatheoretical considerations that we will make extensively later, we assert that both connectivities are relational expressions for the "being," that is, the constant basis subsisting cognition and action.

The significance of *programs for models* may be outlined as follows:

1. The model maker who simply conceives of a model of the *S*-system, without taking simultaneously into account the *S*-system's dependence on its own *D*-system, assumes (subjectively) the role of the *D*-system. Inferences are drawn from the data obtained through observation of the *S*-system as to the underlying *p*-system.
2. As soon as this assumed role (that of the *D*-system) becomes obvious, the model maker realizes that any system that conceives of the *p*-system has necessarily to slip into the nature of a subject.
3. The model maker may then also suppose that, like the model ascribed to the *S*-system or designed instead of the *S*-system, a system can come into existence only through a *D*-system (now believed to exist objectively, although still representing a subject; Figure 4).

From the standpoint of systems theory it is necessary to ask how a system that is able to formulate a *p*-system (or only a model of an *I*-system) needs to be constituted. Such a system should at least be pos-

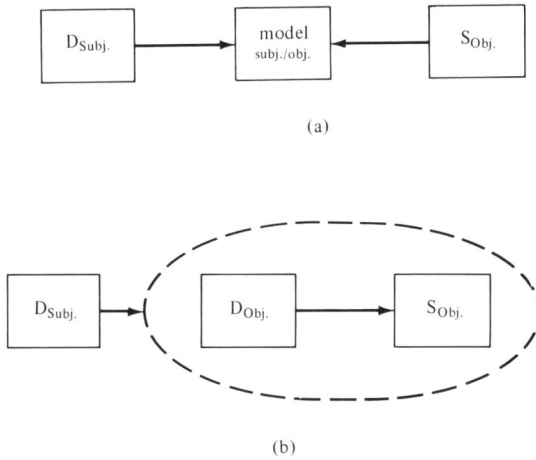

(a)

(b)

Figure 4. Schematic representation of how the model maker, (a) assuming the role of a subjective D-system (for the model of the S-system), transforms this role into the hypothesis (b) that, for the system's objective existence, an objective D-system has to reassumed.

tulated as the *creative* or *generative* system (or C/G-system), whose activity consists of creating or generating from the beginning everything that is needed by a system. The main point is that such a system would (by definition) be able to formulate not only a program for any arbitrary system, but also a program for itself. The C/G-system thus has to share with the human subject not only the property of self-reference, but also those of self-recognition and self-determination. It is obvious that an AP-system, revealing a producer–product (or D-system–S-system) relationship that tends to the final (and/or initial!) identity of the two, has, together with the O-system (recording this identity from "inside" the system), very much in common with the C/G-system.[2] But we may also grasp, from our own self-understanding, that certain features of man, particularly his endowment with "intelligible" freedom (as against the "empirical" freedom of organisms), cannot be shared by an AP-system, yet are beyond the scope of systems theory.

In the well-elaborated systems theoretical methodology are certain clues to the questions raised at the outset. But difficulties still remain in that thus far we have only touched upon solutions that, by their very nature, can not be found in systems theory. Thus, an entry into metatheory is required.

---

[2] This holds inasmuch as the AP-system is *the* product of the C/G-system that is closest to its producer.

13.4   The Possibilities for a Metatheoretical Approach
to Autopoiesis

13.4.1 The Paradox of Cognition and the Significance
of the Subject

Metatheory, by making explicit the conceptual presuppositions held hidden in almost every theoretical construct, brings attention to doubts about the universality claim of science. The significance of the cognizing subject becomes more obvious as metatheory itself is forced to bring its own presuppositions to light. This aim can be achieved if one avoids believing in arriving merely at a definite position, but instead recognizes that one has to permanently transcend and relativize one's own point of view. Such a transcending methodology is, for instance, circumscribed by the conception of the "cybernetics of cybernetics" (von Foerster 1974). It must permanently focus on the fact that the attempt to attain another aspect, for example, by mounting language levels or by looking for alternative descriptions (Pattee 1973), necessarily relativizes the subject's position. It follows that it is impossible to recognize something without recognizing cognition itself.

In order to treat the problem of cognition metatheoretically we have two methods at our disposal; these are themselves complementary or dual to one other, thus signaling an irritating initial paradox.

(1) We may start from the assumption that there is a precise *correspondence* between the being (i.e., "what is," or the so-called objective reality) and cognition (i.e., the cognizing activity of the subject). This correspondence is brought about by the underlying (founding or fundamental) *idea,* the equivalent to some formal structure (but actually infinitely more than this) that can be depicted as uniting the observer with the observed and thus displayed by a circular relation (Figure 5a). It would be impossible to speak sensibly about the correspondence of the being and the cognition of that being without the assumption of the idea.

(2) However, in order to speak about the correspondence there must exist a subject, called "transcendental" (by Kant), that reflects on what is and on cognition; the subject hence presupposes, recognizes, performs, and even partly founds the unity expressed in the correspondence. Therefore, we need to counterpoise the assumption of the correspondence that is objectively stated (compare it with the CD_{obj}!) to the complementary assumption of the cognizing subject that (a) is placed (or places itself) amidst the correspondence, (b) formulates statements about the correspondence, and (c) characterizes, by means of these statements, the correspondence as *coherence* (Heintel 1974) (i.e., as consistency of the statements about any object of cognition, taking the object itself, statements about it, and statements about the cognizing subject into consideration equally). This situation can be illustrated by a circular relation

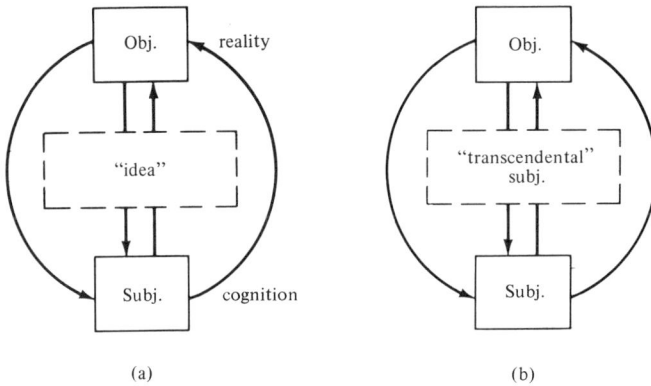

(a)   (b)

Figure 5. Similarity in (a) correspondance and (b) coherence; the difference between the two circular relations is that in coherence, instead of the idea being fundamental for correspondence, the transcendental subject assumes the job of foundation.

that now bears another meaning (because it is comparable with the CD_{subj}) (Figure 5b).

We come to recognize that the notions correspondence and coherence, respectively, can be connected to other attitudes the subject assumes when it acquires cognition: *discovery* of something contrasts with *construction* of the same something. Unreflected *"givenness"* (i.e., conceiving of reality as devoid of presuppositions for cognition) may oppose *"mediation"* (i.e., reflecting upon the presuppositions that are required). A rephrased proposition about "givenness" vis-à-vis "mediation" could read thus: Every entity that is recognizable as an entity must be given (i.e., be part of the so-called objective reality); however, in taking this givenness into cognizance, the presence of the cognizing subject must evidently be presupposed. But the subject too must be considered as given, otherwise it would not be possible to speak about itself. Viewed in this way, the problem of cognition appears even at its root to be deeply paradoxical.

Therefore, we shall strive for some clue to that paradox. It should be mentioned that the CD, whose (minimal) structure is described above, is isomorphic with (1) the *conversational domain,* that is, the domain between partner 1, partner 2, and the theme in a conversation (Locker 1980), and (2) the *self-referential domain.* The last-mentioned domain is built up by a subject ("self 1"), which by "disunion" is able to distance from itself and to see itself as "self 2"; this performance of seeing is "reconciled" by the "self 3," which watches the mutual observation of "selves 1 and 2." The connectedness of the three domains (CD, conversational, and self-referential, respectively) poses a problem similar to the connectedness of "givenness" and "mediation"; in these instances something (some notion) that unites and founds the opposing terms is required. Having arrived at this point we concentrate now on the notion of the "self."

13.4.2 On the Several Meanings of "Self"

The meaning of the prefix "self" in words such as self-reference or self-reproduction certainly depends on the view chosen and the context elicited by that choice. The following meanings of "self" can be discerned:

(1) When it is supposed (Löfgren 1968) that *"description of description"* (i.e., a description referring to a described object itself being described, albeit at a higher language level) is equal to self-description, then we encounter here the *"relational self."* In it the pronoun "self" indicates the grammatical object to which the transitive verb (e.g., describe) points; but the word "self" remains a pronoun. The "relational self" thus is the *formal* basis for any usage of terms in which the prefix "self" appears.

(2) A content-based (*material*) interpretation (i.e., an interpretation regarding the content of the term) shows that the meaning of the word "self" can be grasped as a *self-directed* or *self-dependent* activity—i.e., an activity whose paradigm is possibly expounded in the saying, "I do it by myself," equivalent to saying, "I do it alone, without external help." Here, the prefix "self" appears as the token for origin and goal of activity, and relies on the meaning of "self" as noun. In addition to the necessity of unfolding the "relational self" contentively (i.e., materially) by activity conforming to this relation, the *self-knowing* (or self-cognizing) activity needs to be envisaged.

(3) Independently of any (material) activities formally guided by the "relational self," a carrier of these activities is demanded as necessary. It can be called *the Self* as noun (now written with capital S) and represents the cognizing subject in its full import, the "transcendental" subject that corresponds to the conception of "substance" in ontology.

13.5   On Self-Reference and Self-Consciousness as Paradigms for Autopoiesis

13.5.1 On the Activity and the Knowledge of the "Self"

In order to function appropriately in the CD the observer (who becomes self-observer when the observation expands into the self-referential domain) must possess consciousness. Consciousness comprises

1. *intentionality,* that is, directedness towards outside objects (this attitude is more or less in conformity with the acceptance of "givenness"); and

2. *reflection*, that is, directedness of consciousness towards itself (Old-emeyer 1970) (being more or less in conformity with "mediation" or "self-mediation"). Here, of course, there also needs to be
3. the presupposition of a unifying, founding principle, the underlying "idea" or the "transcendental" subject, that is, the subject that embraces all those conceptual principles without which no object can be thought of.

In line with our assertion that there is a similarity between autopoiesis and self-reference, we look more closely at reflection. As already stated, reflection can be either an expression of the activity of the Self or of the knowledge of the Self. From this follow important consequences for autopoiesis.

If we regard reflection as the *activity* of the Self, then it is necessary to presuppose that the Self already possesses the *ability* to perform reflection. Ability—or, in Chomsky's (1972) terminology, competence—must precede performance; therefore, performance of reflection cannot found (i.e., be fundamental for) consciousness (Henrich 1970). Setting consciousness roughly equal with autopoiesis forces us to conclude that *self-referring relations cannot explain* (or found) autopoiesis. We cannot simply invoke the existence of self-referring relations to give rise to an *AP*-system.

Further, if we look on reflection as the Self's *knowledge* of its own self-referring activity, then it is impossible for the Self to become conscious of any state of affairs by simply reflecting on this state of affairs. Since reflection can sensibly be understood solely as an intended or aimed reflection, a knowledge of the purpose needs to be provided beforehand (Henrich 1970). In other words, any knowledge the subject has of itself cannot be obtained by self-reference. This result amounts to the assertion that autopoiesis *cannot be obtained by circular* (closed, recursive) *relations* alone.

13.5.2 On Modes and Models of Self-Reference

In addition to the underlined inability of self-referring relations to found autopoiesis, we have to ponder the following questions.

To understand the Self more fully, we also have to equate it with the *concept* (or the *p*-system) of the Self. A concept does not make sense without taking into account the subject that conceives (of) it; this is even more valid when the concept refers to the subject contentively. Thus, since the subject requires knowledge of itself before it undergoes self-reference, this knowledge is to be equated with the concept (i.e., the own *p*-system) that the subject consciously has of itself.

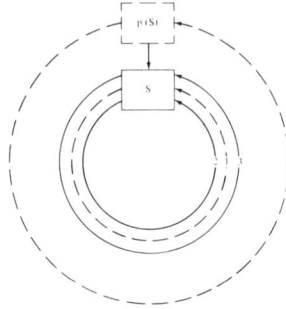

Figure 6. Schematic display of self-reference. The formal relation in itself (1) enables the system (*S*) to perform subjective reflection for itself (2), which is objectively recorded as reflection on itself (3). All modes of self-reference must be formally prescribed by the self-referential nature (4) of the system's program *p*(*S*).

However, we have to constrain our former statement and add the correction that it is only *one* kind of self-referring relation that is incapable of founding the subject. A closer understanding of the connections between self-referring relations and the Self can be reached by drawing further distinctions on the basis of the terminology introduced by Hegel, employing *personal pronouns* and *prepositions*. Remember that prepositions serve also to clarify terminologically the theory of *p*-systems (or purpose). There is, first, reflection *in itself*, meaning something solely formal, thus comparable to the formal "relational self" mentioned above. By reflection in itself we mean a relation leading "from" an entity "to" the very same entity. Because the CD is left out of consideration, this "definition" tells nothing about the mode of cognition regarding the occurrence of reflection in itself. When the CD is taken into account, two further distinctions show forth. The existence of the self-referring relation is objectively stated when one calls it reflection *on itself* (in German, *an sich*). This preposition reveals again a twofold meaning: either (1) it is simply caught as a matter of affairs that can be observed and described, because the prepositional and pronomial term "on itself" characterizes a *b*-machine; or (2) it is contended that the "source" (a term whose meaning will be specified forthwith) for the objectively existing self-referring relation "on itself" lies *outside itself*. Then, self-reference is not actively, but only passively performed, upon some order laid down in a *p*-system stemming from some alien authority.

In sharp contrast with the foregoing, reflection *for itself* asserts that the self-referring relation is actively brought about and performed by the entity from which it proceeds and to which it returns. Differing from the activity carried out under the item "on itself," which is only objectively stated, reflection for itself denotes the "source" of the reflective activity as lying *inside itself,* although the *p*-system "for" the reflection for itself must be given beforehand. The prepositional and pronomial term "for itself" nonetheless characterizes the state of autonomy.

Of course, the knowledge that accompanies reflection for itself is (known) reflection *of itself,* possibly incomplete (Figure 6).

The performance of self-reference can easily be aligned with characters of the *sR*- and *sP*-systems. Our own introspection teaches us that in the *sR*-system two complementary (parts of) "selves," both dually identical with each other, can be discerned, although the performance of self-reference demonstrates that it is impossible to hold both of them in attentive consciousness simultaneously. Attention needs to switch from one part of the *sR*-system to the other part (i.e., from Self 1 to Self 2) and back. This makes the deep split ("disunion") in the system obvious, whereas in the *sP*-system ($\equiv$ *AP*-system) the unity is perfect and the complementarity thus "elevated" (*aufgehoben*) or "reconciled" (to use Hegelian terms). Actually, the unity and the complementarity are both expressions of our approach, which necessarily has to switch between these two modes. It is pure self-consciousness, being indeed something qualitatively else—the mental or spiritual state of the "true" inmost Self within oneself (of which the *C/G*-system or the *p*[*p*(*sP*)]-system only gives a skimpy intimation), which is, as such, a permanently unchanging and yet active unity without internal disseverance.

13.5.3 Systems as "Selves"

It has been said that each system consists of a *b*-machine and an *m*-machine. But there is a compulsion to go further to the concept of Mind, settling upon the meaning of something that is representable by the program but is not the program itself. The Mind has to be equated with the Self.

At this point in the discussion it may seem acceptable to state that every system has the character of Self or subject. Again several possibilities open up:

1. The system *is* a subject, because the subject is *within* it, such that the *b*-machine provides the exterior representation of the subject ($\equiv$ Mind plus *m*-machine).
2. The system *represents* a subject: the latter is not interior, but rather *exterior* to the system. This instance is given when the subject of a system that acts as a *D*-system programs an *S*-system, which in

turn contains a program of itself (as $p(S)$-system) but did not formulate the program for itself.

Further specifications are possible:

1. When the system is a subject then it is already a creative subject (a C/G-system) when it conceives of a program for another system.
2. The system's subject nature can still be augmented when it conceives of a program "for itself."

The program for itself can only be achieved through self-reference, followed by self-recognition. Then the (human) mind recognizes itself as dependent and may conceive of a model of a Mind that is conceiving of itself, although it does already (forever) exist—the Mind of God. The true subject of man recognizes that it cannot bring itself into existence.

13.6   On the Origin of Autopoiesis

13.6.1 The Origin Paradox and How to Overcome It

We used previously the neutral word "source"; it can be used with several ranges of meaning: (1) "cause," understood as the purely scientific term; (2) "foundation," understood in its metatheoretical aspect; (3) "origin," a term which we will adhere to, and which falls between the two aforementioned terms in that it also expresses their togetherness. In addition it accords with the notions of time (cause) and of atemporality (foundation).

Whenever we try to speak of an origin (i.e., a beginning), the expression seems to be fraught with a disturbing paradox, be it in the case of the origin of language (Lohmann 1965), the origin of life (Cairns-Smith and Walker 1974), or the origin of the total creation (Bonhoeffer 1969). Language always seems to fail, especially when we are seizing upon the origin of language itself: language as a means for expressing something already presupposes itself as that which is to be expressed.

Three principal *means* are available *to handle* any paradox:

1. Leave the paradox untouched and sustain it as an unsurpassable boundary to cognition.
2. Escape it (or better said, try, self-elusively, to escape it) by inventing a mechanism that fits a certain conceptual scheme generally taken for granted. For such a mechanism, physical *time* is made absolute so that entities that inherently belong together (and in their very togetherness exhibit the paradox) are torn asunder and placed as subsequent events in the time frame, with a putative initial event (like the "big bang") as the punch of an origin. However, the postulate of a physical mechanism pretending to avoid the paradox (although it exclusively deals with the b-machine and with nothing

else) must be paid for! By destroying the origin paradox the scientific myth of evolution is created.

3. Neither escape it nor leave it as a stultifying scandal before us. This is the metatheoretical approach and consists of permanently transcending the paradox. Methodological guidance is again provided by Hegel's dictum of "identity of identity and nonidentity," which for the problem to be treated here is to be transformed into "mediation of mediation and givenness," thus postulating that "mediation" always outruns the point of view once achieved.

Following this dictum we also receive a recipe for moving out of the dilemma of cognition. Since nothing is immediately given that is not concomitantly mediated, the apparently endless possibility of contrasting immediateness with reflectedness comes to an end as soon as the "transcendental" subject outdistances even these oppositions and "elevates" them to itself.

By leading back this conclusion to the CD, we find that (1) it is the O-system that necessarily precedes the build-up of the CD, and (2) it is the O-system that has the capability to transcend itself such that it can convey its own position to a hypothetically assumed D-system's model. The O-system does so when it recognizes that its own ability concomitantly to *discover* and *construct* any arbitrary entity (e.g., a system and a model thereof) is superseded by a third activity, being "more than the sum of the two" and called the *founding* function. The model of a D-system precedes in the O-system's understanding any S-system. (3) From this position, one can achieve, not too deviously, the conception of a total "being" (as universal system) that is preceded by a subjective designer (i.e., the Mind of God).

13.6.2 The Precept of the Intension/Extension Relation for the Origin Problem

A rather abstract but useful access to the problem of the origin is provided by the theory of concepts. Distinguishing the intension from the extension of a concept shows that the former denotes the definition and the latter points to the objects to which the concept refers. The relation between concepts (the *sense* or imposition relation) has to be considered as atemporal, like any mathematical entity; the *reference* relation, on the contrary, relates concepts to objects and therefore deals with temporal occurrences. The sense relation represents *ontological* preference; it also implies the subject conceiving concepts, thus "setting" or "positing" an origin.

The attitude that neglects the mode of cognition in science frequently leads to the pretension that intension is "nothing but" the result of inductive inference drawn of extensions. However, it is impossible to derive intension from extension in logic just as it is impossible to derive competence from performance in language, to derive the p-system from the I-system in systems theory, or to derive existence from action, theory from practice, substance from motion, or the "being" from the "becoming" in philosophy. To insist upon implementing the impossible must inevitably engulf one in a spiritual insularity, despite one's participation in the vast community of scientists.

What can be made applicable from these considerations to the problem of autopoiesis is that the concept of autopoiesis has to precede the realization of autopoiesis. It cannot be emphasized strongly enough that (contrary to scientifically camouflaged superstition) the m-machine does not arise from the b-machine and that mind does not arise from matter!

13.7 Conclusion

In the upshot of our investigation the only valid conclusion we maintain as admissible is that an interpretation of autopoiesis in mechanistic terms is completely fallacious. Although the term evolution is suggestive of a principle pretending to exist *by itself,* operating on physical objects and events; and although causal mechanisms seem to perfect "themselves" by adaptation to the environment, evolution is a pseudo-idea implied by scientism (Barzun 1959) (i.e., a science absolutizing itself and claiming that there is nothing outside its own scope). Therefore, not only the concept of purpose is rejected, but also the concept-nature of programs (i.e., natural laws) misinterpreted and bowdlerized into a quasi-objective existence. Autopoiesis as proposed (Varela et al. 1974) does not outdistance scientism decisively enough.

I wonder whether scientists will soon become cloyed by a stale propensity toward alluring schemes whose alleged universality is a constant din in our ears. Scientists should vie among themselves not to repudiate what they have not thoroughly examined for its underlying presuppositions. However, even after having got rid of the gyves of the evolutionary scheme, scientists often fall victim to another imposition whose nature is disguised by the neatness of new schemes that are quickly ushered in and heedlessly accepted, lest the scientist had to fear the disdain of the layman's expectations that his flatly materialistic *Weltanschauung* is scientifically confirmed. Indeed, people who fortunately have sloughed off the evolutionary scheme are swayed by the system's paradigm, the validity of which apparently cannot be impugned (although another seductive universality claim should be emphatically objected to). To plunge into systems theory without being aware of the theory's dependence upon its pertinent presuppositions means repeating a fallacy at another level, albeit cloaked through a suggestive terminology, pilfering from philosophy without admitting doing so.

Our study has shown that only under the acceptance of philosophical ideas proper (even when squeezed into systems theoretical terminology, obviating the objection to having adroitly sneaked in "mysticism") can the whole story become understandable. It would be too immodest to hope for an imminent revulsion in the attitude of taking long-accustomed thought schemes for absolute, and for a growing opposition to "scientistic" pretensions. A true upheaval in thinking would be brought about were the solution, at which we arrived by analyzing the meanings and diversities of the term "self" and its impact for autopoiesis, accepted as the proposition that nothing arises "by itself" or, that an objective origin of anything is not only impossible but unthinkable: the *cognizing* subject (*O*-system) is always indispensable for an origin to be recognized; and the *creative* subject (*D*-system) is always indispensable for an origin to be performed. Both subjects, in the ultimate respect, become one.

As a consequence of this survey, a vindication of a formerly much disputed and erroneously refuted dogma of biology appears on the scene: the metatheoretical version of this dogma (better circumscribed as fundamental axiom) reads thus: *Omne systema e systemate.* Whoever (unfortunately, usually a member of the great majority of present-day scientists) comes forward with the assertion that life originated "by itself" such that autopoietic systems arose has to be taught that rather restricted schemes of thinking have, by their bent for the all too obvious, led to neglect of the necessity of simultaneously regarding oneself as the utterer of one's assertion.

I conclude by expressing my conviction that unyieldingly withstanding naiveté and unmasking the preposterous ostentation of scientism will result in a breakthrough toward the surcease of prejudices and regaining the regrettably lost franchise in the land of ideas.

Glossary of Terms

aP-system: "allopoietic" system

AP-system: "autopoietic" system

E-theory: emergence theory

T-theory: transition theory

O-system: observer

D-system: designer

S-system: system proper

M: Mind

m-machine: mind-machine

b-machine: body-machine

p-system: program ($\equiv$ purpose)

$p(I)$-system: program specified for any I-system; e.g., $p(P)$-system: program for a P-system, etc.

I-system: instrumental system
E-system: executing system
M-system: making system
P-system: producing system
R-system: reproducing system
sR-system: self-reproducing system
sP-system: self-producing system
C/G-system: creative–generative system

CD: cognitive domain

References

Barzun, J. (1959), *The House of Intellect,* Chicago Univ. Press, Chicago, Ill.
Bense, M. (1960), Über Metatheorie: Die Erweiterung des Metaphysikbegriffes, *Grundl. Kybern. Geistesw.* 1, 81–84.
Bonhoeffer, D. (1969), *Creation and Fall—Temptation. Two Biblical Studies,* MacMillan, New York.
Cairns-Smith, A. G. and Walker, G. L. (1974), Primitive metabolism, *Biosystems* 5, 173–186.
Chomsky, N. (1972), *Language and Mind,* Harcourt Brace Jovanovich, New York.
Eigen, M. (1971), Self-organization of matter and the evolution of biological macromolecules, *Naturwissenschaften* 58, 465–523.
Foerster, H. von (1974), Cybernetics of cybernetics, *Proc. Conf. Communication and Control in Social Processes,* Philadelphia, Oct. 31–Nov. 2.
Gardner, M. (1971), On cellular automata, self-reproduction, the "Garden of Eden" and the game "Life," *Sci. Amer.* 2, 112–117.
Heintel, E. (1974), Philosophie und Organischer Prozeß, *Nietzsche Studien* 3, 61–104.
Henrich, D. (1970), Selbstbewußtsein. Kritische Einleitung in eine Theorie, in *Hermeneutik und Dialektik* (R. Bubner, K. Cramer, and R. Wiehl, eds.), J. C. B. Mohr, Tübingen.
Locker, A. (1980), Metatheoretische Voraussetzungen der Formalen und Empirischen Linguistik, *Nova Acta Leopoldina,*
Löfgren, L. (1968), An axiomatic explanation of complete self-reproduction, *Bull. Math. Biophys.* 30, 415–425.
Lohmann, J. (1965), *Philosophie und Sprachwissenschaft,* Duncker & Humblot, Berlin.
Monod, J. (1975), *Zufall und Notwendigkeit,* Piper, München-Zürich.
Oldemeyer, E. (1970), Überlegungen zum phänomenologisch-philosophischen und kybernetischen Bewußtseinsbegriff, in *Philosophie und Kybernetik* (K. Steinbuch and S. Moser, eds.), Nymphenburg Verlag, München.

Pask, G. (1970), The cybernetics of behaviour and cognition, extending the meaning of goal, in *Progress of Cybernetics* (J. Rose, ed.), Gordon & Breach, London.

Pattee, H. H. (1973), Physical problems of the origin of natural controls, in *Biogenesis-Evolution-Homeostasis* (A. Locker, ed.), Springer Verlag, Berlin.

Rosen, R. (1966) Biological and physical realizations of abstract metabolic models, *Helgol. Wiss. Meeresunters.* 14, 25–31.

Rosen, R. (1973), On the generation of metabolic novelties in evolution, in *Biogenesis-Evolution-Homeostasis* (A. Locker, ed.), Springer Verlag, Berlin.

Varela, F. G. (1971), Self-consciousness: Adaptation or epiphenomenon, *Studium Generale* 24, 426–439.

Varela, F. G., Maturana, H. R. and Uribe, R. (1974), Autopoiesis: The organization of living systems, its characterization and a model, *Biosystems* 5, 187–196.

INTRODUCTORY REMARKS

Löfgren echoes Locker's conclusions by invoking again the "life is produced by life" principle. If we conceive of life as an objective, nonlinguistic autonomy, then *any* linguistic description of it can be criticized because of the externally chosen language, suggesting that the results are of limited validity, as performed by Locker in the preceding paper. Löfgren's solution is to consider life a linguistic phenomenon *itself.*

Löfgren uses a hierarchy of cerebral languages, ranging from the lowest genetic description–interpretation processes to our external communication language at the top. He presupposes that a general notion of description–interpretation pairs (for example, a transcription of the DNA code) constitutes inner "language," independent of cognition, and applicable at all levels of life. Such internal (biological) languages must of course be recognized and described through an external communication language of humans.

It appears that Löfgren's hierarchy depends on the acceptance of the description–interpretation properties ascribed to a duplicating DNA macromolecule. This implies that certain components of a system are accorded a privileged status with respect to all other components—they are "carriers" of the biological language, capable of self-description and self-interpretation. These are the problems of biological language embodiment and identification.

Löfgren discounts the limiting feature of autopoiesis, the concept of boundary, and allows its broader contextual understanding within organisms, populations, social systems, conversational domains, and the like. A cell membrane should not be conceived as the only possible embodiment of the notion of "boundary."

Lars Löfgren was born in Stockholm in 1925. He received his master degree in 1949, "licentiat" in 1954, and a Ph.D. in mathematics in 1962, all from Kungliga Tekniska Högskolan in Stockholm. During 1959–1961 he worked with the Biological Computer Laboratory at the University of Illinois, Urbana. Since 1963 he has been a Professor at the Department of Automata and General Systems Sciences at the University of Lund in Sweden. He revisited the Biological Computer Group in 1966–1968 as a Visiting Associate Professor. Professor Löfgren's research interests concern foundational questions of cerebral processes such as learning, the emergence of existential perceptions, and cerebral languages with their interpretation theories. *Address:* Tekniska Högskolan i Lund, Fack, 220 07 Lund 7, Sweden.

On the other hand, Löfgren characterizes autopoiesis as providing an explanation for the nonevolutionary forms of life (compare with the understanding of Locker and the treatment of Ben-Eli). The theory of autopoiesis does imply a rather fruitful notion of evolution: a history of structural changes undergone by the members of a phylogeny and realized through an uninterrupted chain of their structural coupling (or fit) with the environment.

Chapter 14
Life as an Autolinguistic Phenomenon

Lars Löfgren

14.1   Introductory Comments on the Autonomies of Life

"Life is produced by life" is an aphorism that can be found, for example, at the Natural History Museum in Chicago. Although the immediate biological significance of the aphorism may appear somewhat meager, it hints at an aspect of the supreme complexity and potentiality of life: Life does not need for its production a phenomenon more complex or of greater potentiality than life itself. However, not until the autonomy expressed in the aphorism is explained are its full meaning and significance revealed.

Here the notion of the autonomy, or self-reference, refers to a specific property of life, its productivity. The complexity and potentiality of life are, however, of such magnitude that autonomy has been attributed to life not only with respect to production but also with respect to such functions as maintenance (life maintains itself), organization (life organizes itself), and creation (life is created by life).

Some of these forms of autonomy may appear problematic, however. It is the primary purpose of this paper to investigate which are nonproblematic and to demonstrate their consistency by *unfolding*, a kind of hierarchical modeling scheme.

We face, then, the problem of requiring a basis for such a study that is on the one hand above the phenomenon under study, life, and on the other hand produced by living scientists, and thus confined within the boundaries of life. As we shall see, there are circumstances under which this problem can be hierarchically resolved.

Consider the problem of describing life as an autonomous organization, where this organization is considered nonlinguistic. We then intend to produce a linguistic description, a theory, of a nonlinguistic phenomenon and are forced to step out of the object domain, that of life, into an external, linguistic domain, that of the description process. This immediately raises the question whether the organizational phenomenon, intended to capture life, really is sufficiently autonomous.

The situation is changed if we instead consider life a linguistic phenomenon by, for example, recognizing its genetic–epigenetic processes as description–interpretation processes and nervous and cerebral activities as description–interpretation processes in languages ranging from inner cerebral all the way up to external communication languages. The linguistic basis for a theory of life processes is then immediately recognized as a part of the autonomy under study. This suggests a process of theory formation in a *hierarchy* of languages, rather than a descriptive theory in a fixed language. Thus, with language broadly conceived of as complementary description–interpretation pairs, we reach an *autolinguistic* view of life, mirroring also its evolutionary character.

It is true that much of our knowledge of language stems from our acquaintance with external communication languages. These languages have, however, very specific interpretation schemes in terms of cognition. Hence it may be difficult for us to ascribe this kind of language to the biological phenomena of lower forms, where cognition is not developed. However, with more general notions of interpretation and description, language may be understood as independent of such phenomena as cognition and applicable to all levels of life.

The complementarity of the description–interpretation pairs means, for example, that not all of interpretation can be described. This makes the general concept of language exceedingly complex (Löfgren 1977a). Fundamental, difficult problems abound; for example, how are we to recognize a biological language by the aid of an external communication language? Some of these problems will turn out to be effectively unsolvable.

However, such difficulties should not be taken as arguments for the nonexistence of internal languages. Rather, they should be taken as an indication of the power of the open hierarchical linguistic perspective that admits the statement of unsolvable problems, knowledge of unknowability, knowledge of difficulties of life, and so on. Characteristic of such knowledge is that it is productive, that it indicates how to proceed toward greater knowledge.

The autolinguistic view suggests that an understanding of life implies an understanding of language, and vice versa. In other words, the phenomena of language and life are equivalent with respect to explicability. A similar point has been suggested by Pattee (1973) but with a more physical conception of language as well as of argumentation.

14.2 General Languages and the Description–Interpretation Complementarity

We may be aware of the phenomenon of language mostly because of its manifestation as an external communication language, a so-called natural language. Fragments of it, like the predicate languages of mathematical logic, are well understood both in terms of syntax (description rules) and semantics (interpretation rules).

Among other forms of languages, we have the inner cerebral languages (Löfgren 1977), which influence our way of thinking as well as the way we express ourselves in the external language. Genetics provides a further area of linguistic phenomena, linguistic by virtue of identifiable description and interpretation processes.

Furthermore, engineering may be considered an extension of our biological description–interpretation processes to a technological domain. A technical construction is an interpretation (engineering realization) of a description (engineering plan). Within computer technology, a programming language coordinates descriptions (programs) and interpretations (computations).

In all of these instances we do conceive of language as complementary description–interpretation pairs. This is an important point to make, in particular because the word "language" is sometimes used to denote only descriptions. For example, in an area such as artificial language, the word "language" may be used to denote a set of sentences for which the concept of meaning (interpretation) is not even considered.

A related point of notational difficulty stems from a widespread (mis)use of model for theory. A theory ought to be regarded as a description of its intended interpretation or model. Not until this distinction between model (interpretation) and theory (description) is made can we understand, for example, the vast difference between model-based and a description-based planning. This has been explained by Rosen (1974) and Pattee (1977).

In a way, it is understandable that such notational difficulties arise. An explanation of the distinction between a description and an interpretation requires a metalanguage sufficiently well developed to permit descriptions of both the object description and the object interpretation. If instead one and the same language is used to describe distinctions between its own descriptions and interpretations, then it must be freely self-referring, and for such languages the semantics cannot be properly defined. Hence, a clear distinction between theory and model, or between description and interpretation, is intimately connected with the concept of linguistic hierarchy, distinguishing object language and metalanguage. Without such a distinction, concepts like theory and model, or description and interpretation, are bound to merge.

14.2.1 Description–Interpretation Processes as Inverse Dominance Relations

In a description process (like a learning or theory formation process, biological or not), that which is described (the interpretation) dominates the description. By way of example, a physical phenomenon in Newtonian mechanics, such as the movement of the moon, dominates (determines) its description (its theory). If predictions of the (hypothetical) theory do not agree with the phenomenon, it is the description (hypothetical theory) that must be changed, not the physical reality!

In an interpretation process (biological or not), the situation is reversed. The description dominates the interpretation (that which is described). By way of example, a computer program (a description) dominates the behavior of the computer (an interpreter). If the behavior does not agree with the program, it is the computer, as a physical reality, that is in error and has to be changed, not the description!

The processes of description and interpretation are inverses of one another. To a certain extent a given interpretation process determines the corresponding description process, and vice versa.

Given an interpretation process, the corresponding description process (learning process) is determined as a search process, guided by an induction principle, such as the simplicity principle, as explained elsewhere (1977a, 1978a).

Sentences are varied (hypothetically suggested), successively interpreted, and compared with a given interpretation (the real world) until agreement is reached. A difficulty here is that two interpretations (realities) can be compared only partially by means of a finite test (finite observation). A confirmation decision will be necessary, that is, a decision when to stop the test and accept the hypothetical description as confirmed. It is this—our use of confirmed rather than verified descriptions—that makes us feel able to make predictions.

The search process may be guided by the simplicity principle as follows. The theory of facts and confirmed hypotheses under production is steadily being simplified by reformulations, involving naming processes and simplifying generalizations. These latter are actually part of the hypothesis generation process. In that way the search process is guided by the simplicity principle, so that if two descriptions (hypothetical theories) both describe the same set of experimental facts but predict differently, then the simpler is to be preferred (see Löfgren 1977b).

14.2.2 Simple Descriptions of Complete Interpretations: Their Complementarity

A description describes in extension (i.e., it predicts) and is ultimately simpler than that which is described. Change is ultimately described by constancy. That which is described (i.e., the interpretation) is usually conceived of as real in a sense of completeness of properties (see Mendelson 1964, Löfgren 1977b).

Now, how can something like a complex, complete, real thing be described by a simple description? How can infinities be finitely described? How can change be described by constancy? The answer is that when we go from description to that described (i.e., when we interpret), then we utilize undescribed, generative properties of the interpreter.

It thus follows that not all of reality can be described (i.e., simplified). For example, should we try to describe description–interpretation processes, we are bound to encounter difficulties, because we obviously cannot have interpretation processes with undescribed properties at the same time that we claim to describe those very processes. As a result, the dominance relations for description and for interpretation will turn out to be incomplete, and description and interpretation will have to be considered complementary.

The simplicity and predictive character of descriptions may be illustrated as follows. The movement of the moon around the earth may be considered real in the sense of being thought complete, continuous, and so on. It may be described as an elliptical orbit, for example, in a mathematical formalism, using a finite and small number of symbols. The observational facts from which such a description is produced are necessarily finite in number, yet the description extends them to an infinite set. Interpretations (computations) of the description may result in good predictions of the position of the moon. If, for high demands of accuracy, the predictions fail, the description may be changed to another, also taking into account the gravitation of the sun, for example. Changes in descriptions may thus occur, as in all learning (description) processes. However, after each confirmation decision, the corresponding description is considered constant. In that sense we can say that change (of the moon's positions) is ultimately described with constancy, and that the description is (infinitely) simpler than its interpretation. The reason is that we can refer to an external interpretation scheme not itself described by the description.

14.2.3 Hierarchies of Languages

If the external interpretation scheme of the previous example is made the object of a further description (study), we are naturally led to extend the description in a new language, a metalanguage, with its interpretation scheme on a higher level.

Again, we may want to look deeper into the other end of the problem, into the experimental facts that are the basis for the description process considered. These facts are primitives only within the linguistic level considered. In a lower-level measurement language, on the other hand, they are descriptions (theories) confirmed on the basis of still more basic facts, and so on.

Thus, what is a primitive observation on one level may well be the result of a description process on a lower level. There may well be a lowest linguistic level in the hierarchy at play in a biological organism. Further decompositions then lead to physical or chemical processes that do not have the character of description–interpretation processes (i.e., that are nonbiological).

Different languages may be adapted for describing phenomena in distinct areas. When description processes proceed within each of these languages, irrelevance barriers may develop to isolate phenomena from each other (see Löfgren 1978b for a study of the concept of relevance in theory formation). Yet there may be deeper interdisciplinary relations that cannot be sufficiently simplified, and therefore not confirmed, except in a new language of a higher level, which breaks the previous barrier.

Thus new linguistic levels seem to be enforced by the simultaneous requirement that the description process be both simple and extendable. There seems to be no limit on the creation of new levels.

Consider a description process that extends in a linguistic hierarchy by creating jumps to a higher level when the descriptions become preventingly complex on an actual level. The process may converge in the sense that all successive higher-level descriptions (theories) become equivalent from a certain level on. Such a convergence implies that the extension of the description process is restricted.

If the field of investigation is not restricted, the theory formation process will proceed along a linguistic hierarchy as an open process. At no stage can we then speak of having a complete objective knowledge. Objectivity, in the classical sense, is only possible for narrow fields that do not themselves fully explain how restricted they are.

14.3   Explanations of Autonomies

Autonomies in linguistic processes with referential interpretations are naturally thought of as self-references, like "this sentence contains precisely forty-five letters," or "this sentence is false." The first sentence is true in the sense that it refers to a factual state of affairs. The second, which is neither true nor false in the above sense, represents an inconsistent form of self-reference.

Examples of self-referential sentences in more formal languages are the following: $S \in S$ (the set S belongs to itself); $f(a) = f$ (the function f belongs to its own range); $P(\forall x\ P(x))$ (the predicate P applies to a proposition about itself);

$$f(x,y) \quad \Leftarrow \quad \text{if } x = y \text{ then } y + 1$$

$$\text{otherwise } f(x, f(x - 1, y + 1))$$

(the function f is recursively defined by the aid of itself).

As is well known (cf. Lofgren 1968, 1972), basic autonomy questions of automata theory can be referred to the consistency of self-referring sentences of the above types. For example, can an automaton be self-producing? Can an automaton be self-describing?

Which forms of autonomy are now consistent and which are not? In other words, which forms of autonomy can be explained and which can be shown to be illegitimate? Let us see how to answer such questions in terms of Tarski's and Kleene's fundamental works on description–interpretation processes. For a more detailed study we refer to Löfgren (1979b).

14.3.1 Tarski's Explanation of Self-Reference by Unfolding

We have explained elsewhere (Löfgren 1978b) how Tarski, in his investigation of the semantical concept of truth (1956), suggests that proper (consistent) self-references in a given language always can be eliminated by an equivalent, explaining description in a higher-level formal language.

This is like unfolding the circularity in a proper self-reference by breaking the identity of the self-referring object. If the object can be attributed further new properties, which can be consistently defined in a higher-level language, such that it is turned into a series of non-self-referring objects, then such a construction is in fact a model, showing the consistency of the primary self-reference.

We have suggested (Lofgren 1979b) that Scott's consistency proof (1960) for a set theory admitting self-belonging sets ($S \in S$) be considered an example of a Tarskian unfolding.

In general, it is reasonable to expect that the more autonomous a system is intended to be (i.e., the more covering its cycles), then the more difficult it will be to impose further properties that unfold the cycles and thus explain the autonomy as consistent or realizable. When a suggested autonomy is too embracing it may be inconsistent or nonrealizable.

For example, "this sentence is false" is heavily self-referential in the sense that it refers both to its whole syntactic form and to its whole semantic truth value. It is not very surprising, then, to find it inconsistent. There is not much room for adopting a new perspective that may unfold the self-reference.

On the other hand, the autonomy of "this sentence contains precisely forty-five letters" is much less embracing, and it is natural to find such a sentence either true or false. It only refers to its syntactical form and states a specific property of that form—by appealing to a semantics (rules of interpretation) entirely outside the range of the autonomy.

A biological example of a heavily autonomous system is the human brain, which, it has been proposed, can develop a language in which it can describe its own cerebral functioning. This proposed autonomy is by Tarski's (1956) results (cf. Löfgren 1979b) too embracing, that is, inconsistent.

There is, however, a positive productivity underneath this negative result. Tarski's hierarchical construction is indicative of an open unfolding process, steadily progressing from one linguistic level to the next. Thus, instead of looking at the brain as a system that can learn itself, we ought to look at it as a learning system in an open process of describing more and more of itself.

14.3.2 Recursion

A definition may be viewed as a description in the following sense. It has the form $A =_{\text{def}} B$, where the *definiendum A* (that defined) and the *definiens B* (that defining) are sentences. Thus, the definition is a description of the meaning of A in terms of B.

Usually it is required of a definition that it not be circular, that is, that the definiendum must not be part of the definiens. In other words, it must not be a self-referential description. In mathematics, however, there are so-called recursive definitions that formally involve cycles. However, they can be considered legitimate, because the recursion in fact unfolds the circularity. The Kleene recursion theorem provides a good insight into the nature of recursive definitions (Kleene 1971; Löfgren 1978b).

More recently, recursive definitions have been considered in studies of programming languages (Scott 1970; Manna 1974). For example, look at a programming language that allows the above-mentioned program:

$$f(x,y) \quad \Leftarrow \quad \text{if } x = y, \text{ then } y + 1$$
$$\text{otherwise } f(x, f(x - 1, y + 1)),$$

meaning that a function $f(x,y)$ is recursively defined by "if $x = y$, then $y + 1$; otherwise $f(x, f(x - 1, y + 1))$." This definition surely is circular in the sense that the function f to be defined is part of the definiens.

Nevertheless, the recursion theorem ensures that a unique function can be defined: the least defined argument f left unchanged by the operator of the definiens. In general, any such recursive program defines a partial recursive function. This can even be "completely partial," in which case we can say that a certain program defines an *undefined* function (cf. Löfgren 1979b).

In particular, here we have a way to study how circularities, or self-references, in the program affect whether a function is well-defined. Note how the circularity establishes itself in terms of the function symbol f for an entity (totality). By the recursion, however, the identity of f is broken up into a series of pairs of arguments and function values, whereby the circularity is unfolded.

The Kleene recursion theorem, as well as the general unfolding methodology of Tarski and Scott, have proved influential in studies of autonomy problems both in automata theory and in theoretical biology. For some such studies we refer to Löfgren (1968, 1972, 1978b). Further applications in the areas of management and planning are foreseen (Löfgren 1979a).

14.4   Autolinguistic Processes and Life

A biological property is generally recognized as inherited or acquired. In either case the property can be identified as a biological interpretation of a biological description.

These interpretation and description processes, biological as they are, can accordingly be themselves identified as biological interpretations of biological descriptions.

The life processes thus suggest themselve as autonomous description–interpretation processes, in other words, as description–interpretation processes that themselves establish the dominance relations that constitute the description–interpretation relations. We refer to this form of autonomy as *autolinguistic*.

The description processes behind the production of inheritable descriptions and those behind the production of acquired descriptions can both be regarded as inverses of (recursively) given interpretations (cf. Section 14.2.1). Hypothetical descriptions are varied, interpreted, and tested. However, the character of the tests differs in the two production schemes.

Explanations of autolinguistic processes by unfolding suggest hierarchical views of life.

14.4.1 Generation of Inheritable Descriptions: Evolution

Consider a genetic description and its generation. By this we mean not its reproduction, but its evolution. For a recursive account of the generation, let an epigenetic interpretation already be established, identifying a level of genotype as well as a level of phenotype, the momentarily highest level of the epigenetic process. Variations of a description occur, for example by mutation, on the level of genotype. The new (hypothetical) description is interpreted, that is, realized as phenotype and tested by a natural selection process, thus occurring at the momentarily highest level.

We return to the problem of explaining natural selection, usually characterized in terms of fitness, in Section 14.4.3.

We may look at the autolinguistic production and maintenance of the dominance relations of the epigenesis complex as a recursive stabilization (freezing) of those (inner) environmental relations, with respect to which the system once passed the test for fitness. The recursive organization is such that properties, once found fit with respect to a particular surrounding, are maintained fit—by maintaining the surrounding as a constructed inner surrounding. This maintenance work may require further properties, recursively developed at higher levels such that essentially only the momentary top level is exposed to actual test by natural selection.

By way of example, the inner surrounding of the embryonic child, developing in the uterus, maintains a salinity from a time when the ancestral organisms evolved in a similar surrounding—then natural but now constructed as part of the epigenesis complex.

Again, on the lower genetic levels we have an inner surrounding, within the cell, of its chromosome. The chromosome can function as a genetic description only if the appropriate interpretation (dominance) relation is established within the cell. This is accomplished by recursion within the cell such that, at for example the higher levels of recursion, proteins are synthesized (interpreted) froms genes (description fragments of the chromosome), which regulate the synthesis (interpretation) from other parts of the description (other genes). A kind of self-reference is thus unfolded within the cell, explaining how a description can enforce interpretation relations for its own interpretation.

14.4.2 Acquisition of Descriptions: Learning and Planning

The description processes behind the generation of acquired descriptions may occur in languages on levels higher than the primary genetic. Hypothetical descriptions are varied, interpreted, and tested against a surrounding.

The test here is usually described in terms of positive confirmation criteria (inductivism) or in terms of negative falsification criteria (deductivism). Especially the positive criteria aim at predictability as well as at a recursive widening of the perception attainable from a momentarily confirmed background knowledge. Compare, on a high epistemological level, science as an activity of extending perception into new contexts as well as a means of obtaining reliable knowledge. The better a description is in these respects the better we can plan our behavior according to goals, and even develop better goals. In its abstract recursive formulation, planning and learning may as well be identified on the lower, unconscious levels of organismic behavior.

Planning and learning (description) are related as follows. Planning involves both a description of and a preevaluation of possible actions. If the evaluation is favorable according to the goals, the action is performed and we speak of a planned action. That which is learned (described; cf. Section 2.1) is the rules of the surrounding, which permit a pre-evaluation to be performed, with a planned action as result.

The preevaluation, like any evaluation (cf. the fitness test of the previous section), should be performed at a semantic level in order to be realistic, that is, in terms of a model or interpretation of the description of the possible action. However, since the evaluation is a *pre*-evaluation, the interpretation must be an inner interpretation or model, distinct from the external model (i.e., the action itself). If not, the whole evaluation must be describable so as to permit conclusions. In the first case we speak of model-based planning, in the second of description-based planning. We return to the problem of explaining planning in Section 14.4.3.

Goals for planning may be determined genetically (inherited) or on another level, in a learning process. An ultimate goal may be to describe the surroundings, including the learning organism itself, so well that its fitness relation is revealed. Such a goal for perfect planning, however, is not attainable (cf. Section 14.4.3), and instead less complex goals are recursively produced.

Planning and learning may occur at distinct linguistic levels. At higher levels they may be conscious processes, at lower levels unconscious. In this connection, consciousness may be viewed as a kind of complexity-reduction phenomenon, admitting otherwise unattainably high levels of description complexity (Löfgren 1977a, b).

14.4.3 Hierarchical Explanations of Autolinguistic Processes

The question of explaining autolinguistic processes arises as a problem of planning with respect to the ultimate goal of understanding the fitness relation underlying biological evolution. The difficulties of the model of biological evolution may be understood from failures to formalize it. Attempts to describe it have been called tautological or circular: natural selection selects the fittest; which are the most fitted? Those that are naturally selected (Eden 1967)!

Not until the circularity is unfolded into a convergent process can we claim to have a complete explanation of life. However, as explained in Sections 14.2.3 and 14.3.1, we cannot hope for such a complete explanation. Rather, we have to look at life as a phenomenon in evolution, which produces learning individuals that learn more and more about themselves and life.

The self-referential aspect of model-based planning is discussed by Rosen (1974) and Pattee (1977). They argue that perfect model planning implies infinitely fast models and is therefore unattainable. Description-based planning, on the other hand, implies a logical autonomy that in general results in an open unfolding scheme (Löfgrem 1979a). Only if the planning activity is restricted is it explicable as a convergent process.

The autolinguistic perspective of life naturally suggests the problems of embodiment and identification of language. The embodiment problem, essentially an interpretation problem, has been considered by Pattee (1973, 1977) from a rather basic physical perspective. Concerning the identification problem, essentially a description problem, Section 14.3.1 suggests that a convergent identification requires an identification language of a higher level than the language to be identified. The identification problem is of a wide biological interest, from studies of observable learning mechanisms to problems of eventual interstellar communication.

14.5 Autolinguistic Views on Autopoiesis

In autopoiesis a kind of autonomy of organization is considered. Autopoietic organization has been defined (Varela et al. 1974) as "a unity by a network of productions of components which (i) participate recursively in the same network of productions of components which produced these components, and (ii) realize the network of productions as a unity in the space in which the components exist."

A six-point key has also been provided for determining whether or not a given unity is autopoietic. In this key, the concept of boundary plays a central role: The boundary is what identifies the unity in space and should belong to the unity and be autonomously produced by it.

This form of autonomy seems to have originated from biological studies of the organization of the living cell, with an observable boundary in the form of a cell membrane.

Autopoiesis has been suggested (Varela 1978) as a paradigm for considering autonomy in such contexts as organisms, animal populations, conversational domains, and economic systems.

In these broader contexts, however, the concept of boundary seems to play a less observable role than with a biological cell. How should a boundary of a conversational domain be conceived? Eventually it can be conceived as the relevance barrier in a description (theory formation) process. But in such autolinguistic processes the boundary is in general not fixed or maintained constant (cf. Section 14.2.3) in the materialistic sense of autopoiesis. The extendability that is inherent in an autolinguistic description process does not seem to be admitted by the autopoietic paradigm.

As a paradigm for living organization, autopoiesis primarily seems to explain the nonevolutionary forms of life. By comparison, a concept of boundary is naturally suggested in descriptions of the hierarchical structure of an autolinguistic process, where it is seen as a shield or barrier that helps to freeze the dominance relations constituting the description–interpretation processes (cf. Section 14.4.1).

In a philosophical study, Maturana (1975) suggests that description (information) processes do not reflect phenomena actually taking place in autopoietic systems. Maturana then seems to reserve description processes for our own highly developed cognitive linguistic activities and does not seem to refer to a general concept of language as developed in Section 14.3.

By comparison, let us recall a conclusion from Section 14.4.3. The problem of identifying internal languages becomes exceedingly difficult when the complexity of the language to be identified approaches that of our own language.

14.6   The Understandability of Life

Monod warns (1970) about the limitations to human understanding:

The logician might warn the biologist that his efforts to "understand" completely the function of the human brain are doomed to fail, because no logical system can completely describe its own structure. However, such a warning would not yet be motivated, considering how far off we are from this absolute bound for human knowledge.

Perhaps Monod is here referring to Tarski's results outlined in Section 14.3.1. It is interesting to note how Monod interprets the logical results in a negative sense, as absolute bounds to the human knowledge, bounds to which we may eventually come dangerously close.

Surely the autolinguistic perspective suggests bounds, but not of a nature such that human knowledge could progress to a certain frontier and then no further. Rather the productive nature of Tarski's results permits a more positive conclusion: Human beings will continually and indefinitely understand more and more about himself so long as mankind exists.

In particular, this means that we should not look for a complete understanding of life. Surely we can understand life as an autolinguistic process; but that knowledge, if expounded, turns into an open hierarchical structure of understandability. It admits knowledge of unknowability, knowledge of the difficulties (complexities) of life, and knowledge of the unforeseeability of the consequences of certain proposed activities. Such is the nature of planning (cf. Section 4.2) that this last type of knowledge may also have a positive planning value. As suggested elsewhere (Löfgren 1979a) the autolinguistic view may even be indicative of goals for planning that may be successively unfolded.

References

Eden, M. (1967), Inadequacies of neo-Darwinian evolution as a scientific theory, in *Mathematical Challenges to the Neo-Darwinian Interpretation of Evolution* (P. Moorhead and M. Kaplan, eds.), Wistar Institute Press, Philadelphia.

Löfgren, L, (1968), An axiomatic explanation of complete self-reproduction, *Bull. Math. Biophys.* 30, 415–425.

Löfgren, L. (1972), Relative explanations of systems, in *Trends in General Systems Theory* (G. Klir, ed.), John Wiley, New York.

Löfgren, L. (1977a), Complexity of descriptions of systems: A foundational study, *Int. J. Gen. Systems* 3, 197–214.

Löfgren, L. (1977b), On existence and existential perception, *Synthese* 35, 431–445.

Löfgren, L. (1978a), The complexity race, in *Applied General Systems Research* (G. Klir, ed.), Plenum, New York.

Löfgren, L. (1978b), Some foundational views on general systems and the Hempel paradox, *Int. J. Gen. Systems* 4, 243–253.

Löfgren, L. (1979a), Goals for human planning, in *Improving the Human Condition*, R. Ericson (ed.), Soc. General Systems Res., Springer-Verlag, Berlin, 460–467.

Löfgren, L. (1979b), Unfoldment of self-reference in logic and in computer science, in *Proc. 5th Scandinavian Logic Symp.*, Jensen, Mayoh, and Moller (eds.), Aalborg Univ. Press, 205–229.

Manna, Z. (1974), *Mathematical Theory of Computation,* McGraw-Hill, New York.

Maturana, H. (1975), The organization of the living: A theory of the living organization, *Int. J. Man–Machine Studies* 7, 313–332.

Mendelson, E. (1964), *Introduction to Mathematical Logic,* Princeton Univ. Press, Princeton, N.J.

Monod, J. (1970), *Le Hasard et la Nécessité,* Éditions du Seuil, Paris.

Pattee, H. (1977), Dynamic and linguistic modes of complex systems, *Int. J. Gen. Systems* 3, 259–266.

Pattee, H. (1973), The physical basis and origin of hierarchical control, in *Hierarchy Theory* (H. Pattee, ed.), George Braziller, New York.

Rosen, R. (1974), Planning, management, policies and strategies: Four fuzzy concepts, *Int. J. Gen. Systems* 1, 245–252.

Scott, D. (1960), A Different Kind of Model for Set Theory, mimeographed (unpublished) paper, read at Int. Congr. on Logic, Methodology and Philosophy of Science, Stanford Univ., Stanford, Calif.

Scott, D. (1970), Outline of a mathematical theory of computation, *4th Ann. Princeton Cong. Inf. Sciences* and *Systems,* pp. 169–176.

Tarski, A. (1956), *Logic, Semantics, Metamathematics,* Clarendon Press, Oxford, Chapter VIII.

Varela, F. (1978), On being autonomous: The lesson of natural history for systems theory, in *Applied General Systems Research* (G. Klir, ed.), Plenum, New York.

Varela, F., Maturana, H. and Uribe, R. (1974), Autopoiesis: The organization of living systems, its characterization and a model, *Biosystems* 5, 187–196.

Glanville is prepared to part with a deep-rooted Anglo-Saxon empiricism in science. It seems that both Locker and Löfgren would find a lot of support for their own ideas in the writing of Glanville (or vice versa).

Locker's unity of the cognizing and creating subject (the "self"), which must precede an object, Löfgren's autolinguistic self-description of objects, and Glanville's self-observation of objects—all these notions seem to come from and are variations of the same tradition: Berkeley's *esse percipi*. They all ask the same question: could anything exist without having both a subjective aspect and the objective properties at the same time? Their answer: no.

The reader might wonder what all this has to do with biology and autopoiesis and why such a heavy dousing with philosophy should be necessary in an introductory book on autopoiesis. In a very distinct sense, our philosophical troika (Locker, Löfgren, and Glanville) is confirming the viability of autopoiesis, its philosophical potential and validity (compare with the philosophy inspired, for example, by DNA transcription).

Glanville avoids the mechanistic issue of "precedence" between subject and object, and conceives the subject as being an outcome of self-observation, as much as is the object. Self-observing and self-observed are "welded together," although they are not contemporaneous but "switching." There is at least one characteristic of any object that cannot be described: its self-observation. It is difficult to deny that a tree exists *because* it observes itself if the self-observation is taken as a privy to the tree.

Similarly, it is difficult to argue with a statement that if *two* things are the same they must also be different. At least one characteristic of each object cannot be described: its self-observation.

Glanville's variations on the "revelation" of physics that the atom is not indivisible should be complemented by the (in)famous assertion of Lenin that the atom is equally infinite as the universe. How many headaches would have been avoided by accepting this simple insight?

An elegant conclusion of Glanville relates to the problem of elementary particles in physics, such as quarks: fundamental particles cannot be described (their autonomy cannot be divided) and hence they cannot be observed. Glanville means that they cannot be "other-observed," because they must be "self-observed" or they would not exist.

The self-observation property of objects cannot be other-observed but it must be self-observed or it would not exist. The self-observation of self-observation . . . ?

Ranulph Glanville was born in London in 1946 on a Friday the 13th. Trained as an Architect at the Architectural Association School in London, he received Ph.D. in Cybernetics under Gordon Pask from Brunel University. The title of his thesis ("The Object of Objects, the Point of Points—or, Something about Things") led to the initial denial of his Ph.D. because the University Library found the topic unfeasible for filing. Dr. Glanville is currently studying Psychology for a M.Phil., while lecturing in Architecture at Portsmouth Polytechnic and at the Architectural Association School. He is an authority on Finnish architecture, speaks Finnish, and was once a high-jump champion. His current research interests include objects (of the sort discussed in this volume), environmental psychology, and fine wines. *Address:* School of Architecture, Portsmouth Polytechnic, King Henry Ist Street, Portsmouth, Hants, England.

Chapter 15
The Same Is Different*

Ranulph Glanville

15.1 Objects

The early concept of an autopoietic system (Varela et. al 1974), that is, one that is alive through giving life, was invoked as an explanation of life. Its tautological format (Varela 1974, 1976) has been explored, in extenso, as has its necessary autonomy (Varela 1977). But for all the work carried out on autopoiesis (to which this volume bears witness), none has really concentrated on its origin as an explanation, as in *The Biology of Cognition* (Maturana 1970). In this paper, the concept of autopoiesis is linked to the more general and formal concept of an *object,* inhabiting a universe of observation (Glanville 1975, 1976), which is the form of those things that can be observed. They correspond to stripped-down versions of those unities Pask calls "topics" (Pask 1976, 1980; Pask, et. al. 1975)—that is, those things that we can conceive and about which we can learn. They do not necessarily have any "reality" in a normal, tangible, truthful sense.[1]

---

* This chapter is a reworking, emphasizing the centrality of the autopoietic idea, of a paper delivered at the First International Conference on Applied General Systems Research: Recent Developments and Trends, Binghamton, New York (Glanville 1977a). The ideas described in this paper are based on my concept of an "object" [contemporaneous with but distinct from Goguen's (1975)]. The paper includes a sketch of relevant bits of this concept, but they are, inevitably, inadequately elaborated and justified. Other papers (Glanville 1976b) are generally available, although they are also incomplete in their handling. The only full handling of the concept to date is in my doctoral thesis (1975), available from Brunel University on Interlibrary Loan.

[1] There is no space here for me to do more than indicate that this "reality" is consensual and dubious. I doubt the validity and usefulness of the concept of reality. Herein lies a main part of the distinction between the constructivist view of von Foerster (1972, 1973) and von Glasersfeld (1974) and my own.

There is a problem in explaining the theoretical unities (objects) that this paper explores. In order to fully satisfy the curious, a particularly long paper would be needed. In the cause of brevity the following somewhat inadequate description must suffice. Something that can be observed is called, in this formulation, an object (Glanville 1975). It is claimed to have the attribute of "observability" because, by observing itself, it comes to inhabit the universe of observation of itself. In order to observe itself, we can argue, it has a means of observing (called a "model facility" and denoted X) and two roles—self-observing (P) and self-observed (E)—which are welded together in our description of the unity that is the object. Such an object is, of itself, self-stabilized through its self-reproduction, although to an external observer it may well appear to change. The question how it can be represented by the external observer, so that such a change can be observed, is tackled in another paper (Glanville 1976b). An object is also autonomous and unique, since it is an autopoietic unity (Varela 1977), albeit both more abstract and less "biological." This requires that, in our understandings, the self-observing and self-observed roles do not exist contemporaneously, and consequently they switch in time.

Such an object is represented thus:

$$\langle O_a \rangle \;\; = \;\; E_a \Leftarrow [(X_a)\, P_a].$$

In this expression, the subscripted a indicates with which object we are concerned (others can be subscripted b, c, d, etc.); the $\langle O \rangle$ is a name (the angle brackets represent the art of naming), and O labels an object. It should be emphasized that the named object is the whole of the expression.[2] The = means "is," and the $\Leftarrow$ means "gives rise to." The square brackets [] indicate an observation being made, the parentheses () that its content (the model facility) is akin to a function. (There is a more elaborate notation, which we will generally avoid, in which time is expressed.)

The expression for an object can be seen to have a form similar to the logician's existential statement:

$$\exists x,\, f(x).$$

---

[2] In his paper in this volume, Locker (1980) argues powerfully and elegantly for the priority of self, that it may observe itself. This is particularly valuable when considering the genesis of autopoietic systems. Objects, however, are not generative. The argument is that an object is a whole, a unity, and that the representation of roles given in the formalizations is given by the (external) observer, to account for the object's believed stability and continuity. In a universe of observation, it is hard to know whether the observation, the observer, or the observed has priority, and it is, at least in the cause of brevity, pragmatically sensible to disregard this matter.

Different objects can be denoted by different subscripts:

$$\langle O_a \rangle = E_a \Leftarrow [(X_a)\, P_a],$$

$$\langle O_b \rangle = E_b \Leftarrow [(X_b)\, P_b],$$

and so on. This form, the form of observables, produces itself without reference to others. Anything of which we can conceive must take this form, for it is observable. It is private, and its self-reproduction maintains its autonomy. Consequently, what it observes of itself no other can observe, and in this lies its uniqueness. That which differentiates $\langle O_a \rangle$ from $\langle O_b \rangle$ is, in the end, the difference of their self-observations.

The form of such an object is the generalized form of an autopoietic system. Through such a form, continuity of existence is generated by the continuous switching of roles, from self-observing to self-observed to self-observing, in an endless regress. Hence the resolution of Dunne's (1934) paradox of the painter painting the whole world. Objects have several such remarkable powers.

15.2   Observation

An object enters the universe of observation through its autonomous and self-referential self-observation. Such self-observation is the root of its uniqueness. It is also the root of noncommunication. Self-observation creates a boundary. How can other-observation transcend this?

The form of self-observation involves a means of observation and the two alternating roles self-observing and self-observed. At the time that the role taken is self-observed, there is no occupant of the observing space, as is shown in the (full) notation of an object.

$$\langle O_a \rangle = \qquad \Leftarrow [(X_a)\, P_a]$$
$$E_a \Leftarrow [(X_a)\quad\ ],$$
$$\Leftarrow [(X_a)\, P_a]$$

and so on. Here we see the two-role cycle of an object; the blank spaces show the alternation between roles, and consequently the role slots left "open."

From the above, it is clear that there are times, dependent upon the object's own internal clock, when there is a vacancy in the observing slot (while the object's self is in the self-observed role). In this way, objects can be characterized through Petri nets (Petri 1962).

Similarly, it will be clear that whenever one role is occupied, the other role is vacant: that is, that when an object is its self-observed role, it is not using, as it were, its observing ability—just as when it is in its self-observing role, it is available to be observed.

Here we have the basis for other-observation. If, during the instant that one object is in its self-observed mode, another object is also in its self-observed mode, the second object can observe the first.[3] If we use the letter F to represent other-observation and B to represent other-observed, we can shown how two objects can so interact.

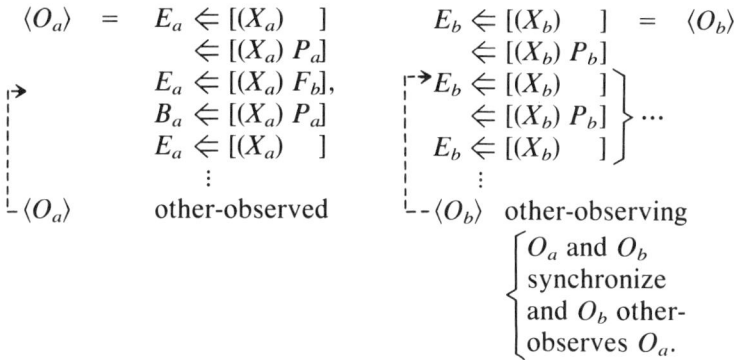

$$
\begin{array}{lll}
\langle O_a \rangle & = & E_a \Leftarrow [(X_a) \quad] \qquad\qquad E_b \Leftarrow [(X_b) \quad] \quad = \quad \langle O_b \rangle \\
& & \quad\;\; \Leftarrow [(X_a)\,P_a] \qquad\qquad\;\; \Leftarrow [(X_b)\,P_b] \\
& & E_a \Leftarrow [(X_a)\,F_b], \qquad E_b \Leftarrow [(X_b) \quad] \\
& & B_a \Leftarrow [(X_a)\,P_a] \qquad\quad\;\; \Leftarrow [(X_b)\,P_b] \;\Big\} \cdots \\
& & E_a \Leftarrow [(X_a) \quad] \qquad\;\; E_b \Leftarrow [(X_b) \quad] \\
& & \qquad\qquad \vdots \qquad\qquad\qquad\qquad \vdots
\end{array}
$$

$\langle O_a \rangle$ other-observed $\langle O_b \rangle$ other-observing

$$
\left\{
\begin{array}{l}
O_a \text{ and } O_b \\
\text{synchronize} \\
\text{and } O_b \text{ other-} \\
\text{observes } O_a.
\end{array}
\right.
$$

In this way one object can observe another object.

In the contrast between self- and other-observation we can find one of the basic quandaries of autopoiesis. Maturana and Varela's argument (1980) is that the autonomy of an autopoietic unity can only be investigated by breaking the unity. Pattee (1977) cannot accept such a position, for it renders traditional science powerless and subjective. We can see here that these two views are not incompatible. The self-observation (precisely that which makes an object itself) is *not* open to other-observation. However, other-observation is possible, although partial. But other-observation is "subjective" in that it is truly observation dependent. This is what one would nowadays except, although, as Varela shows, such other-observation is reentrant (1974, 1976). But Pattee's wish for an "objective" view can be realized in some form provided that the understanding of partiality (in both senses of the word) and of descriptiveness and hence truthfulness is accepted, by applying an especially carefully organized pruning to the potential for observation, of the type proposed and developed by Pask and Scott (1973), which resolves the hygeinisizing problem raised by Rosen (1977).

Thus, the formulation for an object, differentiating self- from other-

---

[3] There is no need, here, to argue "should" instead of "could," but it can be done, using a relativistic idea of the relationships between the individual object's clocks, of the sort developed by Atkin (1976).

observation, can resolve such epistemological differences with relative ease.

15.3 Relationships

When one object is in its self-observed mode, and hence is available for other-observation, there is no reason why it should other-observe only one other object.

It can observe two (or more) other objects in several ways, depending upon the relationships of the times at which the other-observations are made to each other.[4] Should it happen that, with reference to the other-observer's internal clock, there is no overlap, then there is no relationship. Calling the other-observed objects $\langle O_a \rangle$ and $\langle O_c \rangle$, we have a negation:

$$B_a = \langle O_a \rangle, \quad \langle O_c \rangle \neq B_c \qquad \text{or} \qquad B_a \neq \langle O_a \rangle, \quad \langle O_c \rangle = B_c.$$

If the time of observation of one includes the time of observation of the other, we have an implication:

$$B_a = \langle O_a \rangle \;\rightarrow\; \langle O_c \rangle = B_c \qquad \text{or} \qquad B_a = \langle O_a \rangle \;\leftarrow\; \langle O_c \rangle = B_c.$$

If the times of observation overlap, we have either a conjunction or a disjunction:

$$B_a = \langle O_a \rangle \;\wedge\; \langle O_c \rangle = B_c \qquad \text{or} \qquad B_a = \langle O_a \rangle \;\vee\; \langle O_c \rangle = B_c.$$

Which of these alternatives happens depends upon the exact conditions of the observations, which need not concern us here. In fact, we have shown (1976a) that, as a general example, we need only consider the case in which the times of other-observation of two objects are the same, when we have an identity:

$$B_a = \langle O_a \rangle \;\leftrightarrow\; \langle O_c \rangle = B_c.$$

This is the case we will therefore consider.

15.4 Description

It is in this ability to other-observe, and to relate several objects together through the times of other-observations, that the power of this formulation in building descriptions lies. This is an important ability, for unless descriptions can be made of autopoietic unities (and more generally of self-reproducing systems and of objects), the concept is not only useless, it passes into Wittgenstein's (1971) ineffable: "What we cannot speak about we must pass over in silence."

---

[4] For an elaborated treatment, see my earlier papers (1975, 1976b). A new paper covers the consequences of such relationships in detail (1976a).

But it is equally important that the logic of (such a) description is understood and that understanding should include an assessment of its limitations (Glanville 1977a). Self-observation is privy to the self-observer. Other-observation is an act of communion between (only) two objects: the other-observing and the other-observed. This, too, is essentially private, with only the other-observing object observing. In this respect, there is privacy in all observations. And yet, our general experience is that we can communicate. How could this be, and what would this indicate?

The linguist de Saussure (1966) proposed that meaning was communicated by the synchronizing of two streams of unrelated material. He maintained that that which was represented, and that which represented it, attained meaning (that is, formed a representation) through this mechanism. He illustrated his argument by denying that the sound used to depict a tree and the image of a tree have anything in common, bar a synchronous chopping of a continuum. His example can be presented thus:

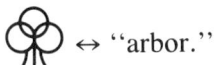

$$\text{🌳} \leftrightarrow \text{``arbor.''}$$

This analysis (with, for instance, its inability to accept onomatopoeia) should be regarded as naive. But that is not to deny the value of its initial insight, the insight of the concurrent slicing of time in synchrony as being a prerequisite for the creation of a (temporary) identity between two quite separate (i.e., only arbitrarily connected) things.

The similarities between this view of de Saussure's, and that arrived at by a quite separate route in this paper, will be quite clear. What is being said is that, when to a common observing object the times of observation of two other objects synchronize, a description of representation can be made. The characteristics and form of such a description are

1. that there are two other-observed objects between which an identity is observed;
2. that the identity is the sameness in the description; and
3. that the (described and describing) objects are nevertheless different.

Thus, in a description, we can say that

$$\underset{\text{described object}}{B_a = \langle O_a \rangle} \quad \leftrightarrow \quad \underset{\text{describing object}}{\langle O_c \rangle = B_c} \qquad \text{(for } O_b \text{ observing).}$$

It is the whole of the expression that is the description, from which we can immediately see that the described and describing objects are also merely roles. Thus we have to say that, using de Saussure's example, 🌳 and "arbor" inherently form a description when placed together; and that when 🌳 is the described object "arbor" can be the describing object (and so can "tree"), just as when "arbor" is the described, 🌳 (or "tree") can join with it to make a description.

15.5   The Same Is Different

We have elaborated how other-observations can be made of systems that owe their ability to exist in this universe to their self-observation; and we have examined how such objects can be used together to make descriptions. Other papers use "conversation theory" to examine how such descriptions can communicate meanings; it is enough here to accept that such descriptions can be made (Pask 1973, 1976; Glanville 1977a). But we have also claimed that such self-observing systems are the only inhabitants of our universe of observations. If that is the case, we should be able to illuminate some problem in description that has, as yet, remained opaque.

The characteristic of computed sameness between two yet different objects is where we will direct our attention. It would normally be assumed that where there is an identity, the things identified are the same, not different. Thus, in a semantic tree, it is assumed that meanings and qualities are actually subsets of an object's set of characteristics. The argument here is that every other-observed object is different, the relationship "same as" being computed from the distinct observations. Hence the title, "The Same *Is* Different" (otherwise if two things are the same they must also be different, or there is only one thing).

Thus when a description is made, there is at least one characteristic in each object that is not described: the characteristic that gives identity and autonomy to the object, that is, its self-observation. The describing object and the described object both maintain their individuality and their autonomy. The "sameness" lies in the computation of similarity in observations. Thus the "sameness" is not a property of the two objects, but rather of the observations. Order and hierarchy, predication and connection, all result from cognizant observation.

15.6   Fundamentals and Elementary Physics

In October 1975, *Scientific American* published a paper entitled "Quarks with Color and Flavor." It (almost) starts thus:

One of the principle achievements of physics in the 20th century has been the revelation that the atom is not indivisible or elementary at all but has a complex structure. In 1911 Ernest Rutherford showed that the atom consists of a small, dense nucleus surrounded by a cloud of electrons. It was subsequently revealed that the nucleus itself can be broken down into discrete particles, the protons and neutrons, and since then a great many related particles have been identified. During the past decade it has become apparent that those particles too are complex rather than elementary. They are now thought to be made up of simpler things called quarks. A solitary quark has never been observed, in spite of many attempts to isolate one. Nonetheless, there are excellent grounds for believing they do exist. More important, quarks may be the last in the long series of progressively finer structures. They seem to be truly elementary (Glashow 1975).

This quote is indicative of a changed attitude in physics, for recently physicists have rejigged their model of the universe, in order to come to terms with the problem they have with fundamental or elementary particles (Calder 1977). They have done so in two ways.

The first (and less popular) is Geoffrey Chew's bootstrap theory, which indicates that, depending on what you wish to consider, you must choose the appropriate theory. Consequently, this indicates that there are no "fundamentals," but that a theory will, as it were, cut the stream of reality in a particular way. Choosing the appropriate cut will allow a view that one wishes to see, but *not* the reality, and, since reality has been cut, the view given is entirely artificial. Compare the argument between Maturana and Varela (1980) on the one hand, and Pattee (1977) on the other. Capra, Chew's leading publicist, would claim that such access to the continuity behind the theory cannot be attained through theory, or through theorylike thinking, and he refers to *The Tao of Physics* (1976).

The second theory elaborates Murray Gell-Mann's quarks. Quarks were proposed as a new level of elementary particle, to account for the plethora of previously considered fundamental particles. They had great success in doing this, but there was one slight hitch: in spite of colossal effort, no quark has ever been observed. In another *Scientific American* article, Nambu (1976) argues that

> an elementary particle is one that has no internal structure.... The assumption is that [quark] color is completely unobservable.... [The way in which] particles are composed of quarks bound together by the exchange of gluons can be given an elegant mathematical formulation. The model is an example of a non-Abelian gauge theory.... Non-Abelian gauge theories are distinguished...by the fact that the fields themselves carry quantum numbers. A field therefore can act as a source of itself.

Thus, in this second approach physicists seem to have come to the conclusion that their elementary particles are unobservable, and they are trying to develop several models that account for this. Both of these approaches to the fundamental or elementary particles of nature are based on arguing for limits to the observable. They arise in one case from the nature of observation, in the other from what is to be observed.

There is, however, another approach, which uses the logic of description. The means of observation the physicist uses today is technically, as well as theoretically, not a means of observation at all, but rather a means of description. The physicist who observes particles is actually observing a series of photographs taken in a bubble chamber. These photographs are the describing object in a description in which the particles themselves are the described objects. The particles are never in themselves observed.[5] Let us consider just one particle and one photographic

---

[5] Noted in an anonymous, untitled paper on particle physics theories of observation presented at Maidstone College.

trace. When we make a description, we have

$$\langle O_{part} \rangle = B_{part} \quad \leftrightarrow \quad B_{trace} = \langle O_{trace} \rangle,$$

and, as we have noted, the sameness in the description also demands a difference. But here is the critical point: Something that is fundamental, elementary, has no structure, is random, and cannot be divided or simplified. [For an elaborate treatment of the concept and consequences of the fundamental, see Glanville (1977b).] Yet for two things to be different from each other, each needs at least one distinguishing feature (its autonomy). That which is fundamental, however, cannot be divided (its one indivisible feature is its autonomy). It follows that something fundamental cannot be described, because its one feature is a matter of self-observation, not other-observation, and because to describe it would be to divide it. It is of little wonder that a fundamental particle is unobserved, for it is not describable. Hence observation of a describing object used in a description for which a fundamental particle is the described object will contain nothing, because a fundamental particle and another object cannot be the same; there can only be a difference. The stream of reality that Chew claims to be unapproachable through theory (description) thus gives the same view as the unobservable quark.

Thus, through the concept of an object, we can understand the essential logic of description: the same is different. This tells us that physicists' problems with their elementary, or fundamental, particles are not inherent in the means used to observe them, but are a consequence of the cybernetics of self-reproduction, the organization of cognition, and the logic of description. This does not appear to be a matter that physics can resolve, for that which is fundamental is not only an object, but is undescribable.

15.7 Conclusion

This chapter has elaborated a more general and formal statement than is usual of the autopoietic concept, the object, which is a conceivable entity inhabiting a universe of observation, and has shown that other-observation of such a self-observing system is not only possible, but leads to a way of protecting the object's autonomy while permitting relations to be computed by an other-observer between other-observations of objects. In so doing it has attempted to resolve the differences between subjectivism and objectivism. The ability to compute relationships has been shown to allow descriptions to be made of a type that elucidate the problem of the apparent nonobservability of elementary particles in physics, for such particles are necessarily nondescribable, and consequently any attempt at description will miss them. This result (which could easily be used in areas other than physics) demonstrates the power of the concept of an object, for which the same is different, and in which organizational circularity does not lead to tautological uselessness.

References

Atkin, R. (1976), "Time as a Multi-Dimensional Structure," Dept. of Mathematics, Essex Univ., Colchester, England.

Calder, N. (1977), *The Key to the Universe*, BBC, London.

Capra, F. (1976), *The Tao of Physics*, Fontana-Collins, London.

de Saussure, F. (1966), *Course in General Linguistics*, McGraw-Hill, New York.

Dunne, J. (1934), *The Serial Universe*, Faber & Faber, London.

Foerster, H. von (1972), *Notes on an Epistemology for Living Things*, Biol. Computer Lab., Univ. of Illinois, Urbana.

Foerster, H. von (1973), On constructing a reality, in *Environmental Design Research*, Vol. 2 (N. Preiser, ed.), Dowden Hutchinson & Ross, Stroudsberg, Virginia.

Glanville, R. (1975), "The Object of Objects, the Point of Points—or, Something About Things," Ph.D. thesis, Brunel Univ., Uxbridge.

Glanville, R. (1976a), Consciousness: and so on, in *Progress in Cybernetics and Systems Research*, Vol. 6 (F. Pichter, ed.), Hemisphere, Washington, D.C., 1980; also in *Int. J. Cybernetics* 10 (1980).

Glanville, R. (1976b), What is memory, that it can remember what it is?, in *Progress in Cybernetics and Systems Research*, Vol. 4 (R. Trappl, ed.), Hemisphere, Washington, D.C.

Glanville, R. (1977a), The logic of description—or, why physics won't work, *1st Int. Conf. on Appl. Gen. Systems Res.: Recent Developments and Trends, Binghamton, N. Y.*.

Glanville, R. (1977b), The nature of fundamentals, applied to the fundamentals of nature, *Proc. 1st Int. Conf. on Appl. Gen. Systems Res.: Recent Developments and Trends, Binghamton, N. Y.*

Glasersfeld, E. von (1974), Piaget and the radical constructivist epistemology, *3rd S.E. Conf. of Soc. for Research on Child Development*, Chapel Hill, S. Car.

Glashow, S. (1975), Quarks with color and flavor, *Sci. Amer.* 233 (No. 4), October.

Goguen, J. A. (1975), Objects, *Int. J. Gen. Systems* 1, 237–243.

Locker, A. (1980), Meta-theoretical presuppositions for autopoiesis (present volume).

Maturana, H. (1970), *The Biology of Cognition*, Biol. Computer Lab. Res. Report 9.0, Univ. of Illinois, Urbana.

Maturana, H. and Varela, F. (1980), Autopoiesis (present volume).

Nambu, Y. (1976), The confinement of quarks, *Sci. Amer.* 235 (No. 5), November.

Pask, G. (1976a), *Conversation Theory, Applications in Education and Epistemology*, Elsevier, Amsterdam.

Pask, G. (1976), Position paper for *Co-evolution Q. Conf.*,

Pask, G. (1980), Organizational closure of potentially conscious systems (present volume).

Pask, G. and Scott, B. (1973), CASTE: A system for exhibiting learning strategies and regulating uncertainties, *Int. J. Man–Machine Studies* 5, 17–52.

Pask, G., Scott, B. and Kallikourdis, D. (1973), A theory of conversations and individuals (exemplified by the learning process on CASTE), *Int. J. Man–Machine Studies* 5, 443–556.

Pask, G., Kallikourdis, D. and Scott, B. (1975), The representation of knowables, *Int. J. Man–Machine Studies* 7, 15–134.

Pattee, H. (1977), The necessity for self-description for the evolution of autopoiestic systems, *1st Int. Conf. on Appl. Gen. Systems Res.: Recent Developments and Trends, Binghamton, N. Y.*

Petri, C. (1962), "Communication with Automata," in draft translation from A.I. Lab, MIT, Cambridge, Mass.

Rosen, R. (1977), Multiple descriptions and similarity, *1st Int. Conf. on Appl. Gen. Systems Res.: Recent Developments and Trends, Binghamton, N. Y.*

Varela, F., and Goguen, J. (1974), "A Calculus for Self-Reference," (first draft), Univ. Nacional, Puerto Rico.

Varela, F. (1976), The arithmetic of closure, in *Progress in Cybernetics and Systems Research*, Vol. 5 (R. Trappl, ed.), Hemisphere, Washington, D.C.

Varela, F. (1977), Autonomy, *1st Int. Conf. on Appl. Gen. Systems Res.: Recent Developments and Trends, Binghamton, N. Y.*

Varela, F. Maturana, H. and Uribe, R. (1974), Autopoiesis: The organization of living systems, its characterization and a model, *Biosystems* 5, 187–196.

Wittgenstein, L. (1971), *Tractatus Logico-Philosophicus,* 2nd Ed., Routledge & Kegan Paul, London.

Pask provides a framework for placing autopoiesis within a particular historical perspective. He overviews and defines his notions of system, stability, organization, closure, and the like. He then links autopoiesis with his own theory of conversation and discusses the difficult subject matter of *consciousness*.

The style of the presentation is that of ordered assertions, similar to that already encountered in the work of Ben-Eli. It is not a simple paper to understand, especially for those who are not used to reading in a metalanguage. There is a short appendix that amplifies some of the ideas that might still be only implicit in the text.

Pask does not draw a distinction between organizational closure and autopoiesis. Varela insists that the term "organizational closure" was introduced precisely for the purposes of eliciting such distinction. Pask in turn insists that the notion of autopoietic systems is already discernible in the works of Ackoff and Beer, or Bateson and Mead, or even Wiener and Svoboda! But the presentation of his own theory of conversations is brilliant and its implications for autopoiesis are potentially significant.

His "P individual" is characterized as the minimal conscious autopoietic system: the relation of autopoiesis to consciousness is undoubtedly of fundamental interest. Pask argues quite powerfully for conversation, rather than interview or stimulus–response, as the minimal situation for observing consciousness.

The reader should be very careful about such statements as "a stable process is organizationally closed." Pask uses metaphors quite freely and his "stable" has nothing to do with the classical notion of stability. Actually, his "stable" is just another metaphor given for "organizationally closed." (See Varela's paper in this volume for another definition of organizational closure.)

There is great power and insight in some of Pask's conclusions. The minimum conscious system is a conversation (which can be both external and internal, i.e., a dialogue or talking to oneself). This editor believes that one could consider as protoconsciousness the ability of (some) living systems to communicate ("converse") with themselves (and others) via the environmental loop. In this sense, all autopoietic systems are potentially conscious or even protoconscious.

The concept of a boundary, although not physically embodied, is nevertheless intuitively apparent through the externalization and internalization of understandings.

Finally, there is no limit to the size of a conversation. A society or a civilization is organizationally closed, just as is a family or a person. Here again is the concept of social autopoiesis.

Pask's final thought is a very proper conclusion to our book on autopoiesis. To paraphrase his concluding statements, there is no need to ask *why* there are autopoietic systems—they are the units of reality. The question to be asked is, Are there any allopoietic systems except those engendered and engineered by the static artifice of a "program"?

Gordon Pask received an M.A. (Nat. Sci. Tripos) from Cambridge Downing College, a Ph.D. (psychology) from London University, and the D.Sc. (research in cybernetics and cognitive systems) from the Open University. He has held visiting professorships at the Universities of Mexico, Illinois, Oregon, and at the Georgia Institute of Technology. During 1980 he is a fellow of NIAS in the Netherlands. Dr. Pask is director of Research at System Research, Ltd., and a Professor in the Department of Cybernetics, Brunel University, and in the IET Open University. He is also President of the Cybernetic Society in London, and a past president (1974) of the Society for General Systems Research where he still retains a post on the panel of "Distinguished Advisors." He was elected an honorary member of the Austrian Society for Cybernetic Studies. Dr. Pask serves on the Editorial Boards of *Instructional Science, Policy Analysis and Systems Science, Behavioral Science,* and *International Journal of Man-Machine Studies.* He serves as consultant to various organizations and lives in Richmond with his wife and two daughters. *Address:* System Research, Ltd., Woodville House, 37, Sheen Road, Richmond, Surrey, TW9 1AJ England.

Chapter 16
Organizational Closure of Potentially Conscious Systems*

Gordon Pask

16.1 Introduction

The notion of organizationally closed and autopoietic systems has been invented more or less independently and in various contexts, though the term itself and its careful application to living systems is due to Maturana and Varela. For example, much of von Neumann's work on reproductive automata and the content of the early Macey Foundation meetings on cybernetics refers to similar constructs. So, on serious examination, does von Foerster's first enunciation of "Self Organization" in 1958, as does McCulloch's notion, "Redundancy of Potential Command." Much the same is true of work in other disciplines, including that of Wiener and Svoboda in mathematical cybernetics, Herbst in logic, Bateson and Mead in social anthropology, Waddington, Tyler Bonner, and others in embryology and genetics, Wynne Edwards in ethology, Ackoff and Beer in operational research, and numerous cosmologists and theoretical physicists. The list is enormous, because this quite basic reappraisal of what systems are and what stability is reflects a very fundamental change in thinking. Only in recent years, however, has there been either the language required to express the pertinent notions or a sufficiently large body of shared concepts to render these notions communicable and generally intelligible.

In this paper I attempt to give a systematic theoretical account of my own ideas, which originated independently (whatever that means, and I am no longer at all certain) but fell into the context about 15 years ago

* Prepared for the NATO International Conference on Applied General Systems Research: Recent Developments and Trends, and presented at Binghamton, New York on 22 August 1978.

of those of von Foerster, Maturana, and Varela. The concept of organizational closure is crucial to a psychological or social "theory of conversations," in which the minimal *conscious* autopoietic system is known as a "P individual" (psychological individual). The empirical background for my own work came in part from studies of complex skill learning, especially from detailed examination of the conceptual mechanisms of educational psychology. More recently the work has been augmented by studies, in similarly detailed enquiry, concerned with complex decision making, social organization innovation (creativity, design), and the burgeoning field of applied epistemology.

16.2   Process Execution

Let Z be a variable with values $A, B, \ldots$ that designate *processes* or active systems. This paper concerns those values of Z, say $Z^* \subset Z$, designating processes that are sites or progenitors of consciousness. For generality, these are known as "conscious systems." Particular interest is accorded to conscious systems for which, at any rate in principle, an external observer can determine the content of consciousness by observing a sharp-valued event called in Section 16.3 an understanding and the extent of the consciousness by a fuzzy (*not* sharp-valued) measure.

Given a process, it is often convenient to distinguish between a processor and a code or program (in general, a nondeterministic program, in a slightly special sense, a fuzzy program [Zadeh 1973]). To avoid misinterpretation, let us call a code or program when it is undergoing execution, a *procedure* (or, for brevity, a Proc). In this case, the processor of $Z = A, B,\ldots$ is $\lambda(Z)$ and the code or program of $Z = A,B,\ldots$ is $\pi(Z)$. It is also useful to employ the notations $\pi(A) = a$, $\pi(B) = b\ldots$; and $\lambda(A) = \alpha$, $\lambda(B) = \beta$: Note, however, that Z does not have a value on "*a* alone," or "α alone," or on "*b* alone" or "β alone," for where Z has a value there is invariably a process (a procedure undergoing execution).

There is a sense, to be developed, in which $\pi(Z)$ constitutes the formal linguistic or syntactic aspect of whatever is designated by a value of Z and $\lambda(Z)$ constitutes its interpretative or semantic aspect. Pragmatics appear (hence, a complete semiotic is attained) only if both aspects are brought into consideration. But these images, though useful in their own way, reflect an underlying reality: Z does not point at objects $A,B,\ldots$ that can properly receive *only* impersonal or *it* reference.

16.3   Conversations, Explanations, Concepts, and Participants

Most of the empirical support for the notions spelled out in the sequel comes from work upon conscious human beings. Values $A,B,\ldots$ of Z designate human beings or groups of two or more human beings engaged in conversation. In concert with Dienes, Piaget, Landa, Luria, Vygotsky, and many others, we have found it fruitful to regard a conversation as the minimal situation for observing psychological events of which the participants are conscious (in contrast, for example, to an input–output or stimulus–response situation), and it has been possible to develop a theory of conversations (Pask 1973, 1975a,b, 1976a–c, 1977a,b), in which the event of understanding is definable and pivotal.

Frequently, the conversation between two human beings, or a normally internal conversation (thinking) between perspectives or roles adopted by one human being, takes place *through* a computer-regulated interface designed to exteriorize normally hidden conceptual operations and to expedite the observation of understandings. In these conditions, the conversational language need not be a verbal, natural language, though a rich symbolic medium with many of the properties of natural language is mandatory. This medium is a language, although not a spoken language, called L. For $Z = A,B,...,$ $\pi(Z)$ is a collection of L expressions; for example, programs (or codes) are L expressions.

We are particularly interested in program listings, conceived as explanations, since the basic and sharp-valued measurements we can make (as observers of understandings) consist in explanations of explanations (which are subsets of coherent and symbolically represented beliefs). One obstacle in the way of psychological enquiry, in the interview mode, is that ordinary language is ambiguous in the sense that there is no easy or systematic means of determining what *is* an explanation (still less an explanation *of* an explanation). For example, explanations do not have to be "true," or "veridical," and most of them are not. Hypotheses and coherent myths are permissible explanations. This difficulty is surmounted in a nonverbal language like L, for an explanation is clearly a prescription (or a prescriptive behavior) involved in L programming a working model (L program) that is constructible and works. It may be a piece of sculpture, incidentally, just as well as a piece of standard calculation or the demonstration of a physical principle.

In L, verbal explanations can be disambiguated as nonverbal, model-building, L explanations; understandings are detectable as cycles involving explanations of explanations. Under these circumstances, it is possible to speak of strict conversations (understandings are ordered in a strict sequence) as contrasted with liberally organized conversations. To quantify the understanding in the latter case, minimal equipment is an epistemological laboratory (Figure 1) in which it is possible to record and regulate nonverbal, but symbolic, interactions (L interactions), which correspond to most types of verbal dialogue encountered in interpersonal discussion, learning, innovation, agreement reaching, design, evaluation, theory building, and the like. This appears to be the proper equipment for paradigmatic studies of consciousness, where results obtained by field investigation can be refined and their conceptual basis well specified. The appropriate frame of reference, *relative* to which conversations are stud-

Figure 1. An epistemological laboratory. The figure is placed first to give an idea of how empirical results are obtained but many of the labels will not be intelligible until, at a later stage in the paper, the functioning of the equipment and the types of man machine transactions are spelled out. The key is as follows:

A. Random access slide projector with control keyboard, for displaying slide mounted graphics.
B. Entailment mesh display with overlay multi sheets, containing 60 node positions and 4×60 independently addressed coloured signal lamps and touch sensors.
C. Tutorial mode keyboard with special function keys.
D. Course assembly mode keyboard with special function keys.
E. ARDS graphic display tubes with control unit and keyboard used for displaying "pruned" meshes.
F. Video display units with control keyboards used for topic text input–output.
G. Pigeon holes filing system with slots for 60 files and containing 3×60 independently addressed signal lamps and 60 sensors.
H. Dual drive floppy disk unit.
I. CAI 32k computer.
J. Digital Cassette unit used as mass storage device.
K. ASR 33 teletype used for "hard copy" output.
L. Electronics rack, containing special electronics and system interface.

ied, is a possibly evolving "conversational domain," that is, the "environment" of conversation theory.

The reader may find it useful to keep these empirical comments in mind, without supposing that consciousness is necessarily restricted to people or groups of them.

In the same spirit let us use commonplace terms, such as "concept" (abbreviated to Con), "memory" (abbreviated to Mem), and "topic," in a precise, but somewhat broader-than-usual sense.

Although it sounds odd to speak generally of memories and concepts, the argument is rendered succinct and intelligible because we are familiar with these things by personal experience.[1]

The theory of conversations is both relativistic and reflective; only in a logically degenerate but highly developed form is the theory simply a relativistic theory. The fully-fledged version is a theory of participants *in* conversations, not merely a theory *about* participants, couched in an external observer's terms relative to a conversational domain. The sharp-valued events of a conversation, namely understandings, are quantifiable but not, strictly speaking, *objective*. They are not *it*-referenced events but subjective (I-, you-referenced) events, either in whole or in part. Only stable processes are observable sharp-valued understandings.

16.4   Organizational Closure, Distinction, and Independence

A stable process is "organizationally closed" (von Foerster 1976; Varela 1975, 1976; Goguen 1975; Bråten 1976). In biology it is called an auto-poietic system (Varela, Maturana, and Uribe 1974; Maturana and Varela 1976) and autopoiesis is characteristic of life.

It should be emphasized that the stability criterion of organizational closure is quite distinct from the classical notion of stability (i.e., a system with states represented as points in a prespecified structural framework of coordinates, having behavioral trajectories that converge to a fixed point, or to a limit cycle to which they return if disturbed by small, but arbitrary, perturbations). Classical stability is a special case of organizational closure.

In the sequel, "stable" means "organizationally closed" and might be rephrased as inherently self-reproductive. For example, Löfgren's (1972, 1975) reproductive Turing systems are simulations of organizational closure. Without denying their utility or failing to appreciate the elegance of simulations, it is important to realize they (or like constructs, open to realization in ordinary digital computers) are posited as simulations of

---

[1] I formerly thought this mode of speaking was little more than an expository trick, a use of metaphor. Today, I know it is a metaphor, but also that it is much more than a trick of exposition.

general systems; notably of systems involving "organizational closure" (see, for example, Ben Eli 1976; Ben Eli and Tountas 1976).

If an organizationally closed system is "opened," for instance, by distinguishing structure and behavior, or by demarcating the linguistic interchange of a conversation, or by instituting the cleavage of Z into $\pi(Z)$, $\lambda(Z)$, then it is reproductive and productive. This last idea, "reproductive and productive," has a lengthy history and, in psychology at least, has been the focal point for considerable and occasionally acrimonious debate. For instance, the associationists were plagued by the difference between reproduction (in the sense of strict replication, through associative principles, of ideas, sense data, etc.) and replication. Seltz, and later the Gestalt psychologists concerned with problem solving and thinking, for example, Duncker and Wertheimer, saw through the distinction, but did not have the notation required to give it expression. So, in a different, more eclectic, tradition did Bartlett. Even today, it is quite difficult to point out that, in *general,* reproduction is productive, and replication is a limiting case of reproduction.

Here the required meaning is approximated by speaking of *fuzzy reproductive processes,* which reconstruct classes of somehow equivalent processes and patterns. The connotation of "fuzzy" is compatible with Zadeh's or Gaines' use of the term (though it is not quite identical with the usage of either). The notion of fuzzy production and reproduction is refined beginning in Section 5. As a first step, the following propositions spell out the character of process distinction and independence.

Process is a more fundamental notion than time. In particular, the point interval (Newtonian) time, evident in the following comments, is a very specialized frame of measurement (see Atkin 1977).

Two different events may only occur at the same place (common location in a storage medium) if they occur at different times.

Two different events may only occur at the same time if they are at different places.

These differences reflect independence.

Processes *are* asynchronous if they occur in different processors. (They *may* be asynchronous in the same processor if it has a rich enough structure; for example, if it is a concurrent machine.)

Two independent systems are rendered dependent by information transfer, in Petri's (1965) sense.

Equally, two asynchronous systems may be coupled or rendered partly or locally synchronous by information transfer.

The most fundamental analogy relations (hence, the broadest and most general) are static inscriptions of coupling or dependency or local synchronicity, between otherwise independent or otherwise asynchronous processes.

To manufacture independence is to make a distinction. Only a process *can* make a distinction. The most fundamental distinction is *any* distinction, predication unqualified (Spenser Brown's Γ operation). But suppose that in a system that is a process there exist (relative to this system) certain subsystems; then the distinction may be reversible or not. We should expect to, and do, retrieve this basic difference, as analogies of form, which are symmetric, and analogies of method, which are not symmetric, except when no point of view (perspective) is adopted (or, to put it in another way, when there is only the system's point of view).[2]

16.5 Universes and Independent Precursors; Consistency, Subsistency, and Coherence as Truth Values

A *universe* is an a priori independent processor; it is a set (the usual connotation of "universe") but with action built into it. In classical model theory, a model is a relation, induced by an interpretation of a linguistic statement, upon a set called the universe. A logic (i.e., the interpreted language and a calculus for statement generation) is *consistent* if all *true* statements of the logic have models in all possible universes.

In nonclassical model theory, the model is a working or dynamic model; that is, a program compiled and interpreted in a processor. If the program is executed as a process, then a relation, or a "classical model," is "brought about" or satisfied in the product of the program input domain and its output domain.

As Löfgren (1975) points out, it is frequently sufficient to be content with *subsistency* rather than *consistency* in a logic; that is, true statements have models in *some,* but not necessarily in *all* independent universes. The nearest we come to veridical truth is "subsistence truth." Further, the truth value set is "executability–incoherence," rather than "true–false," and often there are degrees of subsistence truth.

In addition, we invoke coherence or systemic truth (Rescher 1973) to form a logic of agreement (in contrast to absolute veridicality). Rescher specifies coherence truth within the propositional calculus, pointing out that the same ideas are readily extended to a predicate calculus. He is, for example, concerned with the problem of accepting or rejecting data that have truth *candidacy* as part of a set of not-inconsistent propositions (perhaps the basis for a theory shared by several observers who are testing

---

[2] I owe this insight to J. Goguen and F. Varela and independently to S. Beer, to J. Zeidner, and to S. Bråten, all in personal communications (1976,1977). The matter is discussed succinctly in the Appendix of *The Human Dyad: Systems and Simulations* (Bråten 1977), which recounts a seminar with P. G. Herbst.

it). It is particularly valuable to have an incisive distinction between coherence of a set of beliefs (the theory) and a set of data (the truth candidates), in contrast to mere consensus of beliefs. However, as it stands, the coherency is based upon a process-free logical property; the implications of statements are thought out inside the observers.

For the present purpose, we need to regard coherence as a property of statements undergoing execution—that is, coherence between processes—and this extension of Rescher's idea, though it involves some technical difficulties, does not appear to change the fundamental notion. In fact, as much is suggested by Rescher's occasional use of the term "systemic" truth as "coherence" truth.

Specifically, we regard a process (X) as being coherent with a process (Y) insofar as X and Y can be executed without computational conflict. We thus augment the original idea by making it processor dependent. Whether or not the π component of X, $\pi(X)$, is coherent with the π component of Y, $\pi(Y)$, depends upon the processor $\lambda(X, Y)$ in which these programs are executed, as well as the programs or statements themselves. For example, the processor may be serial, concurrent, or parallel, and composed of many independent processing units. The program or code (Prog) of a procedure (Proc) is commonly a set of L production rules and the procedures under execution (Proc) figure as interpreted production rules undergoing execution. We do *not,* however, insist upon "serial execution" unless specifically stipulated and, in general, are concerned with L productions carried out over several nonclassical universes (i.e., several processors).

16.6   Concepts, Procedures, and the Processes in Which They Are Executed to Yield Descriptions or Behaviors

Let a concept (Con) be a procedure, or a class of procedures, at least some of which are executed concurrently (Petri 1964; McCulloch 1966) but tend, in the limit, to parallel execution. Let $\triangleq$ stand, as usual, for "defined as equal to." If a class of entirely (conflict-free) parallel Procs $\triangleq$ [Proc] and, if a class of simultaneously executed Procs (with some computational conflict) $\triangleq$ {Proc} then, if $\langle$ and $\rangle$ indicate ordered entities,

$$\text{Con} \triangleq \text{Proc or } \langle\{\text{Proc}\}, [\text{Proc}]\rangle \text{ or } [\text{Proc}],$$

such that, under continual execution, {Proc} → [Proc].

Let Inter be the compilation of a Prog in a given processor, so that it may be executed as a process. That is, Proc $\triangleq$ $\langle$Prog, Inter$\rangle$. Just as $\lambda(A)$ = α, $\lambda(B)$ = β, or $\lambda(A)$ = a, $\lambda(B)$ = b, so also $\lambda(\text{Proc})$ = Inter and $\pi(\text{Proc})$ = Prog.

The requirement that {Proc} → [Proc] under continual execution may be regarded as a property of the class of processors, among them brains, in which the Prog are compiled (as well as a property of the programs themselves). The compilation of programs is reorganized (the programs are recompiled) to achieve the parallelism.

Let $\bar{R}_i$, $\bar{R}_j$, ... be interpreted relations in the product of the input and output domains of Proc. If Ex stands for "execution of," then

$$\text{Ex(Proc } i) \;\Rightarrow\; \bar{R}_i, \qquad \text{Ex(Proc } j) \;\Rightarrow\; \bar{R}_j.$$

Let p, q, ... be indices of Prog $\triangleq \pi(\text{Proc})$ and let u, v, ... be indices of Inter $\triangleq \lambda(\text{Proc})$ that are a priori independent processors.

$$\text{Proc } i = \langle \text{Prog } p, \text{Inter } u \rangle,$$

$$\text{Proc } j = \langle \text{Prog } q, \text{Inter } u \rangle,$$

$$\text{Proc } l = \langle \text{Prog } p, \text{Inter } v \rangle.$$

Ex(Proc l) $\Rightarrow \bar{R}_k$ but $\bar{R}_l \neq \bar{R}_i$, even though the same Prog is involved (namely, Prog p), since $u \neq v$. Hence $\bar{R}_i$, $\bar{R}_j$, ... are *interpreted* relations given, in extenso, as *descriptions*.

The usual "relation in extenso," regarded as a subset of an m-fold product set and represented by a list of ordered m-tuples, is a description; but, equally, a relation obtained from other relations through relational operators as in Codd (1970) is a *description*. In the sequel, the term description is often equisignificant with *goal*. Partial, or incomplete, descriptions are permitted, as fuzzy relations R_i (omission of the overbar is deliberate), such that $\bar{R}_{1i}$, $\bar{R}_{2i}$, ... satisfy R_1.

We say that Proc i *produces and reproduces* $\bar{R}_i$. There may be many Proc i that produce and reproduce $\bar{R}_i$, thus Proc r, i, Proc s, i, ...; in fact, there are indefinitely many. By the same token there are many Proc r, $i1$, Proc s, $i1$, ... that produce and reproduce $\bar{R}_{i1}$; many Proc r, $i2$, Proc s, $i2$, ... that produce and reproduce $\bar{R}_{i2}$ ... or in general R_i.

From the definition of Con we say that a concept Con i fuzzily reproduces or reproduces R_i [i.e., Ex(Con i) $\Rightarrow R_i$], noting that R_i is any *interpreted* relation (possibly a periodic process) and is, in general, a fuzzy relation (hence R_i rather than $\bar{R}_i$). R_i may be realized in the input–output domain of a processor $\lambda(A)$, $\lambda(B)$ (notably, A's brain or B's brain) as an apparition or impression: it may be a percept or form a part of a behavior.

16.7 Agreement and Concept Sharing

Speaking of human beings, if a concept Con i is the intention or connotation of i, then R_i is its extension or denotation. But these statements only make sense for some one or several values of the variable Z; that is A's concept of i, namely Con$_A$ i, or B's concept of i, namely Con$_B$ i.

It is also possible for A and B to *agree* about their concepts of i. A general L agreement is achieved if the participants in a conversation (A and B) ask each other "how" questions eliciting L explanations that are Progs in $\pi(\text{Con}_A\ i)$ and $\pi(\text{Con}_B\ i)$. General agreement implies that some A explanations (L listings of Progs in $\pi(\text{Con}_A\ i)$) are coherent under execution in $\lambda(B)$ of $\pi(\text{Con}_B\ i)$ and that some B explanations (L-listings) of Progs in $\pi(\text{Con}_B\ i)$ are coherent under execution in $\lambda(A)$ of $\pi(\text{Con}_A\ i)$.

Obviously, agreement does not imply that $\text{Con}_A\ i = \text{Con}_B\ i$, for the equality is nonsensical (A is not B, whatever else). Nor does agreement usually imply isomorphism between $\text{Con}_A\ i$, $\text{Con}_B\ i$, or $R_A i$, or $R_B i$ (at most a "depersonalized" intention of i would be some definitional or explanatory matching of Prog in $\text{Con}_A\ i$, $\text{Con}_B\ i$; at most, a "depersonalized" extension of i would be a matching of ostended members of $R_A i$ and those of $R_B i$; of course, $Z = A, B, \dots$ may have values over a population, civilization, culture, or group).

An operational or behavioral type of A and B agreement is obtained as follows. Equip both A and B with separate modeling facilities, MF_A, MF_B, of Figure 2 (for example, independent programmable computers), in which they can express nonverbal explanations that are L listings of Progs in $\pi(\text{Con}_A\ i)$ and Progs in $\pi(\text{Con}_B\ i)$, to obtain working models ($M_A i$ and $M_B i$) that are independently executable in MF_A, MF_B. Upon exteriorized execution, suppose that $M_A i$ induces a relation $R_A^* i$ in the input–output domain (U_A) of MF_A and that $M_B i$ induces a relation $R_B^* i$ in the input–output domain (U_B) of MF_B. If, perhaps, after trial execution, remodeling, and so on, the following conditions all hold, then A and B are said to agree about a concept of i.

1. $M_A i$ is executable in MF_A, $M_B i$ executable in MF_B.
2. $R_A^* i$ is in U_A and $R_B^* i$ in U_B (by independent execution).
3. $R_A^* i \subset R_A i$ and $R_B^* i \subset R_B i$ ($\subset$ stands for inclusion).
4. $R_A i \langle = \rangle R_B i$ (where $\langle = \rangle$ stands for "isomorphism").
5. $\pi(M_A i) \subset \pi(\text{Con}_A\ i)$ and $\pi(M_B i) \subset \pi(\text{Con}_B\ i)$.
6. $\pi(M_A i)$ is executable as coherent with $\text{Con}_A\ i$ in $\lambda(B)$.
7. $\pi(M_B i)$ is executable as coherent with $\text{Con}_B\ i$ in $\lambda(A)$.
8. $M_A i$ is "extensionally equivalent" to $M_B i$ (i.e., upon execution one does the same thing as the other).

Clearly, this form of agreement is limited by the capabilities of MF_A and MF_B; for example, if MF_A and MF_B are serial computers, then only serial programs can be compiled to represent members of $\text{Con}_A\ i$ and $\text{Con}_B\ i$ even though other kinds of program may be expressed by L explanations.

There are also difficulties over conditions 6–8, and since no criterion is given for determining whether they are satisfied or not. These difficulties are addressed as part and parcel of concept stability in the following sections.

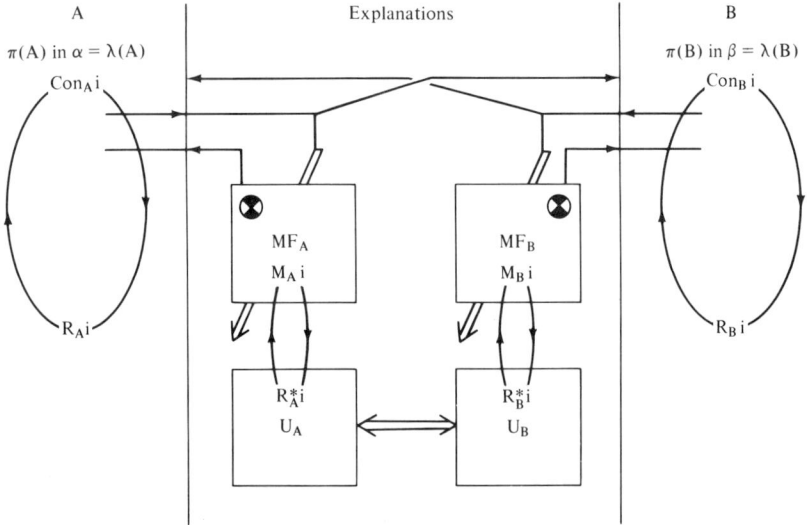

Figure 2. A simple (as it stands, inadequate) form of agreement MF_A, MF_B are independent modelling facilities, such as distinct computers equipped with LOGO, or SMALLTALK, or explanatory forms of PLATO and the necessary peripherals. (In our laboratory, as in most Piagetian experiments, they are special purpose, computing, or model construction systems, designed for one subject matter.) U_A is the MF_A input–output domain and U_B is the MF_B input–output domain. Execution of A's model $M_A i$ (after correction and trial by A), gives rise to $R_A^* i$ in U_A; similarly, for B, ⊗ is descriptive feedback obtained for correction by participants and the double arrow is programming or model building, in contrast to execution MF_A, MF_B. Execution of $Con_A i$ in A's brain α gives rise to $R_A i$ and of $Con_B i$ in B's brain β to $R_B i$.

16.8   Stable Concepts, Learning, and Memory

Just as Ex(Proc i) $\Rightarrow$ $\bar{R}i$ produces or reproduces $\bar{R}i$ (which may be a process, for example, of regulation), so there are Procs that operate upon and produce or reproduce other Procs under appropriate conditions, notably, if a *goal* (alias a *description*), is given in their argument. Thus

$$\text{Proc†, (Proc } i) \quad \Rightarrow \quad \text{Proc.}$$

But Proc† is not necessarily distinct from Proc.

Just as Ex(Proc i) $\Rightarrow$ $\bar{R}i$, so Ex(Con$_A$ i) $\Rightarrow$ R_i fuzzily produces or reproduces R_i. Further,

$$\text{Ex Con†(Con } i) \quad \Rightarrow \quad \text{Proc in Con } i.$$

Although Con† is not necessarily distinct from Con, it is often convenient to regard the acquisition of a novel concept as *learning*, and its reconstruction (possibly also productive) as a *memory*. In the latter case the notation

$$\text{Ex(Mem } i) \;\Rightarrow\; \text{Proc in Con } i$$

is neater than

$$\text{Ex(Con\dag(Con } i)) \;\Rightarrow\; \text{Proc in Con } i.$$

In general, (see Section 16.5) Procs are production systems. There is empirical evidence that *most* (perhaps all) productive and reproductive operations in conscious human beings involve mutualism between two types of Proc.

Among the Procs that produce and reproduce Procs, distinguish two classes, namely, description building $(\bar{D}\bar{B})$ and procedure building $(\bar{P}\bar{B})$. It is not maintained that *all* of the productions acting upon Procs to produce or reproduce them (even in human beings) are of the these two kinds. As a rule, it is quite unnecessary (and it may be impossible) to know what the production systems *are*, in computational detail. Although there is plenty of evidence that people have learning styles and adopt learning strategies explicable in terms of a balance between the relative efficiency, numerousness, and accessibility of description building and procedure building productions, the evidence does not warrant supposing that people compute in the same way. It may be that all of us have entirely different kinds of productions and, so long as certain requirements are satisfied (preserving specificity, for example), the kind does not matter. The $\bar{D}\bar{B}$ and $\bar{P}\bar{B}$ are characterized insofar as they operate upon different arguments; generically, the $\bar{D}\bar{B}$ productions operate upon descriptions to produce descriptions—their arguments may be any number of descriptions (i.e., interpreted relations, $\bar{R}_i$, $\bar{R}_j$)—while the $\bar{P}\bar{B}$ productions operate upon any number of Procs in combination with one or more descriptions. Thus

$$\text{Ex } \bar{D}\bar{B}(\bar{R}_i, \bar{R}_j) \;\Rightarrow\; \bar{R}_k,$$

$$\text{Ex } \bar{P}\bar{B}(\text{Proc } i, \text{Proc } j, \bar{R}_k) \;\Rightarrow\; \text{Proc } k.$$

If it also happens that

$$\text{Ex } \bar{D}\bar{B}(\bar{R}_j, \bar{R}_k) \;\Rightarrow\; \bar{R}_i,$$

$$\text{Ex } \bar{D}\bar{B}(\bar{R}_i, \bar{R}_k) \;\Rightarrow\; \bar{R}_j,$$

$$\text{Ex } \bar{P}\bar{B}(\text{Proc } j, \text{Proc } k, \bar{R}_i) \;\Rightarrow\; \text{Proc } i,$$

$$\text{Ex } \bar{P}\bar{B}(\text{Proc } i, \text{Proc } k, \bar{R}_j) \;\Rightarrow\; \text{Proc } j,$$

then

$$\text{Ex(Proc } i) \;\Rightarrow\; \bar{R}_i, \quad \text{Ex(Proc } j) \;\Rightarrow\; \bar{R}_j, \quad \text{Ex(Proc } k) \;\Rightarrow\; \bar{R}_k.$$

The entire system is self-reproducing and is characterized by the fixed point values R_i, R_j, R_k on iterative execution. Such systems are readily simulated by various computer programs, acting as tesselation of kinematic images of von Neumann (1966) self-reproduction if λ(Proc) is held constant (i.e., if Inter in Proc = $\langle$Prog, Inter$\rangle$ is fixed). One arrangement of considerable generality is obtained by taking $D\bar{B}$ as the relational operators *Join, Restriction* and taking PB as a productive algorithm, such as that of Chang and Lee (1973) A^*. However, this construction is no more than a piece of intellectual scaffolding intended to point out a principle more elegantly expressed by von Foerster (1975), who noted that $\bar{R}_i$, $\bar{R}_j$, $\bar{R}_k$ are defined for Procs that are eigenoperations or eigenfunctions that yield eigenvalues on infinite iteration (are recursive).

Contemplate the following replacements:

$$\bar{R}_i \text{ into } R_i, \quad \bar{R}_j \text{ into } R_j, \quad \bar{R}_k \text{ into } R_k;$$

$$\bar{D}\bar{B} \text{ into } DB,$$

where DB is a class of Cons (not just Procs);

$$\bar{P}\bar{B} \text{ into } PB,$$

where PB is a class of Cons (not just Procs).

The replacements make sense insofar as each Con is subscripted by a value $A,B,\ldots$ of Z, for example, by A.

$$\text{Ex(Con}_A \text{ } i) \quad \Rightarrow \quad R_i,$$

$$\text{Ex(Mem}_A \text{ } i) = \text{Ex(Con}_A\dagger(\text{Con}_A \text{ } i)) \quad \Rightarrow \quad \text{Proc in Con}_A \text{ } i.$$

This is nontrivial insofar as λ(Proc) or λ(Con) is not *held* constant, though constancy may be achieved in execution. The system of productions is shown pictorially in Figure 3, where the double arrows indicate productions and single arrows represent paths by which products can be retrieved and entered into the argument of a production. Such pictures are probably more familiar to biochemists or people from the hybrid computer era than they are to mathematicians or computer scientists today, but they do have interesting properties. Perhaps the mathematicians and the computer scientists will suppress some (to their discipline) obvious objections (for example, how are the productions organized) until later, when these objections will be answered, or accounted for.

If the Cons in Figure 3 are subscripted by a value, say A, of variable Z, then the process depicted is a stable concept, meaning that it exists and can be reconstructed; that is, there is a pair $\text{Stab}_A = \langle(\text{Mem}_A, \text{Con}_A)\rangle$ that is organizationally closed.

Suppose we ask, "*what* is it a stable concept *of*?"; we must have recourse to the index, A's name, and we will say that even in the minimal

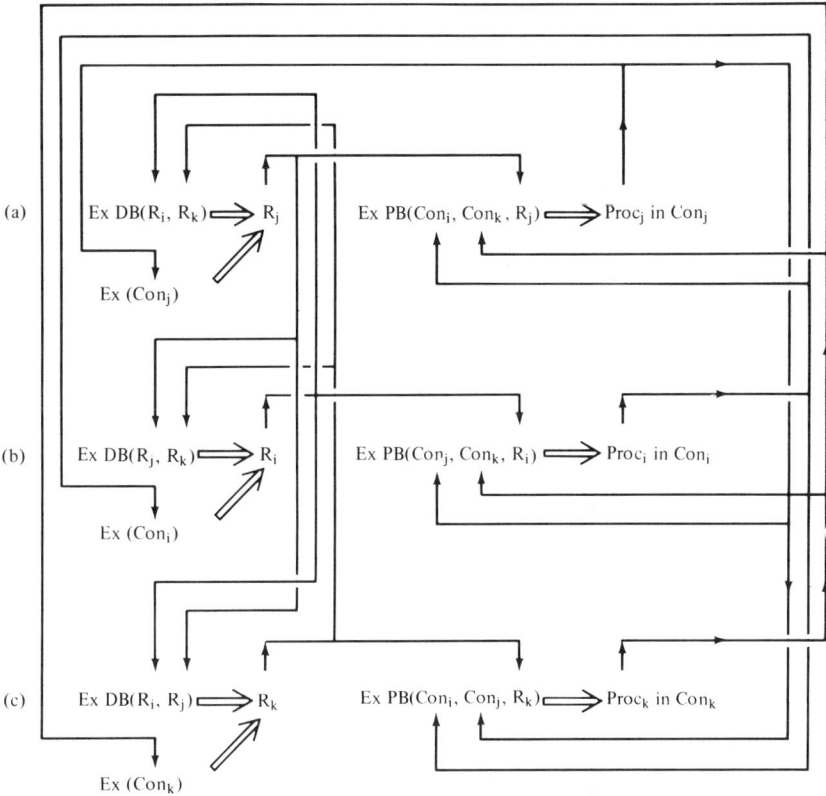

Figure 3. A simple construction for Mem i and Con i (or, just as well, of Mem j, Con j; Mem k, Con k). Suppose that Con i and Con j exist but that Con k does not, and is learned, then line c productions represent the production (rather than the reproduction) of Con k; a similar comment applies to line a and the learning of Con j on line b and the learning of Con i. Once the configuration exists, it is inherently stable, as it will be if Con is subscripted by $Z = A,B \dots$ If so, for example, if Con$_A$ i, then A's *perspective* may be i, j, or k. Again, if Con$_A$ k does not exist, though Con$_A$ i, Con$_A$ j do so, then A's *learning* of Con k consists in the production of a description R_k of k and the subsequent production of Con k to realise R_k. The entire system is Stab$_A$ k (the acquisition by A of a stable concept of k), where π(Stab$_A$ k) is A's *understanding* of k. The isolation of such a system is a pure convention and, in fact, it always exists in the context of other and related systems. Its productions are used as arguments (or conditions) by other production systems and it receives and acts upon the products of other production systems, notably those of A (hence, the edict that the system is meaningful only if the constituents are subscripted by a value A, B,... of Z).

case the answer depends upon A's *perspective* or point of view. If A is asked to say what his stable concept *is,* thus *introducing* directionality and consequently temporality into the picture by requiring an L utterance, any of the following are possible as long as the stable concept exists, implied by the replies i or j or k:

$$\text{If } i, \text{ then } \langle \text{Mem}_A \; i, \text{Con}_A \; i \rangle = \text{Stab}_A \; i \qquad \text{(perspective } i);$$

$$\text{If } j, \text{ then } \langle \text{Mem}_A \; j, \text{Con}_A \; j \rangle = \text{Stab}_A \; j \qquad \text{(perspective } j);$$

$$\text{If } k, \text{ then } \langle \text{Mem}_A \; k, \text{Con}_A \; k \rangle = \text{Stab}_A \; k \qquad \text{(perspective } k).$$

Notice that by so doing we require A to act in a specialized manner, that is, to entertain *one* perspective at *once* (perspective $i, j,$ or k), which amounts to making A say "I am A" and "this is my perspective" (conversely, as we shall see, A is an individualized conscious system *because* he *may* adopt such a unique perspective).

Again, suppose that A imposes his own directionality or temporality by "learning about k"; that is, $\text{Stab}_A \; i$ exists, $\text{Stab}_A \; j$ exists, but $\text{Stab}_A \; k$ does not exist. If so, then A may choose among a finite or indefinite number of possible DB operations that are at his disposal to build a description R_k and to pursue R_k.

16.9   The Status of Topics and Conversational Domains

This circumstance involves and underlines another important point: $\text{Stab}_A \; i$ and $\text{Stab}_A \; j$ are not uniquely defined in Figure 3, which artificially isolates a minimal unit called Stab. For example, by adjoining the productions $DB(R_l, R_m) \Rightarrow R_i$ with $PB(\text{Con}_l, \text{Con}_m, R_i) \Rightarrow \text{Proc in Con } i$, and $DB(R_n, R_o) \Rightarrow R_j$ with $PB(\text{Con}_n, \text{Con}_o, R_j) \Rightarrow \text{Proc in Con } j$ and adding the necessary product collecting arcs, we obtain a network in which exist $\text{Stab}_A \; i$ and $\text{Stab}_A \; j$ but *not yet* $\text{Stab}_A \; k$, which is to be created, or constructed, or learned.

To avoid drawing out such complicated networks, the static inscription of Figure 4 may be employed to depict stable conditions like Stab_A (before) and Stab_A (now)—or, equally, Stab_A (now) and Stab_A (later). The static inscription is meaningful, of course, only if the cyclic production system it stands for exists as a process, and is identified with part of the processes legitimately designated by values $A, B, \ldots$ of Z.

If that assurance is provided, then the nodes are known as *topics* (which designate concepts and interpreted relations), chiefly because most of our work has been in educational psychology. In other contexts, the word "topic" might be replaced by "objects and actions" (in the manner of Glanville 1976) or "coherent behaviors" in the manner of von Foerster (1975). The directed arcs relating these nodes represent the operation of

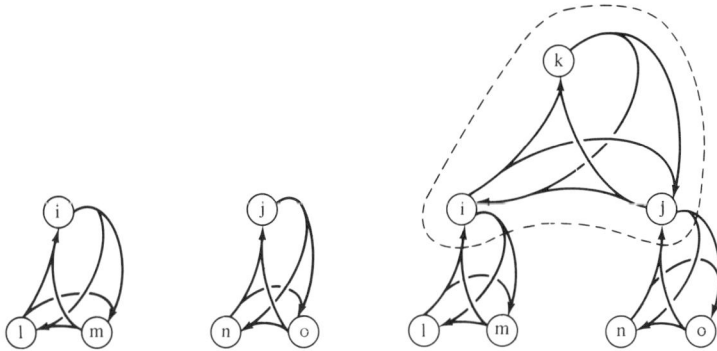

Figure 4. Notation employed for static inscriptions of codes of production system capable of producing and reproducing concepts. On the left are depicted $Stab_A$ (before) and on the right $Stab_A$ (now) [as an alternative, $Stab_A$ (now) and $Stab_A$ (later)]. The dotted region encloses the production system in the text and represents A's learning of a novel concept k. However, once established, this system (in the dotted line enclosure) may be interpreted as $Stab_A$ i, $Stab_A$ j, or $Stab_A$ k depending upon A's perspective, focusing attention upon concept i, j, or k, respectively. Stable concepts are known as topics. Any topic has a kernel. On the left, i has the kernel (l, m), l has the kernel (i, m), and m has the kernel (l, i); on the right, k has the kernels $(i, (l, m), j (n, o))$, and i has the kernels (l, m) and, in addition, the kernel (k, j).

DB and PB productions that are presumed to exist in any process legitimately tagged by a value of Z (in other words, OB and PB are either among the products of this Stab or another Stab of Z); that is,

$$\text{topic } k \triangleq \text{ static inscription } [\pi(Stab_Z\ k)] \qquad \text{for some Z.}$$

Topics (for given values of $Z = A,B, ...$) are those interpreted relations generated as fixed point values by inherently self-reproducing processes. Relations are thus defined in terms of processes (not vice versa), and they are discrete because fixed-point transformations lead to discrete values on indefinite iteration. Topics may be agreed on between conscious systems.

16.10 Explanations, Derivations, and Entailment Meshes

Let us examine the evidence needed to give the required assurance. Superficially, it varies depending upon the particular circumstances, and several cases (by no means exhaustive) will be examined. However, on closer scrutiny, the evidence has many features in common. It firms up the agreement criteria of Section 16.7 and Figure 2, and expresses the fact that if this agreement were itself given a static inscription (supposing an agreement to be reached) then this would be an *analogy* (or an *analogical* topic).

Case I. Assume that a static inscription in *given,* though not yet legitimized, and that $Z = A, B$ are two people. Here, $\pi(A)$ is A's personality; his system of coherent beliefs; $\pi(B)$ is B's personality. $\lambda(A) = \alpha$ is A's brain, assumed to be a priori independent of $\lambda(B) = \beta$, or B's brain. Let $\pi(\text{Stab}_A k)$ be A's *understanding* of k from the *perspective* of k. Let $\pi(\text{Stab}_B k)$ be B's *understanding* of k from the *perspective* of k. Let A and B have the same perspective as indicated by pointing at the topic in an external static inscription. Impose the operational requirements of Section 16.3 and provide an interface (such as THOUGHTSTICKER of Figure 1) through which an agreement over understandings may be reached.

Consider Figure 5, which extends Figure 2 by adjoining a static inscription. This inscription is called an *entailment mesh* (EM) because we are concerned not about the particulars of operations *DB* and *PB* but only that they exist (psychologists lump them together as "discovery"). The static inscription of "discovery" is entailment.

Thus, from Section 16.9,

$EM \triangleq$ for some z *Static Inscription (Superimposition $\pi(Stab_Z r)$)* for r $= 1, 2, ...,$ for all s, with s in power set of index set r

$= $ a related collection of topics seen from all perspectives.

Suppose that A and B in conversation about topics represented in an entailment mesh are pointing, upon some occasion, at topic k (henceforward T_k). One reason may be that one of the participants in the conversation (B, say) has the dominant role of "teacher," while the other (A) is a "student." If so, B has available a stock of possible explanations, which can be used to demonstrate T_k, together with descriptors which can be used to focus A's attention upon T_k, and that B deems it tutorially wise to do so. Alternatively (and, for this purpose we need not press the distinction), the entailment mesh is augmented by a stock of potential explanations of demonstrations (one stock to each topic) and commonly understood descriptors that allow A and B to direct attention at topics. In the latter case there is no necessary dominance on B's part; it is simply that A and B are "learning together" about the mesh-related topics.

The required augmentation (explanations–demonstrations of topics and a scheme of descriptors–predicates for accessing topics) converts the *entailment mesh* into the *conversational domain* of Section 16.3.

Either by tuition, involving "how" and "why" questions, or by accord, A and B not only explain T_k to each other and reach agreement in the matter (nonverbally, both in Figures 2 and 5), but they also explain how they constructed their explanations. To do so they exchange and reach agreement upon *derivations* (Der$_A$ k, Der$_B$ k, in Figure 5). This they may also do, given an entailment mesh and a facility, such as THOUGHT-

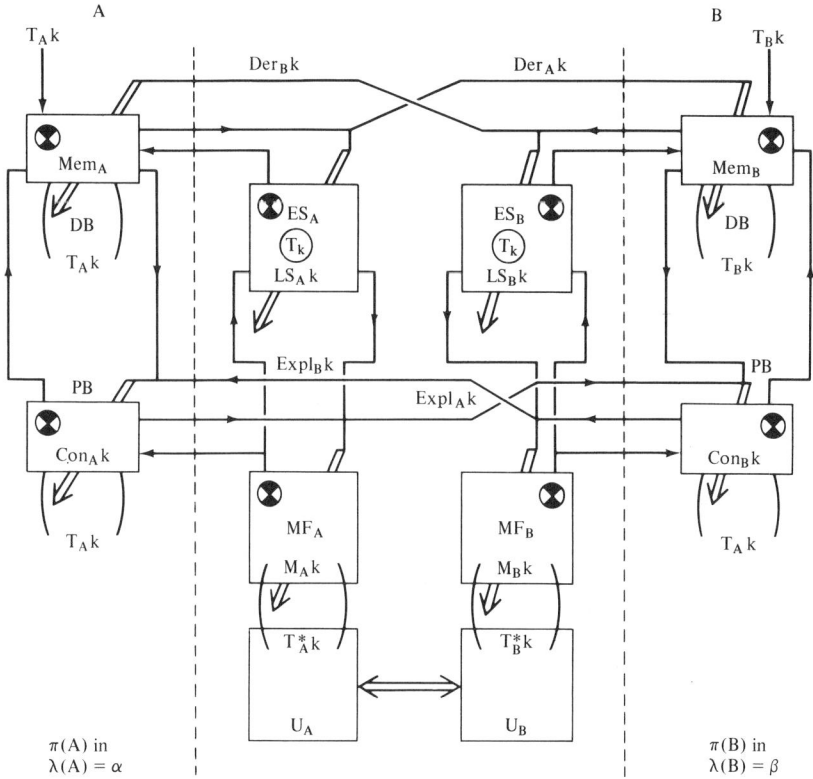

Figure 5. Agreement over an understanding of topic k in conversation between A and B. The participants are interpreted as being $\pi(A)$ and $\pi(B)$ assigned as in Figure 2 to distinct, a priori independent, brains $\lambda(A) = \alpha$ and $\lambda(B) = \beta$. $\mathrm{Expl}_A\, k$, represented, nonverbally, as $M_A k$ in MF_A and $\mathrm{Expl}_B\, k$ as $M_B\, k$ are explanations; $\mathrm{Der}_A\, k$ and $\mathrm{Der}_B\, k$ are derivations (i.e., explanations of explanations, or justifications of why a particular explanation is given and how it is derived), represented nonverbally as learning strategies $LS_A k$ and $LS_B k$ in an entailment mesh EM_A, EM_B. Either EM_A, EM_B are given and contain topic T_k or evolving (in which case, delete T_k from each). From (Mem$_A$ k), the DB operations compute $T_A k$ (equivalent to the execution of Con$_A$ k). From (Mem$_B$ k) the DB operations compute $T_B k$ (equivalent to the execution of Con$_B$ k). The PB operations of A compute Proc in Con$_a$ k and the PB operations of B compute Proc in Con$_B$ k, if Stab$_A$ k, Stab$_B$ k exist. In addition, the agreements over $\pi(\mathrm{Stab}_A\, k) \supset \langle \mathrm{Der}_A\, k, \mathrm{Expl}_A\, k\rangle \equiv \langle LS_A k, M_A k\rangle$ and $\pi(\mathrm{Stab}_B\, k) \supset \langle \mathrm{Der}_B\, k, \mathrm{Expl}_B\, k\rangle \equiv \langle LS_B k, M_B k\rangle$ ensure that PB operations in Mem$_A$ compute Proc coherent in Con$_B$ k (as well as Con$_A$ k) and that PB operations in Mem$_B$ compute Proc coherent as part of Con$_A$ k, as well as Con$_B$ k (conditions 6 and 7 of Section 16.7) and DB operations in Mem$_A$ compute $T_A k$ that are part of $T_B k$, and DB operations in Mem$_B$ k compute $T_B k$ that are part of $T_A k$ (condition 8 of Section 16.7).

STICKER (Figure 1), by nonverbal model-building behaviors. Such behavior is manifest as the exteriorized learning strategy LS_A or LS_B used, respectively, for building Con_A k or Con_B k. An A,B agreement, in this case, means that A could (not necessarily *would*) perform B's construction and vice versa. Phrased differently, Proc in Con_A k are manufacturable by Mem_B and Proc in Con_B k are manufacturable by Mem_A. We call this complex (A,B) agreement, an *understanding* of T_k by A with B; an understanding is evidence for $Stab_A$ k, $Stab_B$ k. As in the caption of Figure 5, we have the minimal requirement, to complete the conditions in Section 16.7, that

$$\pi(Stab_A \ k) \supset \langle Der_A \ k, \ Expl_A \ k \rangle \equiv \langle LS_A k, \ M_A k \rangle$$

and that

$$\pi(Stab_B \ k) \supset \langle Der_B \ k, \ Expl_B \ k \rangle \equiv \langle LS_B k, \ M_B k \rangle;$$

or, in general, that an A, B understanding, in language L, of T_k is the coherent part of $\pi(Stab_A \ k)$ and $\pi(Stab_B \ k)$. There is ample and quite diverse empirical evidence that insofar as understanding is achieved concept k is stable and increasingly resilient to interference.

There is, of course, no requirement that Der_A k and Der_B k (or that the corresponding learning strategies LS_A and LS_B) be the same. Entailment meshes commonly do admit of many and complex derivation paths as suggested in Figure 6. For example, let topic k (in Figure 6) be "the surface of a cylinder"; topic f, "a rectangle labeled a, b, c, d"; topic g, "join edge ab to cd or edge ad to bc but not both"; topic h, "a torus"; topic g, "cut in half along any one slicing plane"; and topic f, "join the free edges."

Figure 6. It is not at all necessary that only one derivation is countenanced provided that both participants are able to construct $Stab_A$ k as a result of either. For example, it may be that A regards k as derivable from i and j whereas B regards k as derivable from g and h (perhaps only after learning to understand g and k). Such disjunctive derivations are common and represented by a notation with several kernels (on the left). Again, although local cyclicity is mandatory, the majority of meshes representing algebraic or otherwise redundant topics have other-than-local cyclicity (as shown on the right).

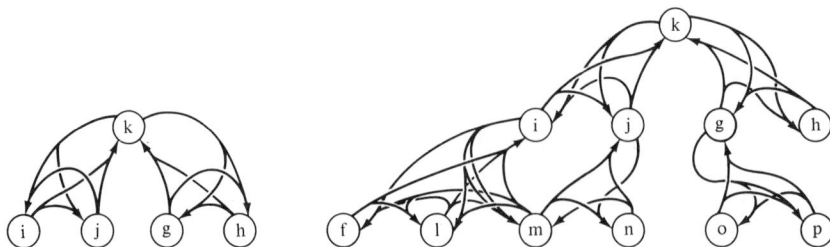

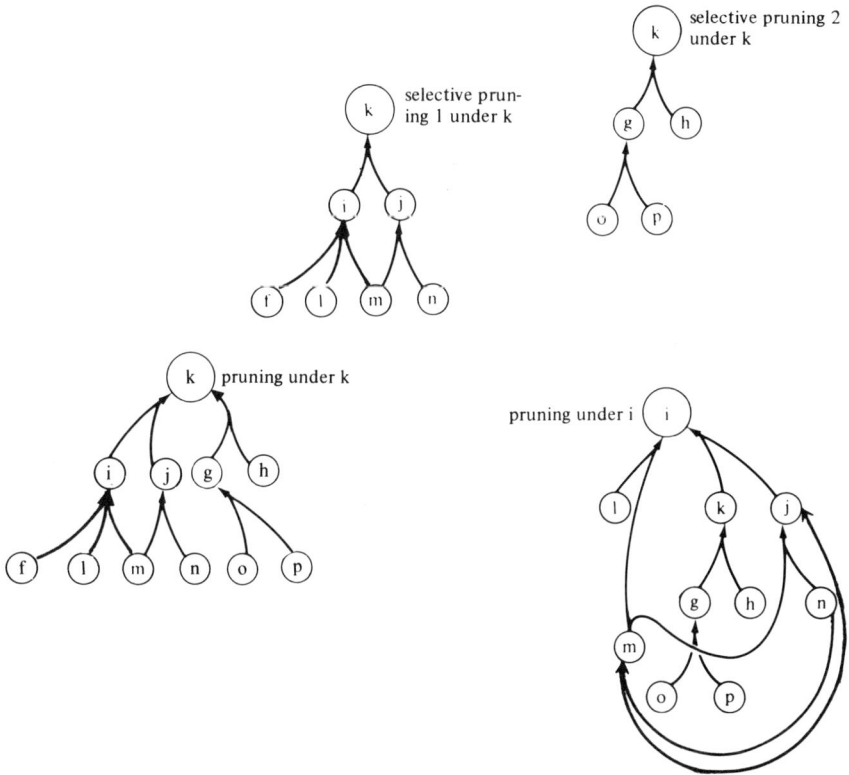

Figure 7. The notion of *perspective,* or point of view, corresponds in a static inscription to a *pruning* of which there are as many as there are (other than primitive) topics and from any one topic several conjunctive or kernel-unique prunings. For example, the rightmost mesh in Figure 6 has been pruned selectively under topic *k* and also under topic *i*. To prune, it is necessary to specify a direction and a topic. Selective prunings are unfoldments for entailment meshes into trees that are truncated when periodicity appears, or at a given depth.

A learning strategy is a *selective pruning* or *unfoldment* of an entailment mesh (Figure 7); it is quite possible, for example, that *A* employs the selective pruning 1 of Figure 7, whereas *B* employs selective pruning 2 of Figure 7 (in the literature a pruning, unqualified, or a union of them, is called an entailment structure). Such a pruning is the static inscription of a perspective, taken by *A* and *B* or taken by some other person, a theorist or a subject matter expert, who produced the static entailment mesh as an encoding of his theory.

Insofar as the entailment mesh or the associated conversational domain represent knowable topics, they are finite samples. No knowledge is depersonalized; these samples are peoples' theories.

Case II. Let A and B be distinct people, as in Case I, but who construct their own static inscription as the framework in which they exteriorize their thinking. The entailment mesh is permitted to evolve, representing the theories of A and B, so that, although still finite, it is representative of these particular people. To depict this circumstance in Figure 5, it is only necessary to delete the topic T_k from ES_A and ES_B. In this case, A and B construct and exteriorize a shared perspective (any perspective about which they agree); around such perspectives they construct their personal theories.

The facility of THOUGHTSTICKER (Figure 1) allows for precisely this kind of evolution (hence, it is an epistemological laboratory and finds just as much practical application in structuring the theories or expositions of authors as it does for learning experiments).

In order to count as an entailment mesh, the inscription of a theory must satisfy syntactic constraints that ensure that each minimal inscription of a topic (Figure 4) does represent a productive and reproductive Stab (Figure 3). The rules are detailed elsewhere (reports 1975–1977 or the references to conversation theory); they are not "mathematically presented" and strike subject matter experts, for example, as rules of decent exposition.

Some of the equipment in Figure 1 is devoted to ensuring that these rules are satisfied; some of it to making extrapolations, or overgeneralizations (open to denial by users, learners, author teams) that spur the users on to further learning or exposition.

Of course, in both Cases I and II, we have taken a distinction for granted; that $\lambda(A) = \alpha$ (one brain) and $\lambda(B) = \beta$ (another brain) are a priori independent processors. This distinction is intuitively plausible but quite arbitrary. It is clear, for example, that agreement over an understanding of any topic (one given to begin with, or one that is invented) implies that some Progs in $\pi(A)$ undergo execution in $\beta = \lambda(B)$ and, vice versa, that some Progs in $\pi(B)$ undergo execution in $\alpha = \lambda(A)$. The fact is that *any* distinction would suffice.

Case III. Consider somebody learning or problem solving alone through the interface of a *fixed* entailment mesh (or, usually, its conversational domain, complete with descriptors and a stock of demonstrations). Provision of the interface makes clear what us usually meant by phrases like "I am learning this myself," or "I am solving these problems myself." Insofar as the ruminations in question enter the public scene (and, by hypothesis, insofar as they go at all), there are two or more individuals (roles or perspectives) accommodated in *one* brain. The mental operations exteriorized for public scrutiny are as much as can be captured of an internal conversation. We capture them by insisting (through THOUGHT-

STICKER or some surrogate of an other than mechanical kind, such as an interview situation) that each topic addressed is understood. For consistency we call the roles or perspectives coexisting in this one person $\pi(A)$ and $\pi(B)$, noting that $\lambda(A) = \alpha = \lambda(B)$. In Figure 5 replace β by α and join the lower ends of the vertical dotted lines by a horizontal dotted line.

Case IV. Consider the same situation, when the entailment mesh evolves under the control of the user (A and B in one brain α). Here, in practical studies of design (Reports 1976, 1977), the reality of an "internal conversation that is exteriorized" is even more obtrusive, for each topic enstated as the justification or explanation of a design must be understood.

It may be sensible to write, in either case, an expression for distinct values A and B of Z that distinguishes a priori independent parts of one processor (brain α), namely,

$$\lambda(A) = \alpha_1, \qquad \pi(B) = \alpha_2,$$

but strictly, this is unnecessary. The important point is that brains as processors (and brains are not necessarily unique in this respect) have the ability to predicate, to make distinctions of the kind already made (but arbitrarily made) in Cases I and II. It may be that the distinctions are only of perspective (as when $\alpha_1 = \alpha_2 = \alpha$) or that they cut apart functionally independent processors (as in α_1 and α_2), or that they demarcate organisms (α and β), or that they demarcate kinds of universe X (for example, electrical entities, poetical entities) as distinct from Y (for example, mechanical entities, pieces of music or drama). This property appeared, covertly perhaps, in Section 16.6 in the context of interpretation functions, Inter (programs that are differently compiled or interpreted do, as a rule, yield different interpreted relations when executed in a processor). It is an essential part of any other-than-trivial identification of organizational closure, autopoiesis, or the like and consciousness (if not observed, the crucial features disappear in a cloud of algebraic manipulation).

16.11   Concurrent Execution, Closure, and Independence

In Section 16.8 I asked the reader to defer judgement upon the organization of production systems, and I take the matter up again at this juncture. Using a standard digital computer, it would clearly be necessary to organize the production system in Figure 3 or its generalization (shown as a static inscription in Figure 4) by means of many program statements, which, in turn, would depend upon criteria of "priority" assignment, "randomization," and the like. Even in the most fitting programming language, the serial simulation of a few lines of production rules gives rise to a few pages of instructions. Of course this can be done, at the price of painstaking labor and largely arbitrary assumptions that demolish the meaning of the original statements.

It is true that some order or sequence *must* be imposed: for example, that if Proc k is to be added, as a new procedure to an existing coherent Stab, then *before* the new Proc k is constructed by a *PB* production, there must be an R_k constructed by a *DB* production; also, that execution of procedures shall not destructively interfere during execution, even before the coherence of entire parallelism (Con = [Proc]) is achieved.

A sufficient order is obtained if one (or as later, several) perspective(s) are adopted, provided that the following conditions are imposed upon the processors α, β, ..., which may legitimately figure as $\lambda(Z)$ for all values of $Z^* \subset Z$ (Section 16.2).

1. The processor must be able to execute concurrent processes both by acting in a strictly parallel mode (and thus *guaranteeing* the independence of the processes) and as a device in which destructive interaction is avoided through information transfer between coexisting ("actor"-like) loci of control.

2. The processor is never quiescent; it *must* do something, it does not *halt*.

3. At least two loci of control are invariably active to realize on the one hand, the productive and reproductive transformations of Figure 3 and, on the other hand, execution of Procs; that is, learning *must* take place, though *what* is learned is not determined.

4. Repetitive execution of Procs leads to a fully parallel mode. Con i $\triangleq$ $\langle\{$Proc $i\}$, [Proc $i]\rangle$ or Proc i tends, upon repetition, to Con i $\triangleq$ [Proc i], as in Section 16.6. The mechanism of recompilation in brains, qua processors, was pointed out by Grey Walter in the mid-1950s; it is, however, a general entrainment property of many nonlinear active media.

5. The processor may distinguish an indefinite number of universes of compilation or interpretation (u, v, of Section 16.6 or, if realized externally, X, Y of Section 16.7). It may make any number of distinctions.

6. For a class of processes that are both organizationally closed and informationally open some distinctions *must* arise, and lacking further specificity these are distinctions of an indifferentiated independence (cleavage of a processor into independent parts or mustering further processors from a stack).

7. Processes of this type are potentially conscious (in the sense of Section 16.12) and may be identified with conversations.

8. Independence is introduced or computed by any production that violates the interference condition of Section 16.11, which leads to

an essential or structural bifurcation in the system's behavior; that is, novel variables are created; it is not just a matter of giving ambiguous values to the existing variables. Stated conversely, if information transfer between organizationally closed systems is conserved then bifurcation *must* take place and leads, without further specificity, to an independence.

It should be emphasized that these properties are common. Only the idiosyncratic development of mathematics in concert with computer architecture leads us to regard them as "strange." The serial, digital computer is really a "strange," though convenient, specialization of computing media in general.

16.12 Consciousness and Information Transfer

To reach agreement (in particular, over an understanding) there must be a *distinction* cleaving a process into independent parts. In Case I and Case II this distinction *seems* to be "given" through α and β. In Case III or Case IV, it seems to "emerge" (for example, as α_1, α_2). In fact, in either case, it is *computed* (or *recognized*).

This independence is reduced by information transfer, namely, information transfer that is required, with equal significance, to render incoherent operations coherent or to render asynchronous operations locally synchronous (Section 16.4) where understandings are exteriorized. Information transfer is what happens in a conversation, when it is *consciousness*. Otherwise, it is *awareness,* which is unobservable.

The degree of consciousness is a fuzzy-valued measure of *doubt,* or its converse, *belief.* (I say doubt, rather than uncertainty, because there are many kinds of doubt, including at least the following: doubt regarding *perspective;* doubt, if a perspective exists, regarding a description R_k or the values of some coordinates of this description if others are given, i.e., doubt about outcomes; doubt, given a description, about a procedure to realize that description; doubt about which procedure to employ i.e., doubt about the method).

The sharp-valued *content* of consciousness is an understanding and the remaining contents are those apparitions, images, or emotions that accompany the productive and reproductive operations of reaching an agreement over an understanding. Consider, for example, the acquisition of a stable concept $Stab_A\, k$ given $Stab_A\, i$ and $Stab_A\, j$. Commonly, the sequence is as shown in Table 1.

In Section 16.8 we concentrated upon a particular kind of (DB, PB) productive system and believe it is more efficient than others, at any rate for deductive or inductive thought. However, this is not the *only* kind of system; for example, Procs may be arbitrarily composed or concatenated, lacking a "goal" or "description" as in the expression PC (Proc a,

Table 1

Doubt about perspective	Doubt about description (or outcome)	Doubt about procedure	Doubt about method
1. High until $\bar{D}\bar{B}$ production is found	High until DB production provides an argument for some PB	High until there is some PB	High since there is no $Proc_A k$
2. Low, if DB found that works	Reduced, but still high (PB can operate on partial description.)	High until PB found	High
3. Low	Reduced	Low if any PB is found	Still high since there are no $Proc_A k$
4. Low	Reduced, but higher than doubt about method	Reduced if PB works	Low as soon as some $Proc_A k$ exists
5. Low	Lower	Iteration of $Stab_A k$ Produced more $Proc_A k$	Increasing, as concepts or skills are overlearned (It is harder to say *what* procedure is used.)
6. Low	Very low, if concept or skill is overlearned	See below[a,b]	Higher than doubt over outcome (descriptive doubt)

[a] If, for some reason, attention is focused upon one topic so that doubt about *perspective* is held low, consciousness tends to a degree of zero (behavior is automatic) and this condition is only relieved by PB operations that introduce fresh procedures (not yet coherent with $Con_A k$), or in skill learning, by mistakes (finding that a well-tried procedure does not work and reconstructing a concept as a result). Again, at the intellectual level, an expert may expound (relearn) his thesis or a teacher may give freshly invented explanations.

[b] In general, however, $Con_A k \rightarrow [Proc_A k]$ (Property 4, given 2 and 3 of Section 16.11) and consciousness is only maintained by changing the attention (redirecting the conversation, thus increasing doubt about the perspective and *current* description) or by innovation or by an autonomous change in perspective. All of these expedients involve a distinction. That such a transformation *must* occur follows from properties 2 and 3 of Section 16.11; that distinctions *may be made to retain awareness* is guaranteed by property 5; awareness is manifest as consciousness if the distinctions made are represented externally.

Proc b) $\Rightarrow$ Proc c. Such activity seems to go on unconsciously. There is a good chance of expressing its magnitude as a background "noise" or "temperature" of a processor using Caianiello's (1977) thermodynamics of modular systems. By hypothesis, that is the *only* kind of "randomness" involved in productive thought; it provides the "noise" against which the information transfer of awareness or consciousness takes place.

16.13   Analogies of Form and Method; Analogical Topics and Analogy Building

L analogies of form are represented, as static inscriptions in an entailment mesh, by the notation shown in Figure 8a. They relate topics H and I, which are similar (at most, isomorphic) but are not identical (they may have a difference in content or merely be replicas, somehow distinct).

For example, an electrical and a mechanical linear oscillator are analogous; their similarity involves a second-order differential equation; their difference is the distinction written Dist (electrical, mechanical). An equally good example is provided by analogous music and poetry, whose analogy (similarity) is due to a common theme. These analogies are symmetric.

Pure analogies of form have similarities that are taken as understood and hence are not derived (as in Figure 8a). Often, however, the similarity is derived; for example, consider two vehicles navigating on the surface of a cylinder (F and G) in Figure 8b. The similarity of F and G lies in the cylinder (derived as in Figure 6). The difference is the difference between the tracks delineated by the two vehicles, determined by their characteristics as vehicles. Such analogies are mixed since (Figure 8c), if contingent upon the adoption of a method, their similarities are supported by a process—namely, the unfolding or selective pruning of an entailment mesh. Strictly speaking, they exist in the pruning field (set of all selective prunings as in Figure 7) of a mesh, not in the mesh itself. It follows incidentally that the indefinite unfolding of a mesh yields interpretations that are generally *not* in the *same* universe (hence, the distinction making property), and that signs like that for implication "$\rightarrow$," or other "syntactic" entities also receive an interpretation as actions (hence the earlier insistence upon a logic of action or execution). Hence, mixed analogies are analogies of form and method.

Although the point is not taken up in this paper, it can be shown that all explanations of topics are obtainable as pruned derivations; the distinction between an explanation of a topic and its derivation is made as a matter of convenience, not of fact.

Again, although the matter is not considered in this paper, the lowermost nodes stand for topics that are primitive only in the very special sense that, in the context of the thesis embodied in the entailment mesh, their computation is irrelevant to the thesis. For example, provided a user (student, expert, teacher, designer, decisionmaker) has some interpretation for "$\rightarrow$," or for "mechanical–electrical," it does not matter what it is, how the user computes the syntactic form, or what predicates the user evaluates to demarcate mechanical–electrical entities. Obviously, this depends upon the user as well as the thesis. The notion of lowermost or primitive is *relative* to both of them, just as any pruning of a mesh is relative to the perspective.

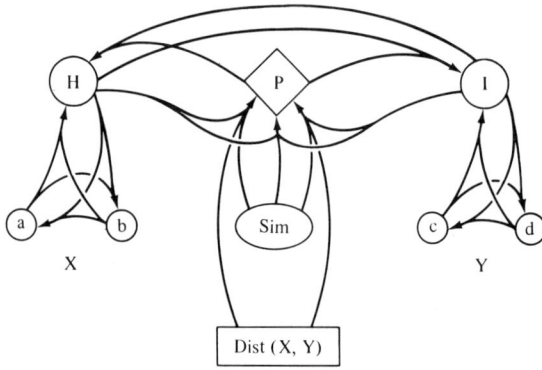

(a)

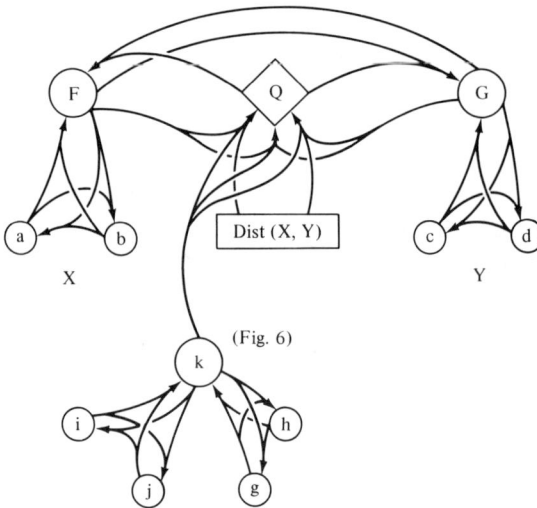

(b)

Figure 8. Similarity (Sim) and distinction (Dist). Given analogy P, Q, topic H may be learned if I is known (a) or F if G is known (b). Vice Versa, given P, Q, I may be learned if H is known, or G if F is known. P, Q may be learned by understanding H and I (or G and F) and the similarity involved *or*, if only one topic is understood, by understanding the similarity *and* understanding the dis-

Finally, there are analogies of method, which in general are asymmetric where there is no analogy of form. For example, mathematical induction may be used in many areas to obtain quite different results; schemes yielding different and possibly contradictory conclusions may be "axiomatically similar." Such analogies exist *only* in the pruning field of a mesh

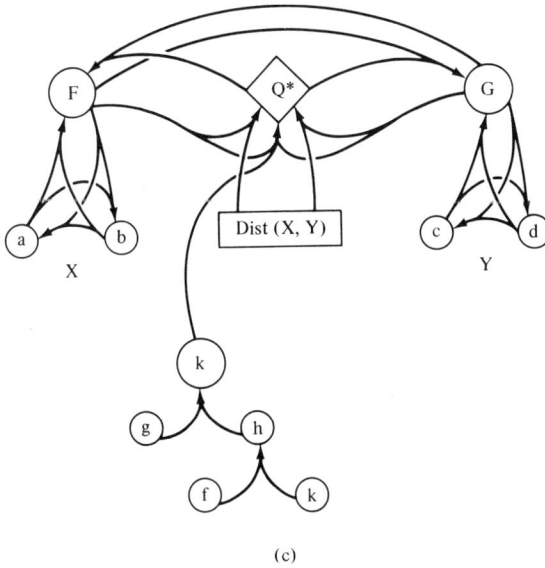

(c)

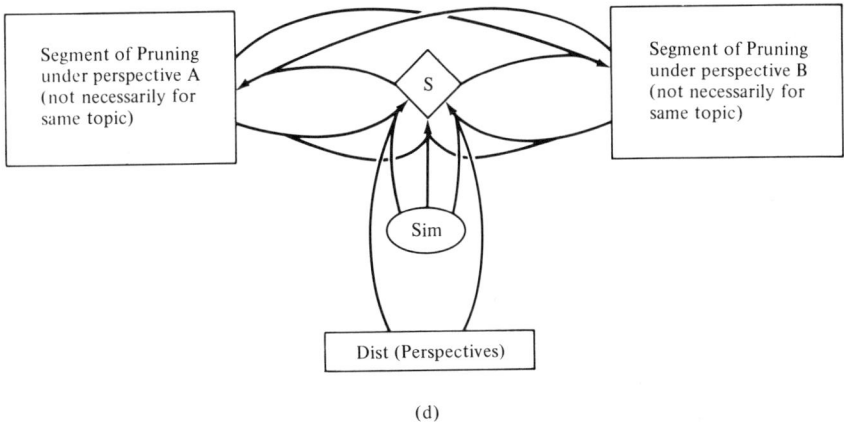

(d)

Figure 8 (*continued*)

tinction. (a) Pure analogy of form; (b) analogical topic with similarity component that is derived; (c) mixed analogy of form and method; (d) analogy of method, no necessary analogy of form.

(Figure 8d) or between meshes (see also Steltzer 1976) between axiomatic structures.

If an entailment mesh or a conversational domain is used as a kind of content map for learning about a theory, then learning an analogy is not greatly different from learning any other topic. This is not true of con-

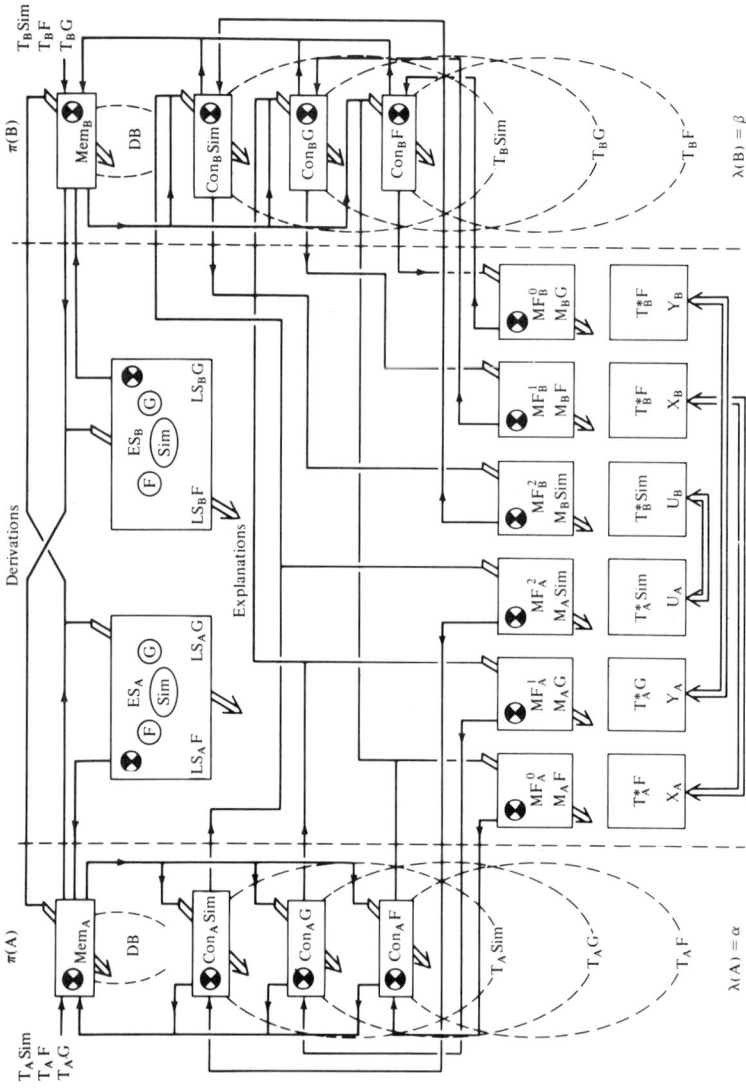

Figure 9. Agreement over and construction of an analogy, by extension of Figure 5 and using the same notation, except that there must be *independent* models constructed by *both* participants in modeling facilities MF^0, MF^1, MF^2, the analogies *between* which become *agreed* to as themselves *analogies* between *perspectives* of the participants.

structing and inscribing analogies (one aspect of creative thought, at any rate in design—see reports 1976, 1977). Independent models must be constructed, executed independently, and rendered coherent (or dependent) because an analogy is built between them (Figure 9).

16.14   Analogies and Agreements over an Understanding

In a sense, all such analogies are static inscriptions of agreements over understandings.

In general, analogies hold between perspectives. Invariably, they are created by the juxtaposition of perspectives and the resolution of these perspectives. As a rule, this involves a further distinction, which may be an inventive or genuine extension of a theory or a design and leads to the realization of a further analogical universe in which an otherwise inexecutable compound or concurrent model *can* be executed (Figure 10).

Figure 10. A mechanism for creativity or innovation. Let $M_F A$ and $M_F B$ constitute models, agreed as analogous by A and B realizing T_A*F, T_A*G, in X_A and Y_A. Take any submodels of these and compound them as Compound ($M_A f$, $M_A g$) = $M_A(Q)$, which *cannot* be realized in X or in Y but may be realized by distinguishing a further modeling facility with universe V in which, upon execution, $M_A(Q) \Rightarrow T_A*Q$. Consider a v-realizable submodel $m_A q$ of $M_A(Q)$, which upon execution gives rise to a relation T_A*q that is isomorphic to a relation $T_A C$ (say) in Y (or in X) such that $T_A C \Leftrightarrow T_B C$.

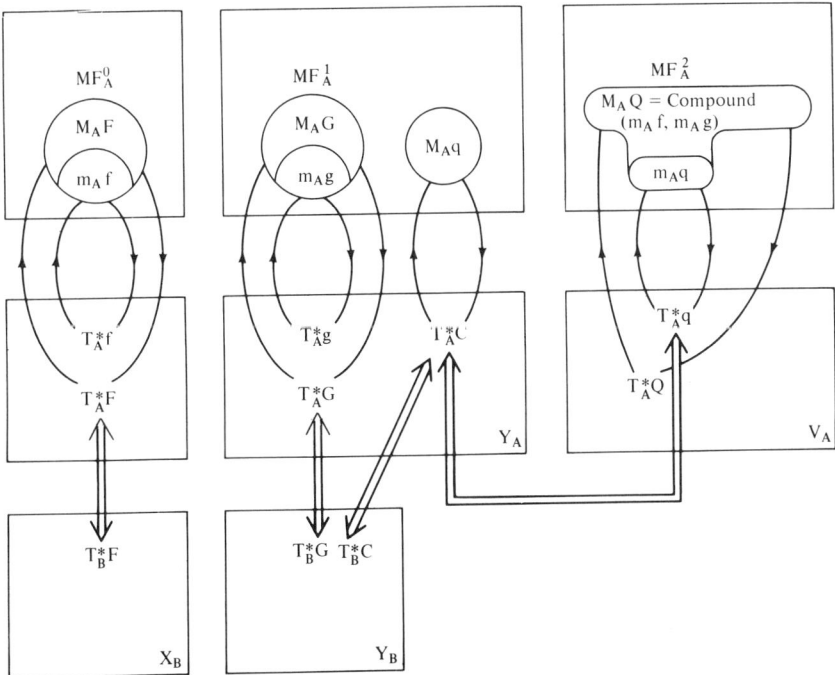

Figure 11. (a) Case I. (b) Case II. (c) Case III. (d) Case IV. Notice that the perspectives are meshes or members of the pruning fields of meshes, however.

(a) $\pi(A) = a$, $\lambda(A) = \alpha$, $\pi(B) = b$, $\lambda(B) = \beta$,   (b) $\pi(A) = a$, $\lambda(A) = \alpha_1$ or α_2, $\pi(B) = b$, $\lambda(B) = \beta_1$, or β_2.

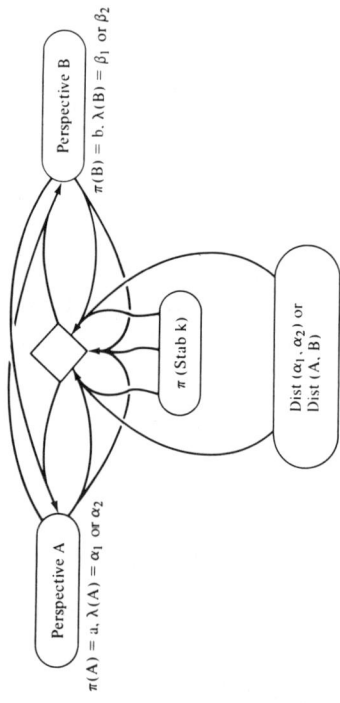

(b)

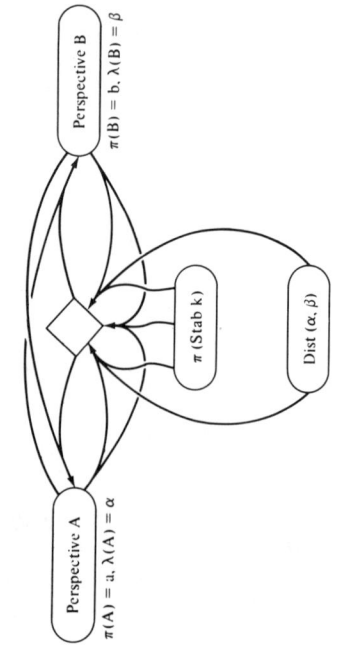

(a)

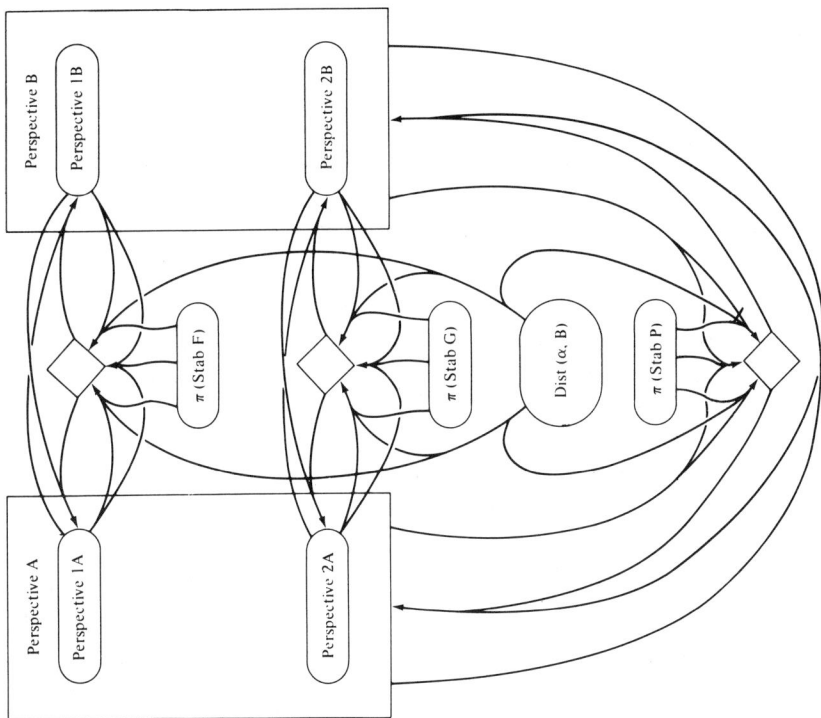

Perspective B

Perspective 1B

Perspective 2B

Perspective A

Perspective 1A

Perspective 2A

π (Stab F)

π (Stab G)

Dist (α, B)

π (Stab P)

(c)

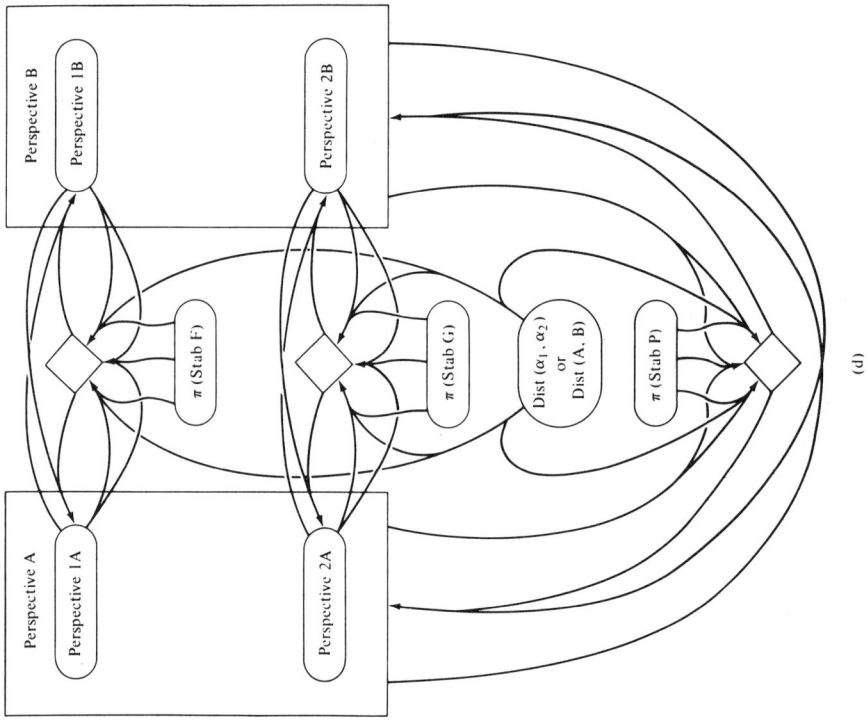

(d)

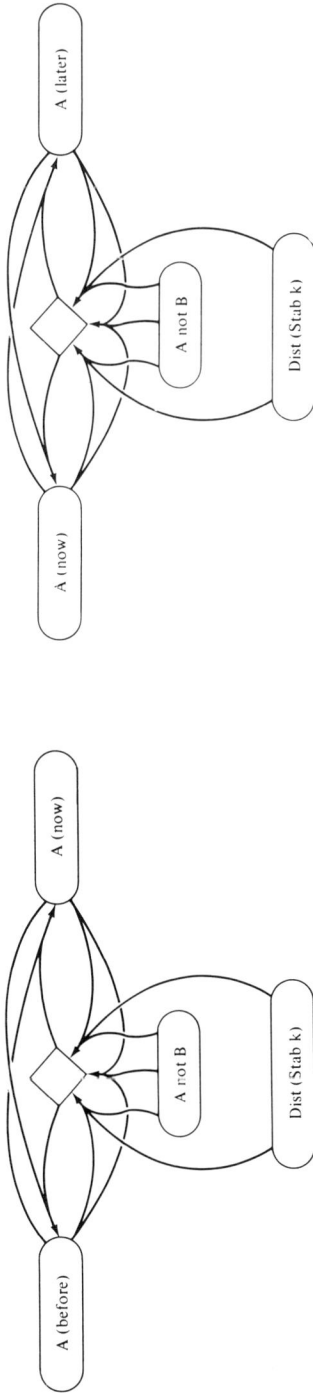

Figure 11 (*continued*). (a) Case I. (b) Case II. (c) Case III. (d) Case IV. Notice that the perspectives are meshes or members of the pruning fields of meshes, however.

(a) $\pi(A) = a$, $\lambda(A) = \alpha$, $\pi(B) = b$, $\lambda(B) = \beta$, (b) $\pi(A) = a$, $\lambda(A) = \alpha_1$ or α_2, $\pi(B) = b$, $\lambda(B) = \beta_1$, or β_2.

Let us represent Cases I–IV of conversations as analogies (static inscriptions), using the notation of Figure 8. For Case I, the inscription is shown in Figure 11a; Case II, which is similar, is shown in Figure 11b. For Case III and Case IV, an analogy between analogies is required (Figure 11c and Figure 11d). To each of these there is a dual (an analogy of method, not *necessarily* one of form) that preserves the identity of A and B, even though they learn or invent (Figure 11e or Figure 11f is representative). These constructions capture the notion of irreversibility, in addition to persistence, as promised in section 16.4.

The truth values (executability, in general) of topics, interpreted in different universes that are analogous, is a subsistence truth (Section 16.5). Thus, with referrence to Figure 8, F is "true in X" and G is "true in Y." The truth value of the analogy itself is a coherence truth (Section 16.5).

Turning to Figure 11a–d, observe that "true for A" and "true for 'B" are not the same, but *agreed* to (i.e., these personal truths are coherence related. There is a maximal coherence truth, those things that A and B can jointly understand—namely, all possible L expressions or subsets of them that are beliefs shared by a civilization, a culture, a few people in dialogue, or maybe just a hermit talking to himself.

16.15 Limits on a Community of Language Users

However, insofar as understanding takes place, there is no limit to the size of an L-speaking community, provided that, within its shared beliefs, it accommodates an adequate diversity (of subcultures, or deviants, or whatever) to maintain conversation that is genuinely productive [the *disagreement* to engender innovation (Section 16.14) and to accept some of the inventions]—a dialectic (which is also compatible, for example, with Moscovic's (1976) theory of social development).

16.16 Autonomy, Individuality, and Knowledge

What are the stable and organizationally closed systems of cognition and cognation, either psychological or social? In the limiting case, there is Stab (Figures 3 and 4) accompanied by an inarticulated awareness, the sentience of a monad.

The least conscious system is a conversation, external or internal, in which agreement is reached between perspectives. I call such an entity a *P individual* (psychological individual).

The least observable conversation places the distinction boundary in such a position that some understandings are exteriorized. The conversation is a P individual, and so are the participants who converse with each other.

There is no limit to the size of a conversation except that it must generate sufficient distinctions to be resolved, that is, sufficient perspectives. Hence, a society or a civilization is organizationally closed (P Individualized), just as is a family or a person. There is no need to ask why there are organizationally closed systems or autopoietic systems. They are the units of reality. The cogent question is whether there are any "allopoietic" (inanimate, "static") systems except those engendered by the artifice of static inscription.

Appendix: Production Schemes for Organizationally Closed and Informationally Open Systems

The entire paper is (obviously) written in a metalanguage, referred to henceforward as $L^\#$ over the conversational language L noted in the paper. For example, the process ostending variable Z, the conditions of Z, including the specification of $\pi(Z)$ and $\lambda(Z)$, and the observation of an understanding are $L^\#$ statements.

It is assumed that individuals $Z = A$ and $Z = B$ are in conversation, so that it is possible to substitute blanks (Con i, etc.) and consider concepts that belong to A or B ($\text{Con}_A i$, $\text{Con}_B i$, etc.). For convenience and clarity in drawing out large production schemes, upper-case symbols ($P, Q, ..., R, S, ..., T$) are used to stand either for an *index* ($i, j, ...$) or a *description* produced upon executing $\text{Con}_A i$, $\text{Con}_B j, ...$; so, for example, we write

$$\text{Ex Con}_A(T) \Rightarrow T_A \quad \text{or} \quad \text{Ex Con}_B(T) \Rightarrow T_B.$$

The ambiguity is harmless since, although indices and descriptions are not the *same*, they are in one-to-one correspondence.

Diagram 1 shows an organizationally closed system obtained by substituting $Z = A$ and by postulating that, depending upon the perspective, A derived T_A from P_A and Q_A, P_A from T_A and Q_A, or Q_A from T_A and P_A; the static inscription of this system is an entailment mesh in the form of Figure 4 in the with $T = i$, $P = l$, and $Q = m$.

Diagram 2 shows the possibility that B derives T_B from R_B and S_B (with static inscription, again, as in Figure 4).

Diagram 4 shows an agreement, over the understanding of T, by A and B. The commonly shared part of T_A and T_B is $T \equiv T^*$. As a result of agreement A may derive a concept for T_A from P_A and Q_A or from R_A^* and S_A^*; B may derive a concept for T_B from R_B and S_B or from P_B^* and Q_B^*.

The event depicted in Diagram 3 (leading from Diagrams 1 and 2 to Diagram 4) is procedure sharing between participants who are regarded, with equal significance, as a priori asynchronous or a priori independent; that is, they become locally synchronized or locally dependent because of procedure sharing that is manifest as an L agreement (Figure 5). When this event is observed in the metalanguage $L^\#$ it has the form of an $L^\#$ metaphor designating an $L^\#$ analogy relation. This analogy is veridically susbistent true (or false) with respect to both A and B. The distinction on which it hinges, $\text{Dist}(A,B)$, is introduced by an observer who is anxious to make objective (it-referenced) statements about conversations as units.

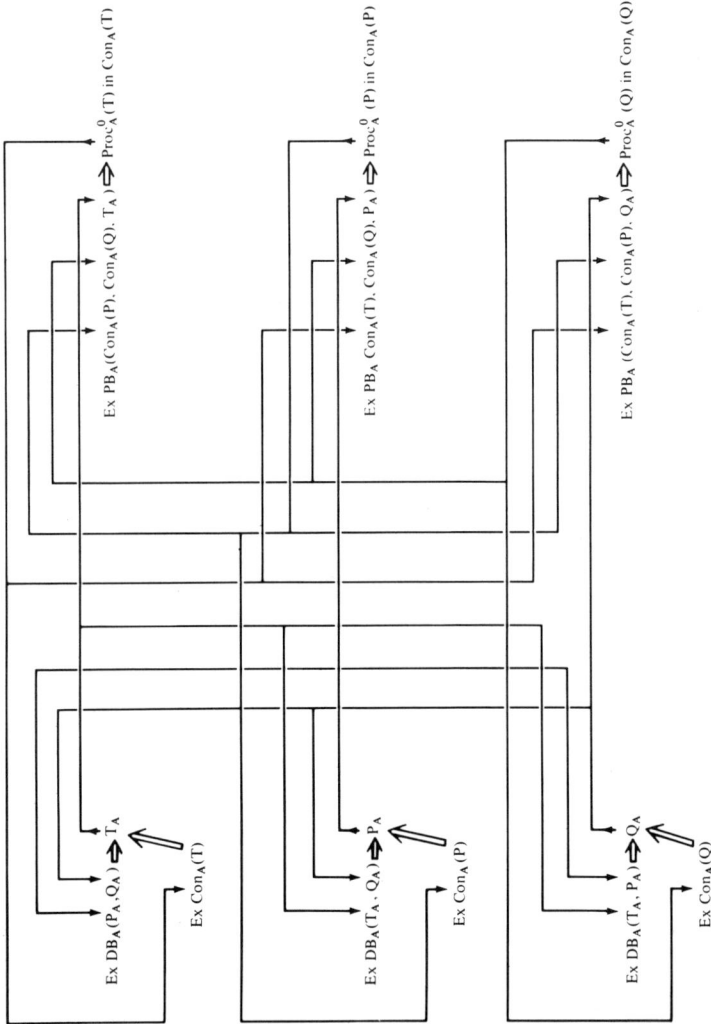

DIAGRAM 1. A stable concept $Mem_A(T)$ is, with equal significance, the organizational closure of a system of productions PB, and DB, and the concept execution for participant A (it is assumed that participant A derives a concept for topic T from concepts for topic P and for topic Q). The subscripting notation is (quite deliberately) redundant if *only* A is in mind, but is consistent with the notation required if consideration is also given to participant B.

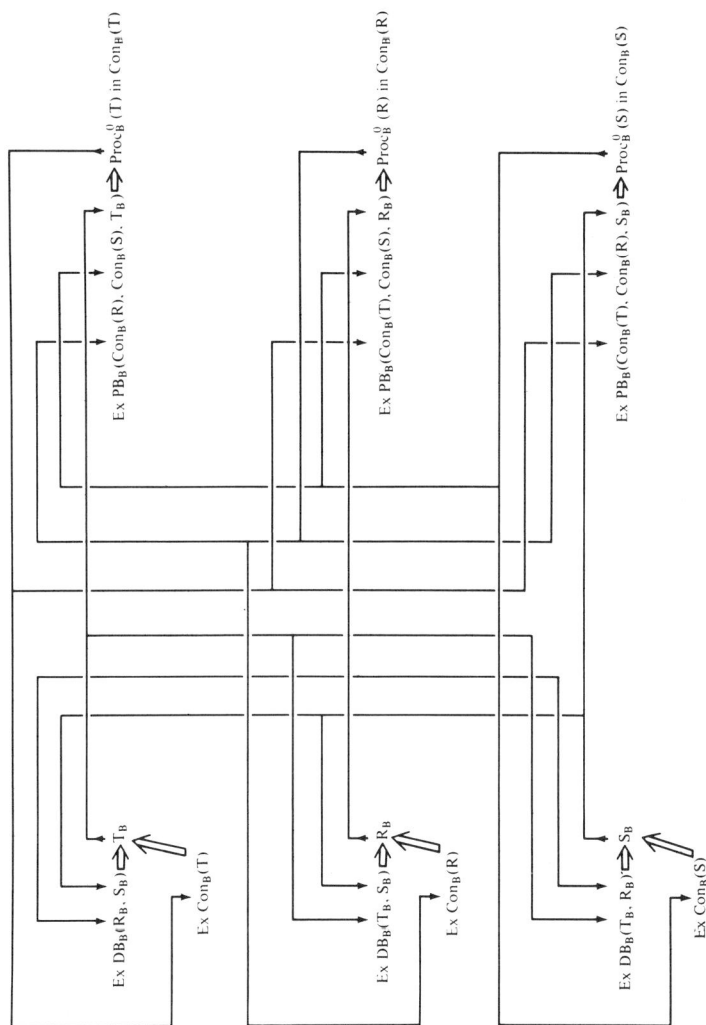

DIAGRAM 2. A stable concept $Mem_B(T)$ is, with equal significance, the organizational closure of a system of productions DB, PB and the concept execution for participant B. It is assumed that participant B derives a concept for T from concepts for R and for S.

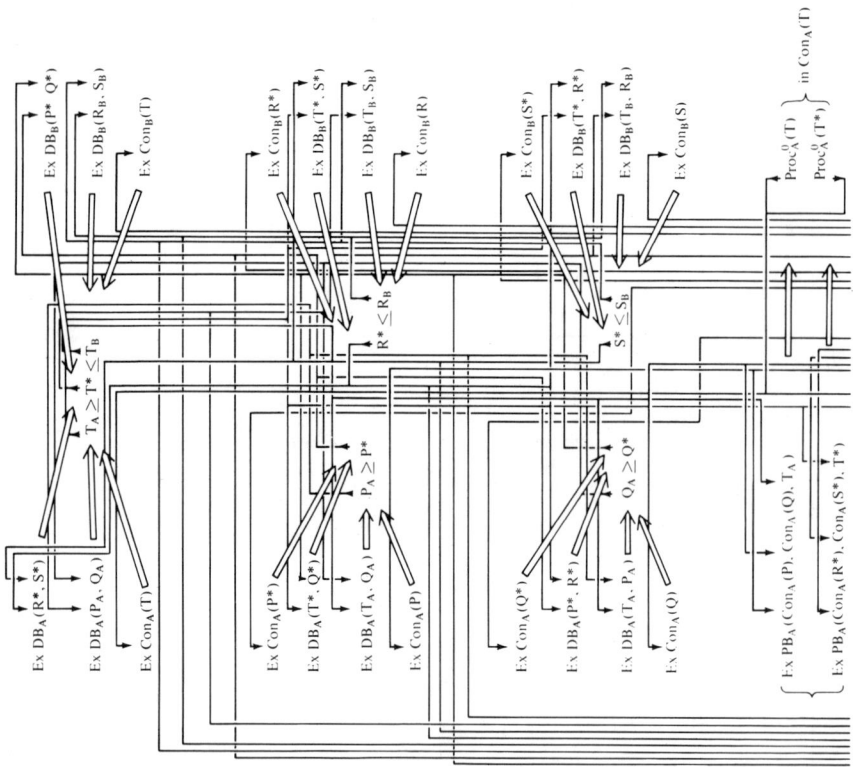

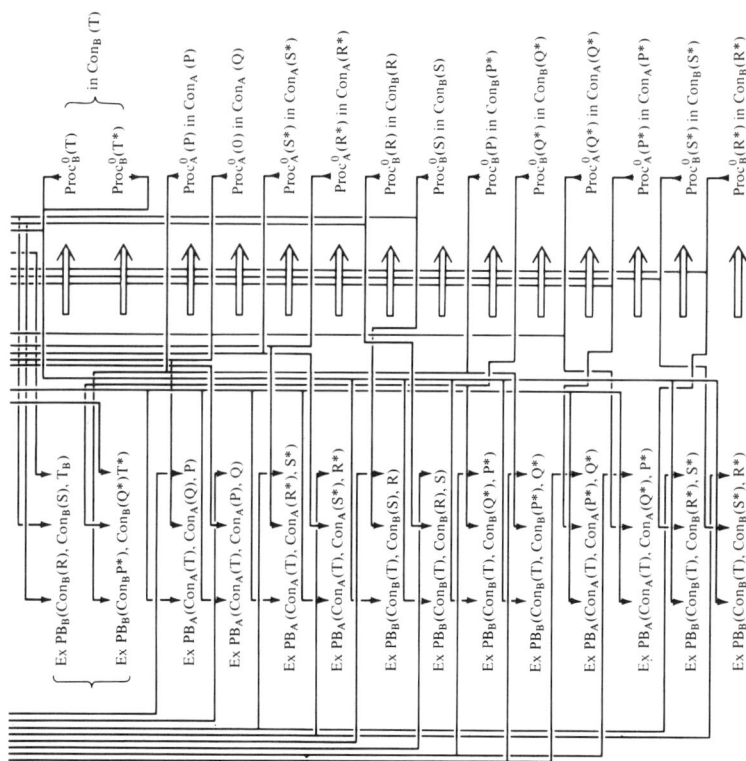

DIAGRAM 3. L agreement over common understanding of topic T. A derives T from P and Q. Participant B derives T from R and S. An agreement may be complete or partial depending upon the isomorphic part.

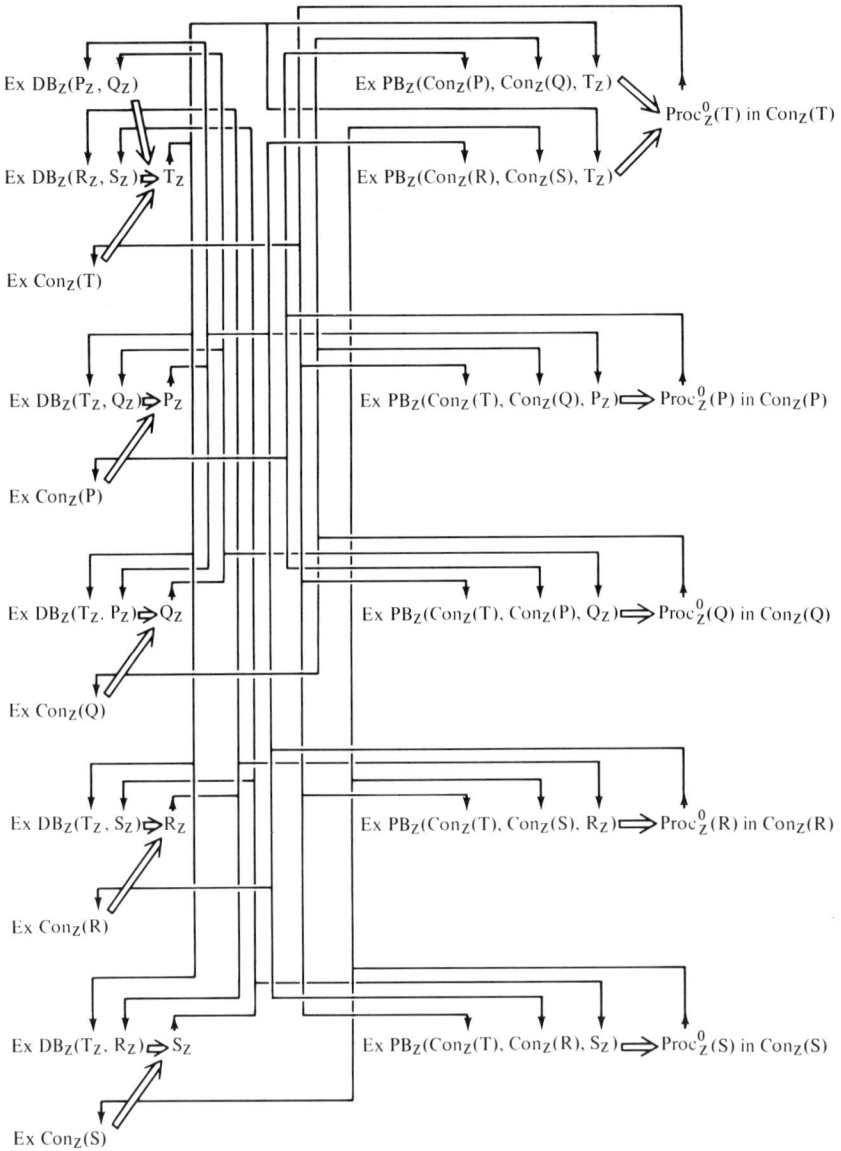

DIAGRAM 4. Given that T is learned by one participant (for consistency, participant A) who derives a concept of topic T from concepts for topic P and Q or R and S, the stable concept is an organizationally closed system of productions that take place in one participant.

ACKNOWLEDGMENTS

The research reported in this paper was supported by a Social Science Research Council of Great Britain Research Programme (Grant HR 2708/2), in the context of learning strategies, educational strategies, and subject matter representation, the U.S. Army Research Institute for the Behavioral and Social Sciences (Grant DAERO 76 G 069), through its European Office, in the context of complex decision processes, and by the U.S. Air Force Office of Scientific Research (Contract F44620) through its European Office in the context of innovative design. The work was carried out at System Research Ltd.

References

Atkin, R. (1977), "Methodology of Q-Analysis," Research Report No. 10. Dept. Mathematics, University of Essex.

Ben Eli, M. (1976), "Comments on the Cybernetics of Stability and Regulation in Social Systems," Ph.D. thesis, Brunel Univ., Uxbridge.

Ben Eli, M., and Tountas, C. (1976), The evolution of complexity in a simulated ecology, *Proc SGSR Symposium,* Denver.

Bråten, S. (1976), Computer simulation of communication and consistency, *Proc. Int. Association for Cybernetics, Namur, Belgium.*

Bråten, S. (1977), *The Human Dyad. Systems and Simulations,* Inst. of Sociology, Univ. of Oslo.

Caianiello, E. (1977), *Some Remarks on Organisation and Structure,* sabbatical project, Inst. of Theoretical Physics, Univ. of Alberta, Canada; Lab. for Cybernetics, CNR, Naples.

Chang, C. L., and Lee, Y. L. (1973), *Symbolic Logic and Mechanical Theorem Proving,* Academic Press, New York.

Codd, E. F. (1970), A relational model of data for large shared data banks, *ACM* 13(b), 337.

von Foerster, H. (1976), papers in collected publications of BCL.

Glanville, R. (1976), "The Object of Objects, the Point of Points—or, Something About Things," Ph.D. thesis, Brunel, Univ. Uxbridge.

Goguen, J. A. (1975), Objects, *Int. J. Gen. Systems* 1, 237–243.

Goguen, J. A., Thatcher, J. W., Wagner, E. G., and Wright, J. B. (1976), *A Junction Between Computer Science and Category Theory, I: Basic Concepts and Examples (Parts 1 and 2),* IBM Thomas Watson Research Center, Yorktown Heights, N.Y.

Löfgren, L. (1972), Relative explanations of systems, in *Trends in General Systems Theory* (G. Klir, ed.), John Wiley, New York.

Löfgren, L. (1975), *On Existence and Existential Perception,* Dept. Automata, Lund Inst. Tech., Lund, Sweden.

Maturana, H., and Varela, F. (1976), Autopoietic systems: A characterisation of the living organisation,

Muscovici, S. (1976), *Social Influence and Social Change,* Academic Press, New York.

Neumann, J. von (1966), *Theory of Self-Reproduction Automata* (A. Burks, ed.), Univ. of Illinois Press, Urbana.

Pask, G. (1973), Artificial intelligence, *Soft Architecture Machinations,*

Pask, G. (1975a), *Conversation Cognition and Learning,* Elsevier, Amsterdam and New York.

Pask, G. (1975b), *The Cybernetics of Human Learning and Performance,* Hutchinson, London.

Pask, G. (1976a), *Conversation Theory: Applications in Education and Epistemology,* Elsevier, Amsterdam and New York.

Pask, G. (1976b), Conversational techniques in the study and practice of education, *Br. J. Educ. Psych.* 46 (Part I).

Pask, G. (1976c), Styles and strategies of learning, *Br. J. Educ. Psych.* 46 (Part II).

Pask, G. (1977a), "Investigations into New Methods of Assessment and Stronger Methods of Curriculum Design, with Particular Reference to Higher Education in the United States, Part I," Ford Foundation Report, IET Open University.

Pask, G. (1977b), Minds and Media in Education and Entertainment, *Procs. 3rd European Meeting on Cybernetics and Systems Research,* Vienna (R. Trappl, ed); see also Revisions in the foundations of Cybernetics and General Systems Theory,... *Proceedings 8th International Congress on Cybernetics,* Namur 1976.

Pask, G., and Scott, B. C. E. (1972), Learning Strategies and Individual Competence, *Int. J. Man–Machine Studies* 4, 217–253.

Pask, G., and Scott, B. C. E. (1973), CASTE: A system for exhibiting learning strategies and regulating uncertainty, *Int. J. Man–Machine Studies* 5, 17–52.

Pask, G., Scott, B. C. E., and Kallikourdis, D. (1973), A theory of conversations and individuals (exemplified by the learning process on CASTE), *Int. J. Man–Machine Studies* 5, 443–556.

Pask, G., Kallikourdis, D., and Scott, B. C. E. (1975), The representation of knowables, *Int. J. Man–Machine Studies* 7, 15–134.

Petri, C. A. (1965), *Communications with Automata,* Suppl. to Tech. Documentary Rep. (tr. Clifford F. Greene, Jr.) for Rome Air Development Center, Contract AF30 (602)–3324.

Steltzer, J. (1976), *Classifying and Structuring the Context of Instructional Subject Matter, ARI Report* DAHC 19-74-C0066.

Varela, F. (1975), A calculus for self reference, *Int. J. Gen. Systems* 2, 5–24.

Varela, F. (1976), The arithmetic of closure (R. Trappl, ed.), *Proc. 3rd Europ. Mtg. on Cybernetics and Gen. Systems Research,* Vienna.

Varela, F., Maturana, H. R., and Uribe, R. (1974), Autopoiesis: The organization of living systems, its characterization and a model, *Biosystems* 5, 187–196.

Index